T0319346

Voltage Stability in Electrical Power Systems

IEEE Press

445 Hoes Lane
Piscataway, NJ 08854

Voltage Stability in Electrical Power Systems

Concepts, Assessment, and Methods for Improvement

Farid Karbalaei
Shahid Rajaee Teacher Training University
Tehran, Iran

Shahriar Abbasi
Technical and Vocational University of Iran
Kermanshah, Iran

Hamid Reza Shabani
Aalborg University
Esbjerg, Denmark

Published by John Wiley & Sons, Inc., Hoboken, New Jersey.
Published simultaneously in Canada.

Library of Congress Cataloging-in-Publication Data:
Names: Karbalaei, Farid, author. | Abbasi, Shahriar (Assistant Professor), author. | Shabani, Hamid Reza, author.
Title: Voltage stability in electrical power systems : concepts, assessment, and methods for improvement / Farid Karbalaei, Shahriar Abbasi, and Hamid Reza Shabani.
Description: Hoboken, New Jersey : Wiley, [2023] | Includes index.
Identifiers: LCCN 2022041340 (print) | LCCN 2022041341 (ebook) | ISBN 9781119830597 (hardback) | ISBN 9781119830641 (adobe pdf) | ISBN 9781119830658 (epub)
Subjects: LCSH: Electric power system stability. | Electric power systems–Control.
Classification: LCC TK1010 .K367 2022 (print) | LCC TK1010 (ebook) | DDC 621.319–dc23/eng/20220919
LC record available at https://lccn.loc.gov/2022041340
LC ebook record available at https://lccn.loc.gov/2022041341

Cover Design: Wiley
Cover Image: © Serg Myshkovsky/Getty Images

Set in 9.5/12.5pt STIXTwoText by Straive, Pondicherry, India

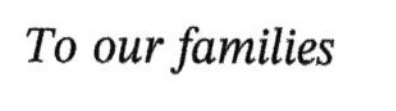

To our families

Contents

Author Biographies

Farid Karbalaei received BSc degree in power engineering from K. N. Toosi University of Technology, Tehran, Iran, in 1997, and MSc and PhD degrees in power engineering from Iran University of Science and Technology, Tehran, in 2000 and 2009, respectively. Currently, he is an associate professor in Shahid Rajaee Teacher Training University. His research interests are power system dynamics and control, voltage stability and collapse, reactive power control, wind power generation, and optimization methods.

Postal Address: Faculty of Electrical Engineering, Shahid Rajaee Teacher Training University, Lavizan, Tehran, Iran. Phone: +98 912 44 45 325, Email: f_karbalaei@sru.ac.ir (Corresponding author)

Shahriar Abbasi received BSc and MSc degrees in power engineering from Shahid Rajaee Teacher Training University Tehran, Iran, in 2008 and 2011, respectively. He received PhD degrees in power engineering from Razi University, Kermanshah, Iran, 2018. Currently, he is an assistant professor in Technical and Vocational University of Iran, Kermanshah Branch. His research interests are power system planning, uncertainty modeling, voltage stability and collapse, wind power generation, and optimization methods.

Postal Address: Technical and Vocational University of Iran, Kermanshah Branch, Kermanshah, Iran. Phone: +98 916 98 48 928, Email: shahriarabasi@gmail.com

Hamid Reza Shabani received the MS degree in power electrical engineering from Shahid Rajaee Teacher Training University (SRTTU), Tehran, Iran, in 2014. Also, he received the PhD degree from Iran University of Science and Technology (IUST), Tehran, Iran, in 2021. His thesis title in the PhD program was "evaluation of large-disturbance rotor angle instability in the modern power systems, with high-Penetration of wind power generation." He currently works as a postdoctoral researcher at Aalborg University (AAU Energy) in the Esbjerg Energy Section. His main research interests include power system stability, power system dynamic and control, and renewable energies.

Postal Address: The Faculty of Engineering and Science (AAU Energy), Aalborg University, 6700 Esbjerg, Denmark. Phone: +45 52 78 06 90, Email: hmdrzshabani94@gmail.com

Preface

Voltage instability has been considered about 60 years ago and is still a major cause of blackouts in electrical power systems. So far, extensive studies have been conducted on this topic and the result of which is publication of thousands of articles and a few number of books. The articles cover a wide range of subjects related to voltage stability, including proper system modeling for voltage stability studies, online and offline voltage stability assessment methods, and methods to prevent voltage instability, which include two sets of preventive and emergency methods.

Being familiar with the all above-mentioned subjects is necessary for power engineers because effective prevention of voltage instability necessitates timely detection of it. Timely detection also requires knowledge of the mechanism of voltage instability and proper modeling of the system. Since voltage instability is a local phenomenon, contrary to rotor instability, many methods of detecting and preventing voltage instability are performed locally. Therefore, in addition to the engineers working in the system control center, all power engineers in local control centers should be fully familiar with this field.

Good understanding of issues related to voltage stability requires reading hundreds of articles. Due to the fact that articles do not have educational purpose, it is difficult to fully understand, summarize, and relate them together. Therefore, a book that presents all the above subjects in a complete, arranged and comprehensible way is needed; so that it first explains the necessary fundamentals (which is not done in articles) then presents the related subjects in an appropriate classification and sequence. The aim of this book is to present all the voltage stability subjects so that readers with the level of bachelor information can use it well. Of course, the state-of-the art on voltage stability is introduced in it to be useful for university professors, master and doctoral students.

This book consisted of three parts. The contents of these parts can be summarized as follows:

The Part I: *Concept of Voltage Stability, Effective Factors and Devices, and Suitable System Modeling* consisted of four chapters. In Chapter 1, the concept of voltage instability and its types are first described. Then how long-term instability and voltage drop occur due to the activities of tap changers and thermostatic loads are illustrated. The occurrence of short-term voltage instability due to the presence of induction motors is also described. In this chapter, by simulating on a simple system, the importance of loadability limit increase in maintaining voltage stability is shown. The concept of exiting from attraction region and the importance of timely performing of emergency measures are also explained by simulation. The purpose of this chapter is to familiarize the reader quickly and in general (not in full detail) with the concept and causes of voltage instability as well as how to prevent it from occurring.

In Chapter 2, different dynamic and static load models used in references to analyze voltage stability are introduced. These models represent the behavior of integrated loads seen from different buss of the power system. In short-term voltage stability analysis, the dynamic behavior of load is simulated as the dynamic model of induction motor. Hence, a part of the third chapter is devoted to presenting the algebraic and differential relations of induction motor. Another part of this chapter introduces the types of tap changers and modeling transformers with variable tap. In references, there are two models to represent variable tap transformers. The difference between these two models is in the side that the equivalent impedance of transformer is seen from it. The simulations verify that these two models lead to different values for the system loadability limit. Given the importance of determining the correct (real) loadability limit in voltage stability studies, selection of the proper model is very important. This is discussed in this chapter.

Chapter 3 deals with modeling of synchronous generator and two types of distributed generation sources (FSIG- and DFIG-based wind turbines). The modeling degree should be chosen according to the intended type of study. In this chapter, suitable models for studying each type of voltage instability (long term and short term) are presented.

Chapter 4 explains the importance of concurrent modeling of distribution and transmission networks in assessing voltage stability. This is shown that sometimes, separate modeling of distribution and transmission networks causes a significant error in determining the voltage stability limit. Also in this chapter, the effect of the presence of distributed generation (DG) sources on voltage stability is investigated. These sources, which are mainly connected to distribution networks, up to the condition have different effects on voltage stability.

The Part II: *Voltage Stability Assessment Methods*, includes the Chapters 5–9. This Part of the book is dedicated to voltage stability assessment methods. Voltage stability assessment is performed for several purposes. One of them is determination of the voltage stability margin, which is calculated for both the current (no-contingency) system and probable contingencies. What is important in calculating the voltage stability margin is speed and accuracy of the calculation. Since the determination of the stability margin must be repeated every few minutes, its calculation for a large number of probable contingencies is possible only if the calculation time be very short while maintaining the required accuracy. For this, the methods of continuation power flow (CPF) and PV-curve fitting are presented, which are discussed in Chapters 5 and 6.

Voltage stability margin shows the level of system stability in the face of various contingencies. It does not directly provide information about the vulnerable points of system as well as the important elements in instability occurrence. This information, which helps operators to decide about taking the voltage instability preventive methods, is obtained by voltage stability indices. A number of voltage stability indices are calculated based on the system model. These indices help a lot in defining and ranking the critical contingencies, as well as in determining the necessary actions after each contingency occurrence. Another set of indices uses only variables measured at different points of system and does not require the system model. These indices can be used to quickly identify the current status of system and early detection of voltage instability. Also, when voltage stability assessment requires dynamic analysis and time simulation, the voltage stability indices can be used to reduce the required simulation time. These indices are all introduced in Chapters 7 and 8, and the advantages and applications of each one are stated.

In Chapter 9, the machine learning-based methods to assess and monitor voltage stability of power system are introduced. General topologies of these methods and their capabilities were introduced. These methods are categorized in five methods. For each method, a table including input(s), output(s), used technique, and case study is presented.

In the Part III of this book: *Methods of Preventing Voltage Instability*, the methods to prevent voltage instability are discussed. All actions used to prevent voltage collapse are divided into two categories: preventive and emergency. The purpose of preventive actions is to increase the voltage stability margin of the power system. Increasing the voltage stability margin is considered for both the current (no-contingency) system and probable contingencies. Therefore, these actions are applied when the system is stable, but there is a small distance between the current operating point and the voltage stability limit. The purpose of determining preventive actions is to improve system stability with the least measures (especially with the minimum load shedding).

The emergency actions are performed when the system becomes unstable due to one or more contingencies, and if these actions are not applied, a voltage collapse will occur in a few moments or minutes. In determining the emergency actions, the speed of calculations is very important because the later these actions are applied, voltage stability maintenance is possible with more actions.

Some of the voltage stability studies are devoted to methods for determining preventive and emergency actions, which are discussed in Chapters 10 and 11.

Farid Karbalaei, Shahriar Abbasi and
Hamid Reza Shabani
Tehran

Part I

Concept of Voltage Stability, Effective Factors and Devices, and Suitable System Modeling

1

How Does Voltage Instability Occur?

1.1 Introduction

The phenomenon of voltage instability is one of the major problems of today's power systems. According to the Institute of Electrical and Electronics Engineers (IEEE)/The International Council on Large Electric Systems (CIGRE) definition, voltage stability is the ability of a power system in maintaining an acceptable steady-state voltage at all buses when subjected to a contingency. The consequence of voltage instability is voltage collapse. Unlike rotor angle instability, which is more related to generator operation, voltage instability depends on the amount and characteristic of loads. For this, voltage instability is also called load instability [1]. Voltage instability is divided into two categories; long term and short term. In the long-term type, voltage collapse occurs during a process of a few tens of seconds or a few minutes, but in the short-term voltage instability, the voltage collapse occurs rapidly and within a few seconds.

In general, the reason of voltage instability is the presence of devices whose power consumption is not much dependent on voltage. Voltage drop initially reduces the input power of these devices, but after a few moments or minutes, the reaction of these devices causes their receiving power to increase to a value close to the value before the voltage drop. Power recovery may be done for active power only or for both active and reactive powers. Figure 1.1 conceptually shows the process of recovering the active power of a device when voltage drops. It is observed that at first the power is reduced but in the steady state its value become close to the initial value. Therefore, it is said that this device has the characteristic of constant steady-state power. In this figure, it is assumed that a constant reduced voltage is applied to the device. Also, the fluctuations of power are ignored when it is recovering. In practice, the voltage of a device decreases as its power consumption increases. In steady-state conditions, this voltage drop is small, but when voltage instability occurs, power recovery causes a severe voltage drop. A necessary

Voltage Stability in Electrical Power Systems: Concepts, Assessment, and Methods for Improvement, First Edition. Farid Karbalaei, Shahriar Abbasi, and Hamid Reza Shabani.

Published 2023 by John Wiley & Sons, Inc.

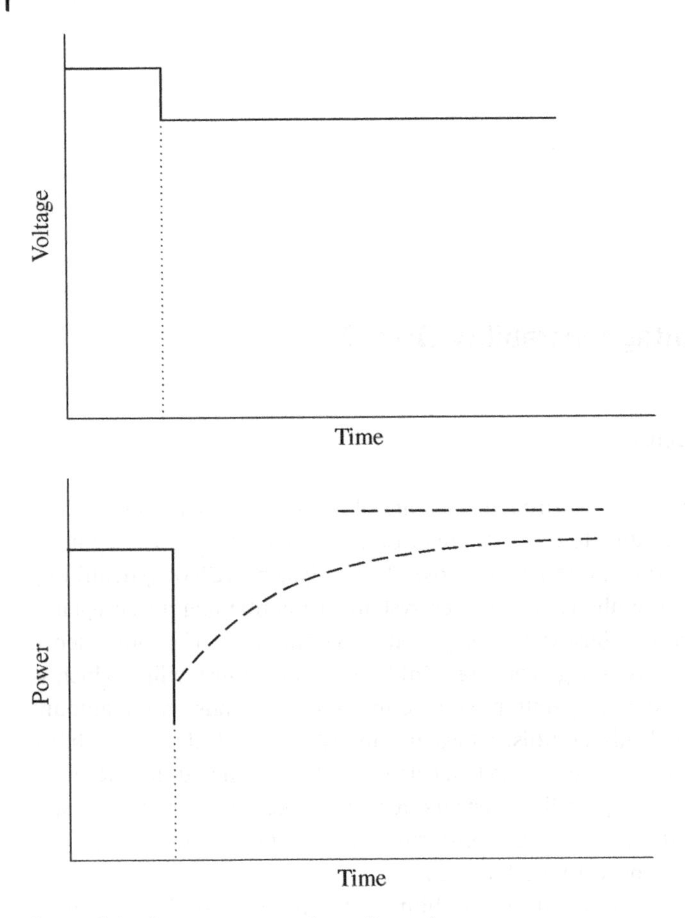

Figure 1.1 Power recovery after voltage drop.

(not sufficient) condition for maintaining voltage stability is that the transfer of the required power to constant power consumers is possible. Otherwise, long- or short-term voltage instability will occur, depending on to the power recovery time constant.

One of the most important devices that create constant power characteristic is load tap changer (LTC) transformer. In most cases, the variable tap is located on the high voltage (HV) side of the transformer. This is due to less current in the HV side, which makes it easier to change the tap. Another reason is the high number of winding turns in the HV side, which makes the voltage regulation more accurate [1]. The tap control system usually controls the voltage of low voltage (LV) side, which has lower short-circuit level. These devices fix the voltage of LV side, independent of the HV side voltage. By keeping this voltage fixed, the power consumption at the LV side also remains constant. Hence, assuming the transformer losses remain constant, the power received from the HV side is independent of its voltage. Therefore, the load seen from the HV side of the transformer has the characteristic of constant steady-state power. If the system is stable, voltage and power recovery will be done by tap changer in a few tens of seconds. In addition to tap changer, thermostatically controlled heat loads (TCLs) can also create constant power characteristic. When voltage drops, these loads will remain in the circuit longer because they must produce the necessary heat energy. Therefore, as the voltage decreases, the impedance of a set of these loads decreases. As a result, when the voltage drops, the power consumption by these loads does not change much in the steady state. It is clear that in the early moments of voltage drop, TCLs have an impedance characteristic and a constant power characteristic is created during a process of several minutes.

The main reason of short-term voltage instability is the presence of induction motor loads. Speed reduction and stalling of these motors lead to sudden increase in their reactive power consumption by them and voltage collapse. After a voltage reduction, initially, the power consumption of an induction motor decreases, but due to decreasing speed and increasing slip, the active power consumed by the motor gradually will increase. The amount of active power increment depends on the type of mechanical load supplied by the motor. The power recovery in induction motors is done quickly in a few seconds. Motor stalling happens when the motor is unable to supply its connected mechanical load. In this chapter, using simulation on simple networks, the procedure of voltage instability occurrence due to the above-mentioned factors is illustrated.

1.2 Long-Term Voltage Instability

In this section, using an example, the procedure of voltage instability due to operation of LTCs is shown. Also, the reason of this occurrence is illustrated.

1.2.1 A Simple System

To simulate the occurrence of voltage instability, the simple system of Figure 1.2 is used. In this system, a generator supplies a static load with voltage-dependent characteristic through two parallel lines and an LTC transformer. For the load, the exponential model according to Eqs. (1.1) and (1.2) is used. In these equations, P_0 and Q_0 are, respectively, the demanded active and reactive powers of this load at the voltage 1 pu. A complete discussion about load modeling is presented in Chapter 2.

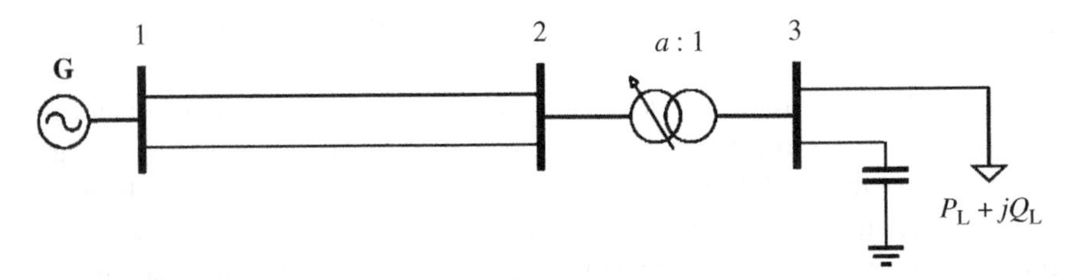

Figure 1.2 Single-line diagram of a simple system with LTC transformer.

$$P_L = P_0(V_3)^{1.5} \tag{1.1}$$

$$Q_L = Q_0(V_3)^{2.5} \tag{1.2}$$

The transformer is modeled as a series leakage impedance and an ideal transformer. Next to the load, there is a compensating capacitor with the admittance 0.45 pu. In this example, the generator is modeled as a constant voltage source, and the voltage V_1 is assumed to be 1 pu. Therefore, the dynamic behavior of this system is only due to the operation of LTC. It is assumed that the LTC is installed at the HV side and its value (a) can vary from 0.85 to 1.15 pu in the steps of 0.005 pu (0.5%). The duty of the LTC is maintaining V_3 between 0.99 and 1.01 pu. This range is called the LTC dead band. The dead band is always considered larger than LTC steps, usually twice of LTC steps. Otherwise, the LTC changes will not converge to a steady-state value.

1.2.2 Voltage Calculation

To calculate V_3 for different tap values, the equivalent circuit π is used to model LTC transformer [2]. By doing so, the equivalent single-phase circuit of this system is as shown in Figure 1.3. By changing the tap value, the impedance of the equivalent circuit branches of the transformer changes. In this system, the bus 1 is slack

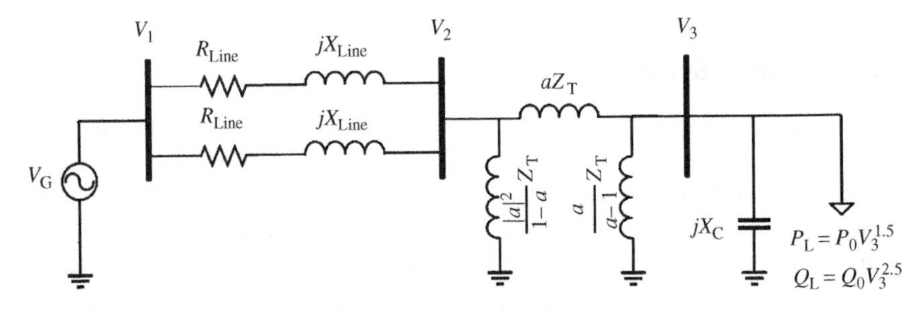

Figure 1.3 Single-phase equivalent of the system of Figure 1.2.

and the buses 2 and 3 are the PQ type. To calculate the voltages, the load flow equations are written and solved with voltage-dependent load at bus 3. Assuming P_0 and Q_0 are, respectively, equal to 0.5 and 0.2 pu, the steady-state tap value becomes 1.00 and the magnitude of voltage V_3 is equal to 0.992 (a value between 0.99 and 1.01 pu).

Assuming the outage of one of the parallel lines, the voltage of bus 2 and consequently the voltage of bus 3 decreases. After this voltage drop, the LTC reduces the tap value installed at the HV side to recover the voltage. The LTC and voltage changes are shown in Figure 1.4. It can be seen that by a few changes of tap value, the voltage of bus 3 is recovered to the desired range.

1.2.3 Illustration of Voltage Collapse

Now with the two parallel lines, the values of P_0 and Q_0 are increased to 0.83 and 0.33 pu, respectively. In this condition, in steady state, the tap value and voltage of bus 3, respectively, converge to 1.05 and 0.991 pu. Now, with these values for P_0 and Q_0, and outage of one of the parallel lines, the LTC control system will start reducing its tap value again to recover the voltage of bus 3. But in this case, as shown in Figure 1.5, voltage recovery is not achieved. At the first few steps, the voltage of bus 3 increases by any reduction in the tap. But after these initial steps, any decrease in the tap leads to a voltage drop, which means that the voltage instability has occurred.

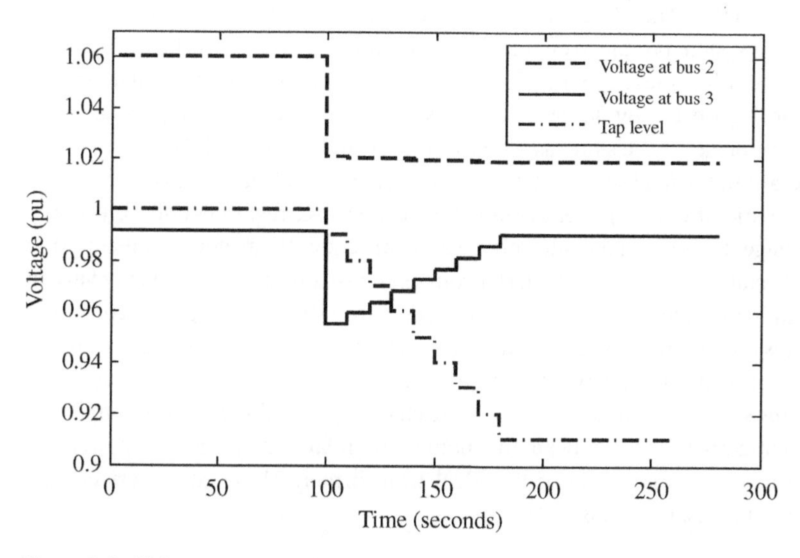

Figure 1.4 Voltage recovery after line outage.

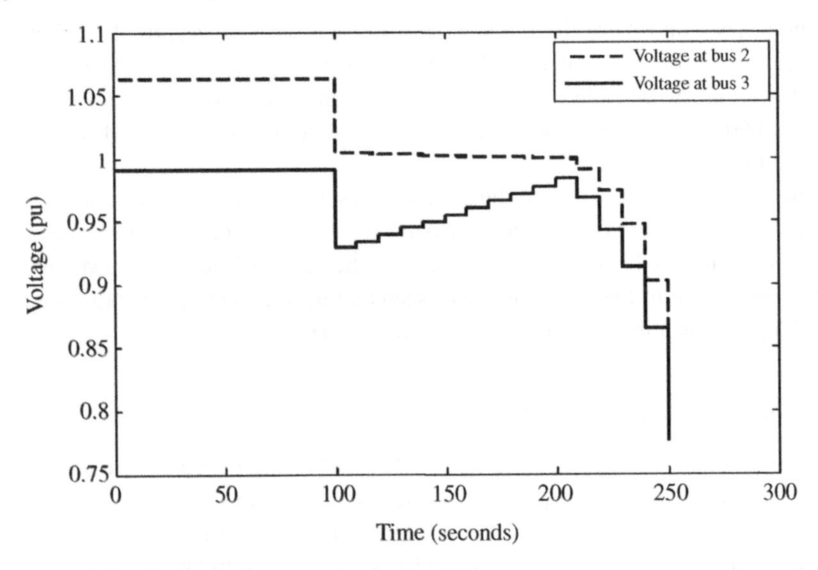

Figure 1.5 Voltage instability after line outage.

1.2.4 The Reason of Voltage Collapse Occurrence

The reason of impossibility of voltage recovery is that voltage recovery means the recovery of active and reactive powers to values close to P_0 and Q_0 (0.83 and 0.33 pu in this example). Therefore, voltage recovery is achieved only when after a line outage, the system be able to deliver these amounts of power to the bus 3. To better understand this subject, the transformer impedance is shifted to its primary side as shown in Figure 1.6. By doing so, the power delivered to the primary side of the ideal transformer is the same as the power delivered to the load. Now, due to the tap operation, the load seen from the primary side of the ideal transformer has the characteristic of constant steady-state power. Because, regardless of the primary side voltage, the secondary side voltage and consequently its power consumption in steady state is almost constant. (Of course, the steady-state voltage may vary in the dead band range, causing slight changes in the steady-state power consumption.) It is obvious that the constant power characteristic will be obtained during a few tens of seconds to a few minutes process, depending on the time delay of the tap changes. In addition to the steady-state characteristic, there are also some transient characteristics that indicate the momentary relationship between power and voltage in the primary side of the ideal transformer. These characteristics are according to Eqs. (1.3) and (1.4).

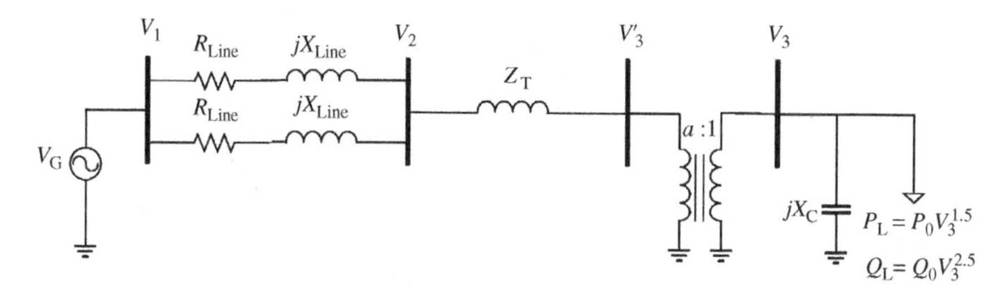

Figure 1.6 Transformer impedance transferred to the primary side.

$$P_T = P_0 \left(\frac{V'_3}{a} \right)^{1.5} \tag{1.3}$$

$$Q_T = Q_0 \left(\frac{V'_3}{a} \right)^{2.5} \tag{1.4}$$

A load that has a transient characteristic in addition to the steady-state characteristic is called a dynamic load. In the above-mentioned example, the load seen from the primary side of the transformer is a dynamic load. The power variations of dynamic loads due to voltage change, in addition to the steady-state term, have a transient term. Now, due to the constant steady-state power characteristic, a stable operating point is achieved only when it is possible to deliver power to the demanded load. For this reason, in many references, the maximum loadability of a power system is introduced as the long-term voltage stability limit.

Occurrence of voltage instability implies that the active and reactive powers of 0.83 and 0.33 pu are definitely more than the maximum powers that can be transferred to the primary side of the ideal transformer. Figure 1.7 shows the voltage variations versus the active power variations received at the primary side of the ideal transformer before and after the line outage. In drawing these curves, the variation of the transformer series impedance due to the tap changes is neglected. The error of this approximation is very small since the transformer impedance is connected in series with the line impedance. These curves, which are widely used in power system voltage stability studies, are called power voltage (PV) curves [3]. Usually, these curves are drawn with constant power factors. The power factor in Figure 1.7 is chosen based on the active and reactive powers of the load and the reactive power produced by the capacitor at voltage 1 pu. It can be seen that the maximum transferable active power after line outage is equal to 0.75 pu, which is less than the power required to voltage recovery (0.83 pu). In Figure 1.7, along with

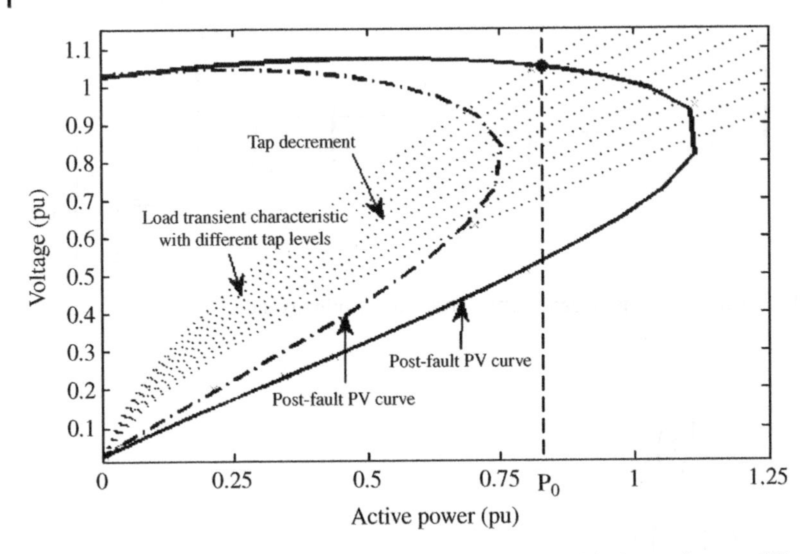

Figure 1.7 The PV curves seen from the primary side of the ideal transformer of Figure 1.6.

the PV curves, the transient characteristics of the active power seen from the primary side of the ideal transformer are plotted. These characteristics are obtained for different tap values. The intersection points of the transient characteristics with the PV curves are the active power and voltage values for different tap values. As can be seen, as the tap decreases, the intersection point moves toward the point of maximum transferable power (the nose of PV curve). Until the intersection point reaches the nose of PV curve, any tap reduction will increase the received power by the load. Due to the voltage-dependent characteristic of the load, increasing the received power means increasing the voltage V_3, although the voltage V_3' decreases. After crossing the nose of PV curve, the power and voltage V_3 will reduce after each tap reduction. Therefore, the nose of the PV curve is also called the collapse point.

Figure 1.8 shows the PV curves with different power factors of the load. It is shown that the more lead characteristic of the load, i.e. the more reactive power compensation, the higher maximum loadability, and also the higher collapse point voltage. If the reactive power compensation level is not high, the voltage of collapse point will be low (even about 0.5 pu). However, if the reactive power compensation level is high, the collapse point voltage may even reach values higher than 0.9 pu. Given that the operating voltage of the system is usually in the range of 0.95–1.05 pu, if the compensation level is low, then there will be a large distance between the operating range and the collapse point. Under these conditions, the probability of long-term voltage instability is very low. But if the level of

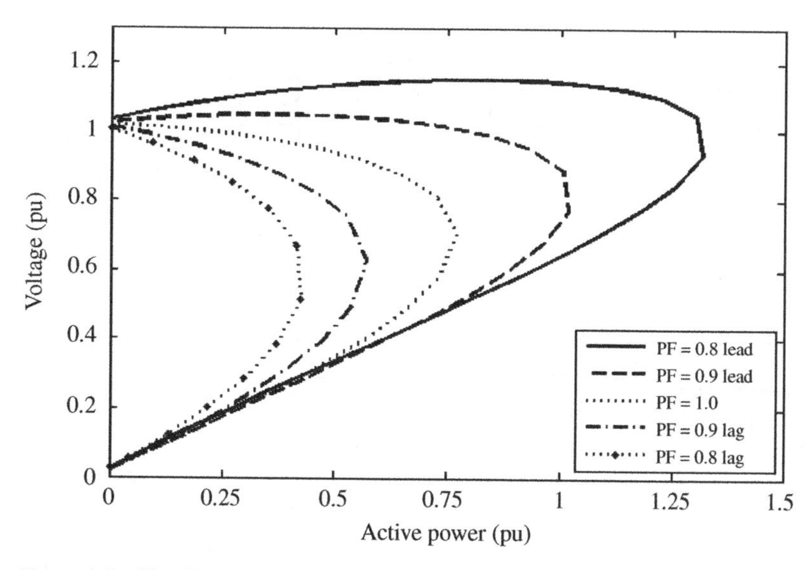

Figure 1.8 The PV curves with different power factors of load.

compensation is high, the operating range is close to the collapse point. Hence, it is found that the long-term voltage instability is the problem of power systems that have been highly compensated.

1.2.5 The Importance of Timely Emergency Measures

The Figure 1.7 illustrates that the inability of LTC in maintaining voltage appears when it is not possible to transfer the needed power to a bus whose voltage is controlled. Therefore, as this will be shown in future chapters, the basis of all voltage instability preventing methods is increasing the maximum loadability level of the power system. This increase is usually done by measures such as increasing the voltage of the generators, connecting the capacitors, and removing the reactors. If increasing the loadability level be not sufficient, the last measure is to reduce the system load. The important point is that the above measures can be effective when applied in a timely manner. The later the measures, the more measures must be taken to maintain voltage stability. To explain this, let us assume that there is another capacitor with susceptance 0.20 pu in bus 3 that connects when needed (Figure 1.9). By connecting this capacitor, it is possible to maintain the voltage stability of the system. Because, as shown in Figure 1.10, the maximum transferable power of the system when the capacitor is connected and one of the parallel lines is removed becomes more than the demanded power. In this figure, similar to Figure 1.7, the dashed-line curves are the load transient characteristics seen from

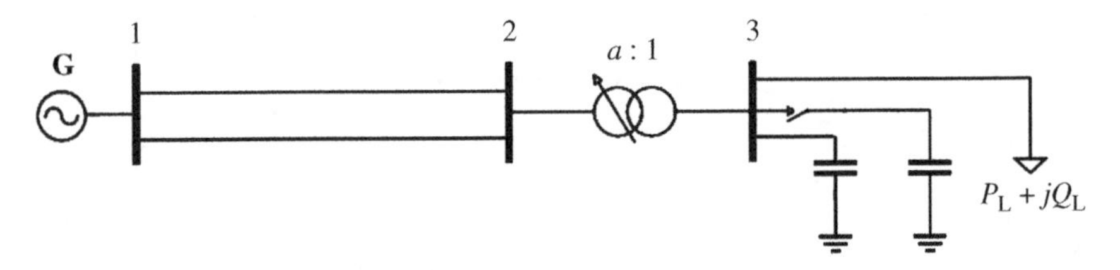

Figure 1.9 Capacitor connection as an emergency measure.

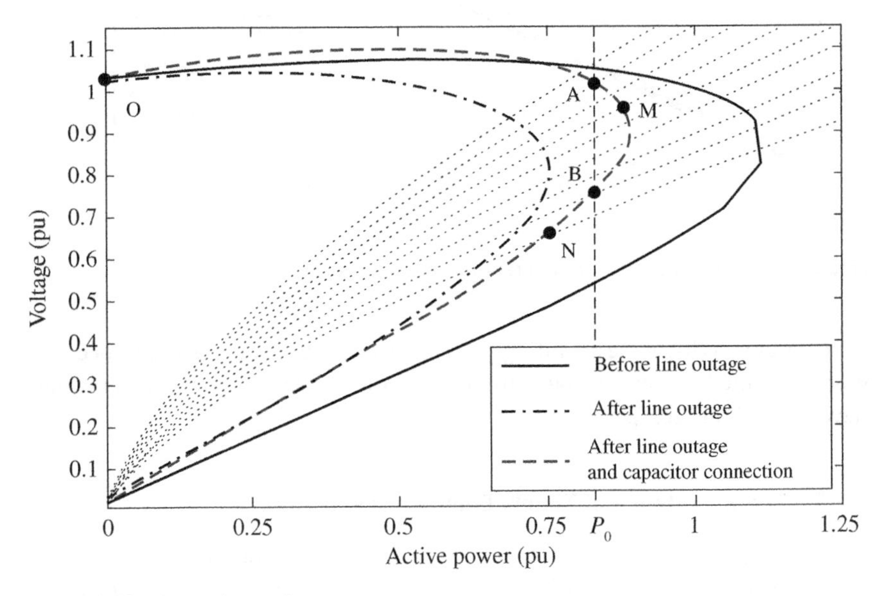

Figure 1.10 Attraction region concept.

the primary side of the transformer, which have changed due to the tap changes. To illustrate the importance of taking timely measures, two scenarios are considered. In the first one, the capacitor is connected when the intersection of the load transient characteristics and the new PV curve (the PV curve after the capacitor connection) is at the point M. In this moment, the received power by load is more than P_0, which implies that the voltage of bus 3 is higher than 1 pu. This causes the tap value to increase and the load characteristics to move toward the point A, which is the equilibrium point after capacitor connection. In this scenario, the capacitor is connected in time and consequently the voltage stability is maintained.

In the second scenario, it is assumed that the capacitor is connected with a longer delay so that the load transient characteristics collide with the new PV curve at the point N. In this moment, the load active power is less than P_0, which

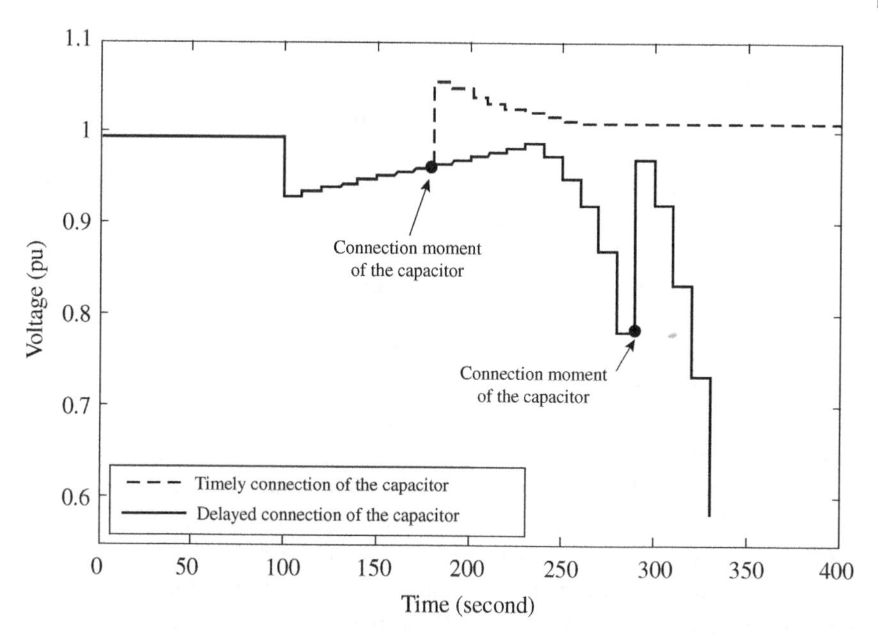

Figure 1.11 Importance of timely measures.

implies that the voltage of bus 3 is less than 1 pu. In this situation, the tap reduction continues and, as shown in Figure 1.10, the load power and voltage decrease more. According to this figure, voltage stability is achieved when the intersection of the load transient characteristic and the PV curve is in the OAB region. This region, the dashed line with gray color, is called the attraction region of the equilibrium point A. The attraction region refers to the region in which, if the operating point is located in it, the trajectory of system changes moves toward the equilibrium point. Figure 1.11 shows the response time of the system for the above two scenarios. It is illustrated that delayed connection of the capacitor cannot prevent voltage instability.

1.3 Short-Term Voltage Instability

Voltage collapses due to the short-term voltage instability that occurs rapidly and during a few seconds. The reason of this instability is induction motors stalling. The induction motors stalling causes a fast increment in the received reactive power of these motors and in consequence an unacceptable voltage drop at its input terminal.

1.3.1 The Process of Induction Motors Stalling

An induction motor stalls when it fails to produce the required torque to its mechanical load. Impossibility of supplying torque means impossibility of transferring the required active power to the motor output. The power transfer limitation is due to both system impedance and motor impedance. Increasing in the system impedance, which originates from occurring a fault in it such as outage of a line, reduces the input voltage of the motor. This voltage drop initiates a process that can lead to stalling of motor. In Figure 1.12, the process of induction motors stalling due to voltage drop is simply analyzed. In this figure, with different voltage levels, the torque-slip characteristic of the induction motor and the motor load characteristic which is assumed to be constant torque (T_L) is shown. C_0^- and S_0^- are in respect the torque-slip characteristic and the slip value of the induction motor before voltage drop. The value of this slip depends on the intersection point of the motor torque-slip and load characteristics. After a fault occurrence in the power system, the voltage drops and the motor torque-slip characteristic changes to C_0^+. Since the speed and consequently the motor slip cannot change instantly, immediately after the voltage drops, the output torque of the motor becomes T_0^+, which is the torque related to the slip S_0^- on the curve C_0^+. Now, since the produced torque by motor is less than the load torque, the motor speed decelerates and its slip increases. As the slip increases to S_1, the motor current increases and its supplying voltage drops more. This voltage drop changes the motor torque-speed characteristic to C_1. Again, it is observed that the motor torque related

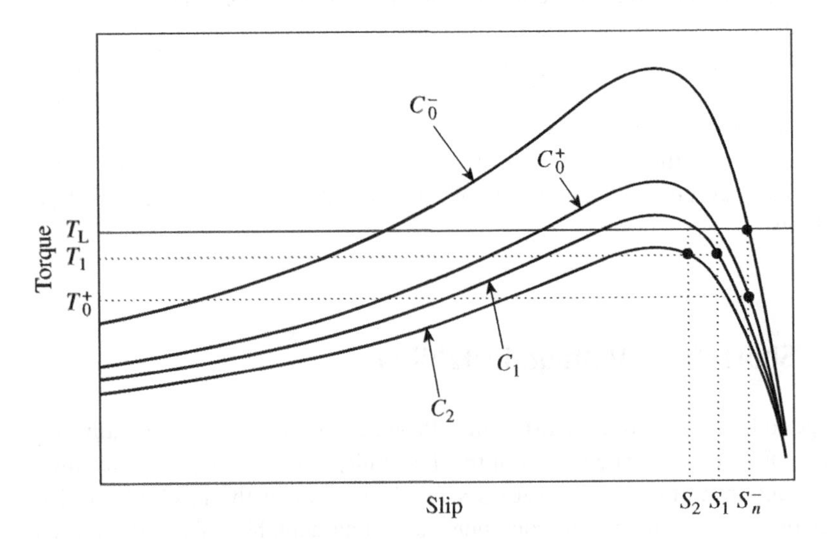

Figure 1.12 The process of induction motors stalling.

to the slip S_1 on the curve C_1 (i.e. T_1) is less than the load torque, and therefore the motor slip will increase more to S_2. With this slip increment and consequently voltage drop, the torque-speed characteristic of the motor re-changes to C_2. As can be seen, the maximum producible torque by the motor on the characteristic curve C_2 is less than the load torque, i.e. it is no longer possible to reach a stable operating point, and increasing the slip will eventually cause the motor to stall and the voltage to collapse.

What was mentioned illustrates the process of induction motors stalling. But in many cases after a fault occurrence, depending on the motor load amount and the system conditions, the motor slip and voltage reach to stable points. To achieve a stable point, the amount of voltage drop when the motor current is increasing is very determinant. Of course, the higher the level the motor loading, the more likely of its stalling. Accurate evaluation of short-term voltage stability requires dynamic analysis and calculation of time response of variables, although sometimes static analysis can be used [4].

1.3.2 Dynamic Analysis

To obtain the time response of variables, dynamic modeling of the induction motor is necessary. Here, the simple system of Figure 1.13 and the third-order model of the induction motor are used for simulation. The motor specifications are as shown in Table 1.1. Accurate analysis of power system stability requires dynamic model of all devices, including generators, their controllers, and limiters. However, some simplifications are considered in the models, depending on the type of study considered. In this simulation, since the focus is on analyzing the stability of the induction motor, similar to Figure 1.2, a simple constant voltage source is used to model the (Infinite bus) generator. Chapters 2 and 3 describe the dynamic models required for various voltage stability studies.

It is assumed that the mechanical load of the motor is of the constant torque type. Similar results are obtained for variable torque loads. The simulated fault is a three-phase short circuit that occurs at the moment $t = 1$ second at the receiving end of one of the parallel lines and is cleared by removing that line. Figures 1.14–1.17 show the motor voltage response for different values of load torques and fault clearing times (t_{cl}). Note that, when t_{cl} is not more than 0.1 second,

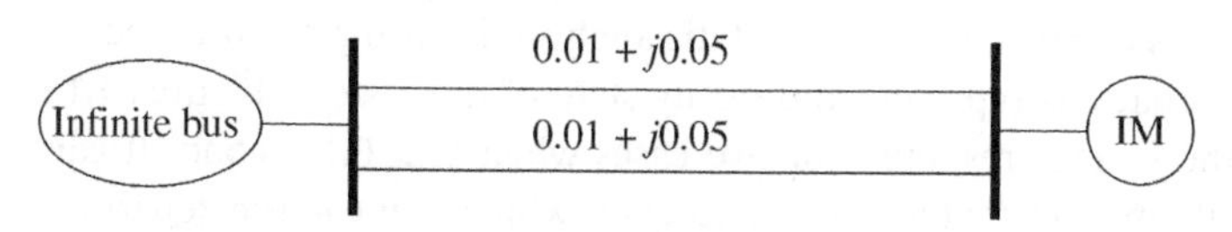

Figure 1.13 A simple system.

Table 1.1 Characteristics of the induction motor.

Parameter	Symbol	Value
Frequency	F	50 Hz
Inertia constant	H	2 s
Stator resistance	R_1	0.0040 pu
Rotor resistance	R_2	0.0040 pu
Stator reactance	X_1	0.0435 pu
Rotor reactance	X_2	0.0392 pu
Mutual reactance	X_m	2.9745 pu

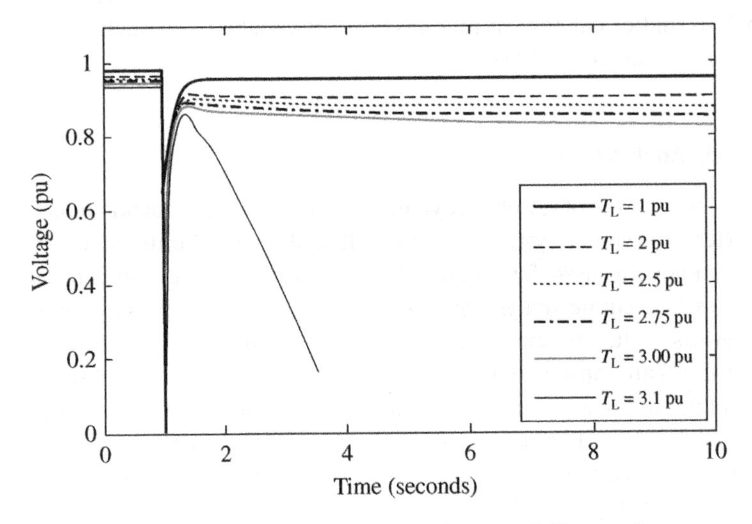

Figure 1.14 The motor voltage response when t_{cl} = 0.05 seconds.

until $T_L = 3$ pu, voltage instability and collapse will not occur. Nondependence on fault clearing time (until the fault clearing time of 0.1 second) verifies that the instability occurs due to the absence of a steady-state operating point (not exiting from the attraction region). In other words, in the steady state, it is not possible to transfer the required active power to the motor output to produce some torque more than 3 pu. This can be illustrated by static analysis, i.e. using algebraic equations. Algebraic equations represent the steady state of the system. Figures 1.16 and 1.17 show the voltage response for two cases when $t_{cl} > 0.1$ second. It can be seen that instability occurs at lower load torques, which is due to the departure from the attraction region.

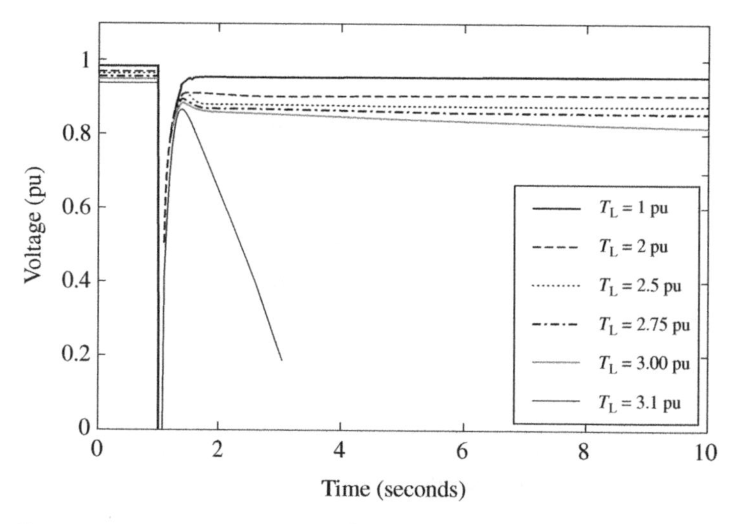

Figure 1.15 The motor voltage response when t_{cl} = 0.10 seconds.

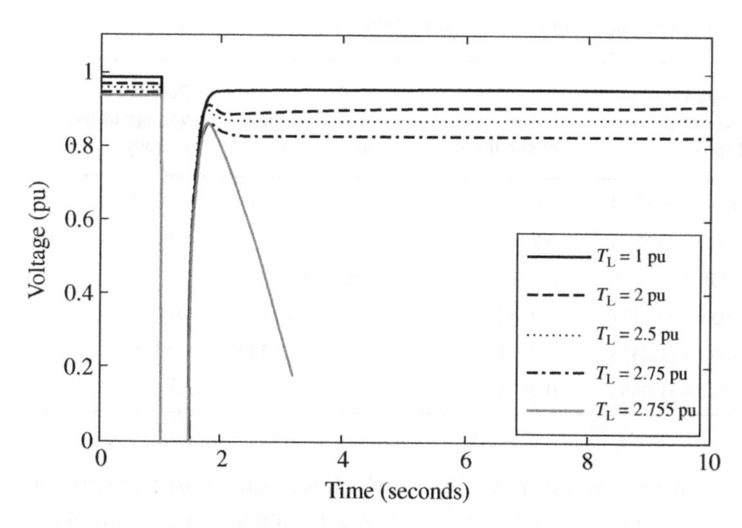

Figure 1.16 The motor voltage response when t_{cl} = 0.25 seconds.

1.3.3 Static Analysis

In this analysis, the differential equations of induction motor are replaced by its equilibrium equations. Table 1.2 shows the results of static analysis. It can be seen that the divergence of the calculations, which means a steady-state operating point

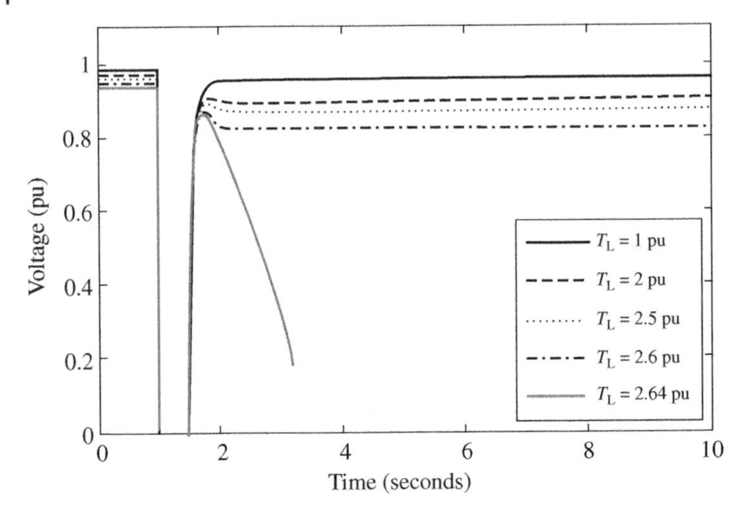

Figure 1.17 The motor voltage response when t_{cl} = 0.50 seconds.

Table 1.2 The input voltage and power values of the motor in steady state.

Torque (pu)	Pre-fault input power of the IM (pu)	Pre-fault voltage of the IM (pu)	Post-fault input power of IM (pu)	Post-fault voltage of the IM (pu)
1	$1.0558 + j0.3790$	0.9792	$1.0875 + j0.3980$	0.9552
2	$2.2588 + j0.6892$	0.9637	$2.3266 + j0.7581$	0.9059
2.5	$2.8235 + j0.8615$	0.9557	$2.9082 + j0.9907$	0.8774
2.75	$3.1058 + j0.9477$	0.9542	$3.1990 + j1.1183$	0.8533
3.00	$3.3882 + j1.0338$	0.9533	$3.4898 + j1.2406$	0.8185
3.10	$3.5011 + j1.0683$	0.9525	Calculation divergence	

is not available longer, occurs at $T_L = 3.10$ pu, which is exactly consistent with the results obtained in Figures 1.14 and 1.15. In the second and fourth columns of the table, the steady-state input powers to the motor are given. These powers are related to the states before and after the line outage. It can be seen that the active power input to the motor does not change much, but the changes in reactive power, especially with increasing T_L, are quite significant. Also in this table, the steady-state values of the input voltage to the motor in both before and after line outage are presented. It is shown that with increasing T_L, the line outage leads to more voltage drop.

1.3.4 The Relationship Between Short-Term Voltage Instability and Loadability Limit

According to Table 1.2, at the beginning of voltage instability, the power received by the motor is 3.4898 + j1.2406 pu. In order to investigate the distance between this power and the maximum power that the system can deliver to the motor, the PV curve at the bus 2 is plotted (Figure 1.18). The power factor used to plot this PV curve is equal to the power factor of the motor at this load level, which is equal to 0.943 lag. It is shown that short-term voltage instability occurs before reaching the maximum transferable power (i.e. 4 pu). The reason of this can be understood from the steady-state equivalent circuit of the system and the induction motor that are shown in Figure 1.19. The output active power of the motor, which provides the needed torque for the mechanical load, is the power that reaches the variable resistance $R_2(1-s)/s$. Due to the impedances existed between this resistor and

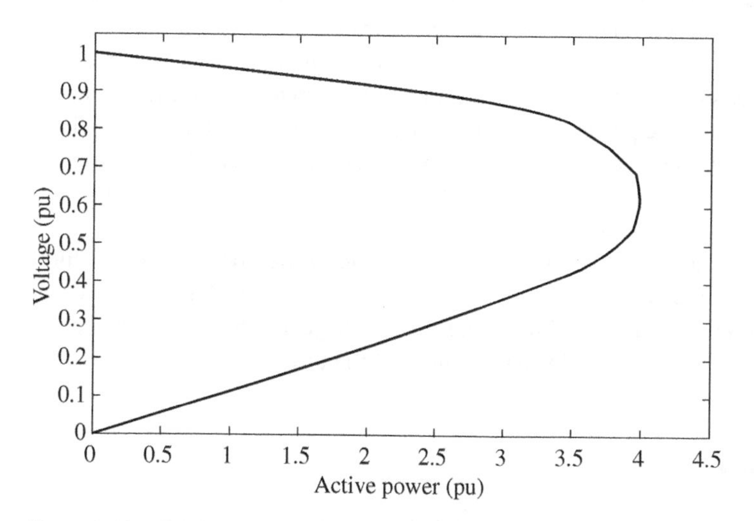

Figure 1.18 The PV curve seen from the input of induction motor.

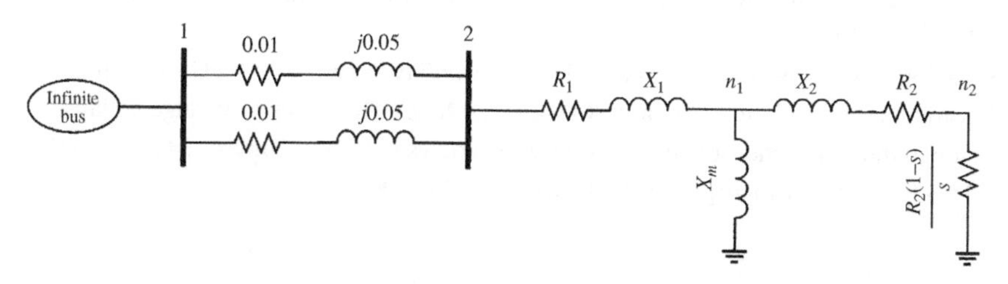

Figure 1.19 The steady-state equivalent circuit of the system and the induction motor.

the bus 2, the maximum transferable power to this resistor is reached before reaching the maximum transferable power to the bus 2. The same is true of long-term voltage instability, when the distribution system impedance does not allow the LTCs of distribution transformers to recover voltage and power. This issue is discussed in Chapter 4. It should be noted that the power transmission limitation can cause the system to have no steady-state operating point, but as shown, instability can occur even when a steady-state operating point is available.

1.4 Summary

This chapter described how long- and short-term voltage instabilities occur. Long-term voltage instability is caused by devices that create constant power characteristic, such as LTC transformers and TCLs. The necessary condition for long-term voltage stability is that the power system should be able to deliver the demanded power by these devices. Hence, distance to the maximum loadability limit of power systems can be a good criterion for long-term stability margin. After occurrence of a contingency such as outage of a generator or a line, measures for increasing the maximum deliverable power of system must be taken in a timely manner. These measures, which include, for example, connecting capacitors and increasing the voltage of generators, if applied with a delay will not have much effect on the prevention of voltage instability.

Short-term voltage instability occurs mainly due to the presence of induction motors. Speed reduction and stalling of these motors cause a sudden voltage drop. Unlike long-term voltage instability, short-term voltage instability can occur at an operating point far from the loadability limit of the power system.

References

1. Van Cutsem, T. and Vournas, C. (1988). *Voltage Stability of Electric Power Systems*. Norwell, MA: Kluwer Academic Publishers.
2. Grainger, J.J. and Stevenson, W.D. (1994). *Power System Analysis*. New York: McGraw-Hill.
3. Taylor, C.W. (1993). *Power System Voltage Stability*. New York: McGraw-Hill.
4. Karbalaei, F., Kalantar, M., and Kazemi, A. (2008). Diagnosis of voltage collapse due to induction motor stalling using static analysis. *Energy Convers. Manag.* 49 (2): 151–156. https://doi.org/10.1016/j.enconman.2007.06.024.

2

Loads and Load Tap Changer (LTC) Transformer Modeling

2.1 Introduction

If model-based analysis of voltage stability is considered, it is necessary to carefully model the system devices, especially devices that play an important role in the occurrence of voltage instability. As mentioned in Chapter 1, load characteristics play an important role in the occurrence of voltage instability. The load here refers to a set of consumers seen from a bus of power system. Figure 2.1 shows a situation in which the goal is to model the aggregate load seen from a transmission bus.

In such a modeling, just the transmission network is modeled in full details, and all downstream devices of the transmission bus are represented by an aggregate load model. The purpose of this modeling is to simulate how active and reactive powers received from the transmission bus change due to its voltage changes. These changes depend on the characteristics of consumers and devices installed at the downstream network of the transmission bus. The used model can be static or dynamic. The dynamic load model, which consists of both algebraic and differential equations, in addition to the steady-state changes of active and reactive powers due to voltage variations, illustrates the transient changes of these powers. The static model, which consists of algebraic equations, illustrates the relationship between the power consumption of static loads and their voltage. Unlike the dynamic loads, following a voltage change, the power consumption of static loads does not have a significant transient period and quickly reaches a steady-state value. Dryers, kitchen ranges, incandescent lamps, and electronic equipment are examples of static loads [1, 2]. When the static load model is used for the dynamic loads, the aim is to model the steady-state behavior of the load, not the changes during the transient period. Dynamic analysis of voltage stability, which is done for various purposes such as finding trajectory of voltage changes, assessing impact of timely control measures, and detecting possibility of leaving attraction region, requires the use of dynamic load models.

Voltage Stability in Electrical Power Systems: Concepts, Assessment, and Methods for Improvement, First Edition. Farid Karbalaei, Shahriar Abbasi, and Hamid Reza Shabani.

Published 2023 by John Wiley & Sons, Inc.

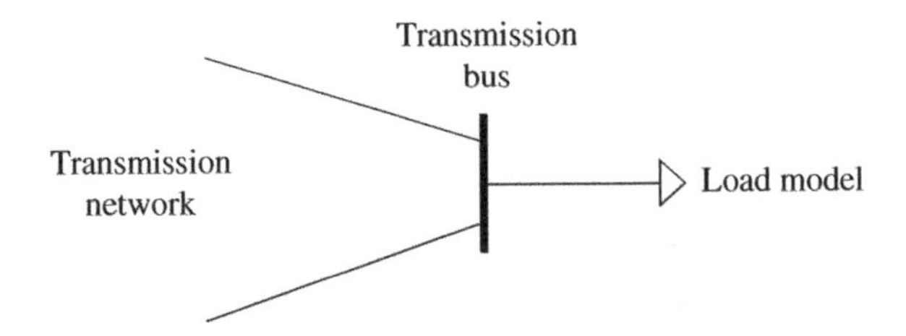

Figure 2.1 Aggregate load model at transmission buses.

Load modeling from the view of transmission buses is commonly used for power flow calculations as well as transient stability analysis. It can also be used to calculate the loadability limit of the transmission network, which is an important indicator in voltage stability assessment. However, accurate analysis of voltage stability is not possible with this model. Because, as shown in Chapter 1, the necessary condition to maintain voltage stability is the ability to transfer the required power to the buses that have steady-state constant power characteristic. The modeling used in Figure 2.1 is capable in predicting voltage instabilities that occur due to the impossibility of transferring the requested power to the transmission bus. This is shown in Chapter 4, and the voltage instability may occur due to the operation of the load tap changer (LTC) of distribution transformers and the impossibility of transferring the required power to these transformers. This instability can only be predicted by considering the model of distribution network.

On the other hand, the above-mentioned modeling that tries to simulate the performance of all downstream devices of transmission bus is reliable only for limited voltage changes because these devices will not have the expected performance when the voltage changes increase. For example, one of the important characteristics of load in voltage stability assessment is the recovery of power consumption after a voltage drop (Figure 1.1). A reason for this characteristic is the operation of transmission and distribution LTC transformers, which causes voltage recovery at buses under their control and, consequently, power recovery. This is the reason of choosing the constant power characteristic for loads in power flow calculations [3]. Even if stable, voltage recovery is possible only if the required values of transformers' taps are within the allowable range of their changes. The greater the voltage change at the transmission bus, the power recovery needs more tap changes that may not be possible always. Accurate analysis of this situation is possible only by separate modeling of LTC transformers.

A more accurate model of the power system is obtained when load modeling is performed at buses with lower voltage, and upstream devices such as LTCs at extra high voltage (EHV)/ high voltage (HV) and HV/medium voltage (MV) transformers are modeled accurately and separately. Figure 2.2 shows an example of this

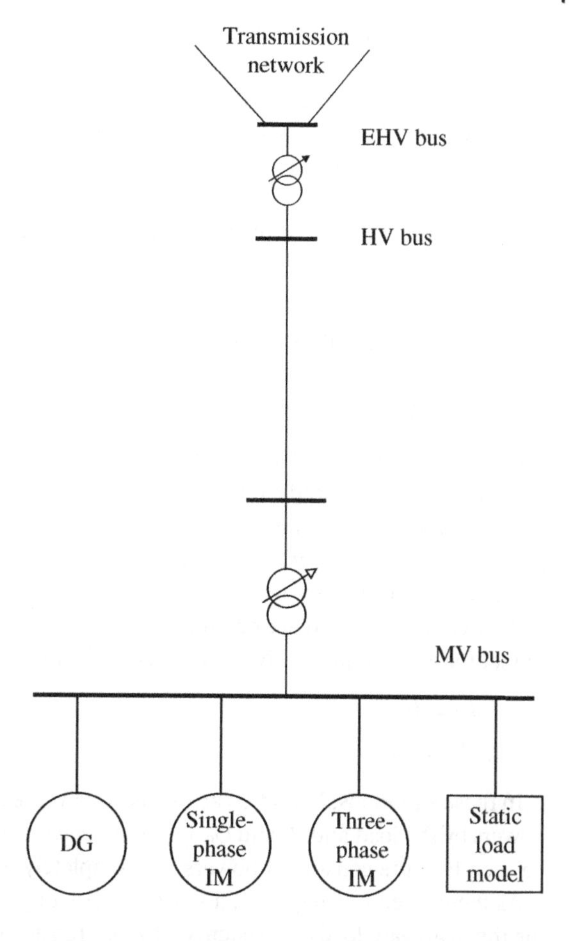

Figure 2.2 Component-based aggregate load model at an MV bus.

modeling type in which the component-based aggregate model is used for loads. In this modeling, some aggregate models are used for three-phase induction motors, single-phase induction motors, static loads, and DG units. In some references, a separate aggregate induction motor model is considered for each type of mechanical loads [3, 4]. Also for electronic devices a separate model can be used [4]. The model used in Figure 2.2 provides an accurate assessment of both long- and short-term voltage stability. However, this model leads to a system with large number of algebraic and differential equations that are very difficult to solve simultaneously. So, proper methods to solve this system of equations easier are required. Solving the equations of distribution and transmission networks separately is one of these methods [5, 6]. Another method could be the separate analysis of fast and slow dynamics [7]. Long-term voltage stability is mainly related to state variables that have slow changes. Therefore, in analysis of changes of these variables, state equations that have fast dynamics can be replaced with their equilibrium equations. For example, the dynamic model of induction motors in Figure 2.2 can be replaced with the static load model. Also, in analyzing the fast dynamics that play an important role in creating short-term voltage instability, the slow state variables can be assumed to be constant. In the second section of this book, this issue is explained in more detail. In the following, the static and dynamic load models and then the LTC transformer model are presented.

2.2 Static Load Models

2.2.1 The Constant Power Model

The constant power model is the most common static model that is more used for the both steady-state studies (Such as power flow calculation) and dynamic studies of the power systems than other static models [8]. Of course, the use of constant power model for stability studies is conservative [3]. Because, in the static constant power model, the transient period of power recovery that can be an opportunity for system controllers to react or take emergency measures is not considered. This model for active and reactive powers is as shown in Eqs. (2.1) and (2.2):

$$P_{\mathrm{d}} = P_0 \tag{2.1}$$

$$Q_{\mathrm{d}} = Q_0 \tag{2.2}$$

In these equations, P_{d} and Q_{d} are, respectively, the consumed active and reactive powers by the load and P_0 and Q_0 are initial powers (before voltage change). It can be seen that the consumed powers are completely independent of voltage.

As mentioned in Chapter 1, a reason of creating constant power characteristic for the aggregate loads is existence of LTC transformers. When the primary side voltage changes, the secondary side voltage and consequently the power of the secondary side the transformer are kept constant using the LTC operation. Hence, the power passing through the transformer is independent of its primary side voltage. Therefore, the load characteristic seen from the buses in which the downstream network has automatic voltage control devices such as LTC can be constant power. Of course, creation of constant power characteristic, depending on the speed of LTC response, occurs during a process of several tens of seconds to a few minutes. Another reason of creating constant power characteristic is the presence of thermal loads that are controlled by thermostat. These loads stay in the circuit longer when their voltages drop. Therefore, the total impedance created by a set of loads decreases with voltage drop. As a result, when the voltage drops, the steady-state power consumption of these loads does not change much. In view of the above, the constant power characteristic created by the operation of LTCs as well as the thermostatic loads is in fact the steady-state characteristic of a dynamic load. In addition to the above-mentioned cases, the active power consumption of some electronic devices such as monitors and computers also has constant power characteristics, which of course are not notable in the aggregate load model [9].

2.2.2 The Polynomial and Exponential Models

Another static load model is the polynomial model in the form of Eqs. (2.3) and (2.4). This model, known as the ZIP model, consists of three parts: constant impedance, constant current, and constant power.

$$P_d = P_0 \left[Z_p \left(\frac{V}{V_0} \right)^2 + I_p \left(\frac{V}{V_0} \right) + P_p \right] \tag{2.3}$$

$$Q_d = Q_0 \left[Z_q \left(\frac{V}{V_0} \right)^2 + I_q \left(\frac{V}{V_0} \right) + P_q \right] \tag{2.4}$$

where

$$Z_p + I_p + P_p = Z_q + I_q + P_q = 1 \tag{2.5}$$

where Z_p, I_p, and P_p are, respectively, the coefficients of the parts of constant impedance, constant current, and constant power of the active load. There is a similar definition for Z_q, I_q, and P_q. The P_0 and Q_0, respectively, denote the load active and reactive powers before the voltage change, and V_0 is the initial voltage (before the change). The P_d and Q_d are the consumed active and reactive powers by the load at the voltage magnitude of V. If $I_q = P_q = I_p = P_p = 0$, the load is converted to constant impedance load. In the same way, it is possible to create constant current and constant power types.

Another static load model is the exponential model whose equations (2.6) and (2.7) are as follows:

$$P_d = P_0 \left(\frac{V}{V_0} \right)^{np} \tag{2.6}$$

$$Q_d = Q_0 \left(\frac{V}{V_0} \right)^{nq} \tag{2.7}$$

where np and nq are exponents that indicate the load type. The P_0 and Q_0 are the initial active and reactive powers with the initial voltage magnitude V_0 (before the change). The value of nq is usually larger than the value of np because the reactive power is more dependent on the voltage. By changing the values of np and nq, it is possible to create constant power, constant current, and constant impedance characteristics.

Both polynomial and exponential models are used for modeling static consumers as well as modeling the steady-state characteristic of dynamic loads. When these models are used to model the steady-state characteristics of the induction motors, their parameters change by changing the nominal values of the motors as well as changing their mechanical load value [10]. The changes of the exponential model parameters are less than those of the polynomial model parameters [11]. Both models can be used to model the aggregate loads at high voltage levels. The aggregate load model can be obtained by the weighted integrating the model of all components or by a measurement method. The first method is easier to implement when using a polynomial model for different consumers [12, 13]. In a number of power system buses whose downstream network has a large number of LTC transformers, np and nq may have negative values in the aggregate load model. This is owing to keeping the power of consumers constant by LTCs when the bus voltage changes. In this situation, increasing the bus voltage reduces the current in the downstream network and, consequently, reduces the losses. This decreases the power received from the bus.

2.3 Dynamic Load Models

The dynamic load models must be able to show the transient behavior of loads in addition to the steady-state characteristics. What is important about the behavior of loads in voltage stability studies is the changes of their power consumption when voltage drops. Accurate modeling of a device that causes power consumption to be recovered to its previous values after an initial reduction (Figure 1.1) is of great importance. The most important of these devices are induction motors, LTC transformers, and thermostatically controlled thermal loads.

When according to Figure 2.2, the modeling is done from the point of view of an MV bus, it is possible to model these devices separately and accurately. But, if the modeling is done from the point of view of a transmission bus, we need a model that represents as much as possible the aggregate effect of the set of loads and the downstream devices of this bus. However, for the reasons mentioned in Section 2.1, this model at most is able to predict the approximate changes in power received from the transmission bus even during limited voltage changes. Given that the power recovery is done almost exponentially [14], the presented model is also based on the simulation of the same behavior, which is called the Exponential Recovery Model [14–16]. First, this model is described and then an overview of the induction motor model is presented.

2.3.1 Exponential Recovery Model

As mentioned, the important characteristic of the dynamic loads is the recovery of their power consumption after a voltage drop. In this characteristic, which is again shown for the active power in Figure 2.3, the power recovery is done almost exponentially. There is a similar response for reactive power. In this figure, P_d is the load power consumption at each instant. It is assumed that at the instant t_0, the load voltage jumps from V_0 to V_+. ΔP_t indicates the initial decrease in power consumption and ΔP_s is its steady-state change. The length of the transient period

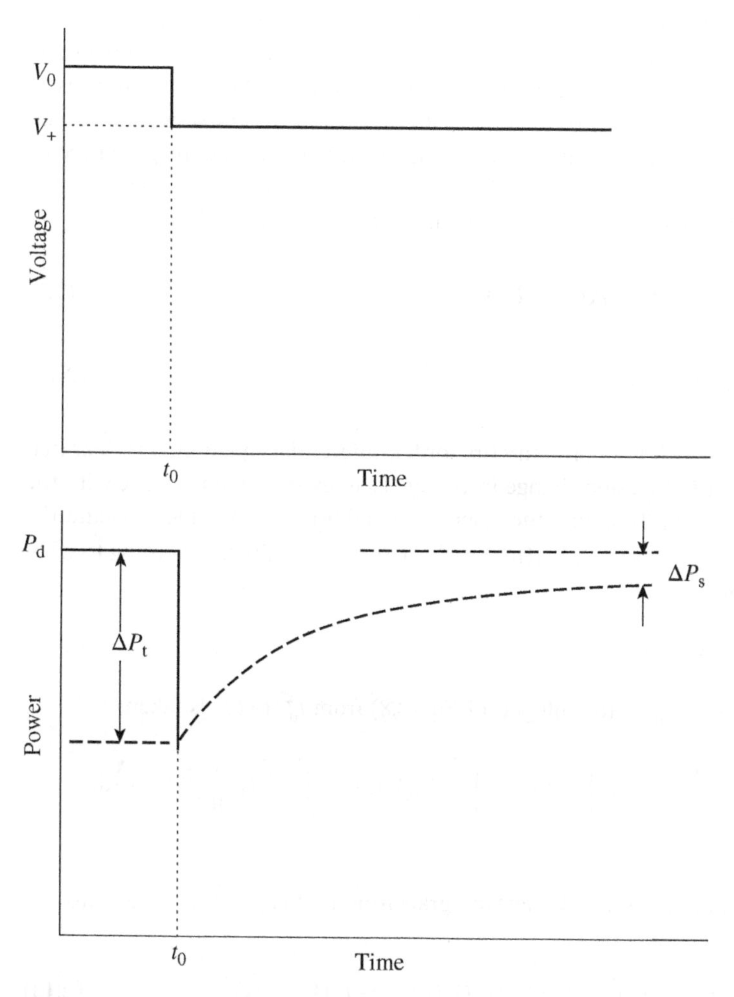

Figure 2.3 Exponential recovery of power consumption after an initial reduction.

(or the recovery constant time) depends on the predominant devices in the load. If the power recovery is due to the presence of induction motors, the recovery is done in a fraction of a second. However, when power recovery is due to the operation of LTCs or thermostatically controlled loads, the recovery period may last from a few seconds to a few minutes.

To achieve the characteristic of Figure 2.3 in [15], a first-order differential equation as (2.8) is presented.

$$T_p \frac{dP_d}{dt} + P_d = P_s(V) + T_p \frac{d}{dt}(P_t(V)) \tag{2.8}$$

By equating the derivative terms to zero, it is observed that P_d is equal to $P_s(V)$ in the steady state. It will also be shown that $P_t(V)$ specifies the value of initial jump of P_d. To solve (2.8), its equivalent which is in the form of (2.9) and (2.10) is used. The equivalence of these two equations with (2.8) can be simply shown by extracting x_p from (2.10) and placing it in (2.9). x_p is a state variable that the power recovery process is done by changing it. Here also it can be seen that, P_d which is the instantaneous value of power will be equal to $P_s(V)$ in the steady state.

$$\frac{dx_p}{dt} = -\frac{1}{T_p}x_p + P_s(V) - P_t(V) \tag{2.9}$$

$$P_d = \frac{1}{T_p}x_p + P_t(V) \tag{2.10}$$

Given (2.9), which is an ordinary first-order differential equation, it is clear that the response of P_d to a step change in voltage is an exponential response with the time constant of T_p. To obtain the response, in addition to the time constant, the initial and final values of P_d $[P_d(t_0^+)$ and $P_d(\infty)]$ are required. According to what was mentioned:

$$P_d(\infty) = P_s(V^+) \tag{2.11}$$

To calculate $P_d(t_0^+)$, the integral of Eq. (2.8) from t_0^- to t_0^+ is taken.

$$\int_{t_0^-}^{t_0^+} \left(T_p \frac{dP_d}{dt}\right) dt + \int_{t_0^-}^{t_0^+} P_d dt = \int_{t_0^-}^{t_0^+} (P_s(V)) dt + \int_{t_0^-}^{t_0^+} \left(T_p \frac{d}{dt}(P_t(V))\right) dt \tag{2.12}$$

Since P_d and P_s are limited, their integrals from t_0^- to t_0^+ are equal to zero. So, from Eq. (2.12):

$$T_p(P_d(t_0^+) - P_d(t_0^-)) = T_p(P_t(V(t_0^+)) - P_t(V(t_0^-))) \tag{2.13}$$

Therefore,

$$P_d(t_0^+) = P_d(t_0^-) + P_t(V(t_0^+)) - P_t(V(t_0^-)) = P_d(t_0^-) + P_t(V_+) - P_t(V_0) \tag{2.14}$$

Furthermore, this is obvious that

$$P_d(t_0^-) = P_s(V_0) \tag{2.15}$$

Because, before the instant t_0, the power has reached its steady-state value. Now, according to the global equation of (2.16):

$$P_d(t) = P_d(\infty) + (P_d(t_0^+) - P_d(\infty))e^{\frac{-(t-t_0)}{T_P}} \quad t > t_0 \tag{2.16}$$

The response $P_d(t)$ will be calculated as

$$P_d(t) = P_s(V_+) + (P_s(V_0) + P_t(V_+) - P_t(V_0) - P_s(V_+))e^{\frac{-(t-t_0)}{T_P}} \quad t > t_0 \tag{2.17}$$

From Figure 2.3 and Eq. (2.13):

$$\Delta P_t \triangleq P_d(t_0^-) - P_d(t_0^+) = P_t(V_0) - P_t(V_+) \tag{2.18}$$

And,

$$\Delta P_s \triangleq P_d(t_0^-) - P_d(\infty) = P_s(V_0) - P_s(V_+) \tag{2.19}$$

It can be seen that the value of initial jump as well as the steady state change of power is independently determined by the functions P_t and P_s. P_t and P_s are, respectively, called the transient and steady-state functions of active load power. The following exponential functions are usually used for P_s and P_t.

$$P_s(V) = P_0\left(\frac{V}{V_0}\right)^{\alpha_s} \tag{2.20}$$

$$P_t(V) = P_0\left(\frac{V}{V_0}\right)^{\alpha_t} \tag{2.21}$$

In these equations, V_0 and P_0 are the initial voltage magnitude (before the voltage change) and the active power at that voltage, respectively. α_s and α_t are the exponents related to the steady state and transient active powers, respectively. There are similar equations for reactive power, as Eqs. (2.22) and (2.23). The values of α_t and β_t are usually larger than α_s and β_s because dependency of the load active and reactive powers on voltage is greater in the transient period than in the steady state.

$$Q_s(V) = Q_0\left(\frac{V}{V_0}\right)^{\beta_s} \tag{2.22}$$

$$Q_t(V) = Q_0\left(\frac{V}{V_0}\right)^{\beta_t} \tag{2.23}$$

The exponential recovery model of Eqs. (2.9) and (2.10) is called the additive model. Because, in this model, in the instantaneous value of active and reactive powers, the state variable is added to the transient state function. The complete equations of this model, which include algebraic and differential equations for active and reactive powers, are given here again. Given that there is a constant (voltage independent) term in the instantaneous value of powers, the Jacobin matrix may be singular in the transient period and the power flow equations may diverge. This situation is unrealistic and causes inaccessibility of the system response [7].

$$\frac{dx_p}{dt} = -\frac{1}{T_p}x_p + P_0\left(\frac{V}{V_0}\right)^{\alpha_s} - P_0\left(\frac{V}{V_0}\right)^{\alpha_t} \tag{2.24}$$

$$P_d = \frac{1}{T_p}x_p + P_0\left(\frac{V}{V_0}\right)^{\alpha_t} \tag{2.25}$$

$$\frac{dx_q}{dt} = -\frac{1}{T_q}x_q + Q_0\left(\frac{V}{V_0}\right)^{\beta_s} - Q_0\left(\frac{V}{V_0}\right)^{\beta_t} \tag{2.26}$$

$$Q_d = \frac{1}{T_q}x_q + Q_0\left(\frac{V}{V_0}\right)^{\beta_t} \tag{2.27}$$

There is another exponential recovery model called the multiplicative model, in which state variables are multiplied by transient state functions. This improve the shortcoming of the additive model. The equations of this model are as follows [16]:

$$T_p\frac{dx_p}{dt} = -x_p\left(\frac{V}{V_0}\right)^{\alpha_t} + P_0\left(\frac{V}{V_0}\right)^{\alpha_s} \tag{2.28}$$

$$P_d = x_p\left(\frac{V}{V_0}\right)^{\alpha_t} \tag{2.29}$$

$$T_q\frac{dx_q}{dt} = -x_q\left(\frac{V}{V_0}\right)^{\beta_t} + Q_0\left(\frac{V}{V_0}\right)^{\beta_s} \tag{2.30}$$

$$Q_d = x_q\left(\frac{V}{V_0}\right)^{\beta_t} \tag{2.31}$$

In [17], a comparison between additive and multiplicative models is done. It is shown that there is no significant difference between these two models in terms of accuracy and ability in identifying parameters.

The constant time values T_p and T_q depend on the devices and consumers available in the aggregate load. If power recovery is done only by thermostatically controlled thermal loads, the recovery will be done only for active power with the time constant T_p from a few tens of seconds to a few minutes. The power recovery time constant in induction motors is from a fraction of second to maximum of a few seconds. When the reason of power recovery is the presence of LTC transformers, the recovery will be done for both active and reactive powers with a time constant that is in the range of a few seconds to a few tens of seconds.

The values of the load parameters are different in different seasons as well as at different times of a day. For example, if the power recovery is done by thermostatically controlled thermal loads, the value of α_s in winter is less than its value in summer. Also, the α_s value is smaller at night than during the day and may even reach a value close to zero [15]. When power recovery is done by LTC, the value of α_s and especially β_s may be negative. This situation appears when the operation of LTC causes a significant portion of downstream network consumers to attain the characteristic of steady-state constant power. In these conditions, with increasing voltage of the supply bus, the current passing through the downstream lines decreases, which reduces the network losses and consequently the total power received from this bus. In other words, increasing the voltage reduces the received power.

2.3.2 Induction Motor Model

Another important and effective consumers in creating voltage instability are the induction motors. Speed reduction and stalling of these motors lead to a rapid voltage collapse. Hence, they are the main cause of short-term voltage instability. In many dynamic studies, the induction motor model is used to model the load. In the United States, about 30% of power utilities use a composite model for dynamic studies, which includes a dynamic induction model with a polynomial static model [8]. Ref. [18] shows that the instabilities of 1996 and 2000 in the US power system were not predictable using a static model for the load. On the contrary, using the induction motor model with a static model, the results of the happened disturbances can be well simulated. The important point is the considered ratio for the induction motor to the total load. Simulations demonstrate that in the studied system, the best result is obtained when the portions of the induction motor and the static load from the total load are, respectively, 20% and 80%. Of course, this is for the case where the load is modeled at the HV buses. If the load model is transferred to the low voltage (LV) buses, the required portion of the induction motor from the total load will increase to 50%. The following is the equation of the induction motor model. In [19], the method of the determination of the induction motor model parameters is explained.

In an induction machine, the stator has three coils with spatially 120 electrical degree different from each other. The relation between electrical and mechanical angles is given as Eq. (2.32):

$$\theta_e = \frac{P}{2}\theta_m \tag{2.32}$$

where θ_e and θ_m, respectively, are electrical and mechanical angles, and P is the number of poles. There is a similar relation between electrical and mechanical speeds. Using the electrical angle instead of the mechanical angle, the equations of the machine become independent of the number of poles. The rotor structure can be either wound or squirrel cage. In the former, the rotor has the normal three phase windings whose terminals can be connected to an external circuit. But in the latter, the rotor contains longitudinal conductive bars set into grooves and connected at both ends by shorting rings. Since the currents induced in these bars provide a field with the same number of field poles as the stator windings, the rotor bars can be modeled as the balanced three-phase windings.

Due to the dependence of machine inductances on the position of the rotor, it is common to transfer machine equations to the dq0 rotating frame, that is, to convert variables from abc to dq0. Here it is assumed that the axis q is 90° lead of the axis d. A feature of the dq0 transformation is that it converts the balanced three-phase variables into fixed variables. For example, if the three-phase currents are assumed as

$$\begin{aligned} i_a &= I_m \cos(\omega_s t + \alpha) \\ i_b &= I_m \cos(\omega_s t + \alpha - 120^\circ) \\ i_c &= I_m \cos(\omega_s t + \alpha + 120^\circ) \end{aligned} \tag{2.33}$$

Having the bellow dq0 transformation matrix,

$$T_{dq0} \triangleq \begin{bmatrix} \cos(\omega_s t) & \cos(\omega_s t - 120^\circ) & \cos(\omega_s t + 120^\circ) \\ \sin(\omega_s t) & \sin(\omega_s t - 120^\circ) & \sin(\omega_s t + 120^\circ) \\ ½ & ½ & ½ \end{bmatrix} \tag{2.34}$$

The dq0 currents are obtained as:

$$\begin{bmatrix} i_d \\ i_q \\ i_0 \end{bmatrix} = T_{dq0} \begin{bmatrix} i_a \\ i_b \\ i_c \end{bmatrix} = \begin{bmatrix} I_m \cos(\alpha) \\ I_m \sin(\alpha) \\ 0 \end{bmatrix} \tag{2.35}$$

It can be shown that the currents i_d and i_q can be assumed to be phasor components related to sinusoidal currents. Then, the current can be written as

$$I = i_d + j i_q = I_m \angle \alpha \tag{2.36}$$

Therefore, it can be supposed that the phasor variables are obtained using the dq0 transformation on the abc variables.

In connecting machines to grid, a common rotating frame is required to write the equations between the machines' output voltage and current and the grid variables based on it. The speed of this common frame is equal to the synchronous speed, and its transformation matrix is similar to Eq. (2.34). In stability studies, it is assumed that the frequency of variables in the transient period is equal to the steady-state frequency [20, 21]. With this assumption, it is possible to use the phasor equations for the dynamic equations of the grid. The transformation matrix of the common rotating frame can be used to convert variables in time domain into phasor one.

In an induction machine, the rotor speed depends on the load value. Hence, it is preferred that the speed of the rotating frame for the induction machine be equal to the synchronous speed instead of the rotor speed. Therefore, its transformation matrix can be the same as the transformation matrix of the common rotating frame. The model presented here for the induction motor is a third-order model that includes the dynamics of the rotor fluxes as well as the swing equation. This is the model used for dynamic studies in [18, 21–23]. In this model, the dynamics of stator fluxes, which are very faster than the dynamics of rotor fluxes and quickly reach their steady state, are ignored. Differential and algebraic equations of this model are as follows:

$$\frac{dv'_d}{dt} = -\frac{1}{T'_0}\left[v'_d + (X_s - X'_s)i_{qs}\right] + s\omega_s v'_q \tag{2.37}$$

$$\frac{dv'_q}{dt} = -\frac{1}{T'_0}\left[v'_q - (X_s - X'_s)i_{ds}\right] - s\omega_s v'_d \tag{2.38}$$

$$\frac{d\omega_r}{dt} = \frac{\omega_s}{2H}(T_e - T_m) \tag{2.39}$$

$$v_{ds} = R_s i_{ds} - X'_s i_{qs} + v'_d \tag{2.40}$$

$$v_{qs} = R_s i_{qs} + X'_s i_{ds} + v'_q \tag{2.41}$$

$$s = \frac{\omega_s - \omega_r}{\omega_s} \tag{2.42}$$

where v'_d and v'_q are, respectively, the voltages of d and q axes behind the transient reactance X'_s. Eqs. (2.40) and (2.41) can be combined and re-written in the following phasor form:

$$v_{ds} + jv_{qs} = (R_s + jX'_s)(i_{ds} + ji_{qs}) + \left(v'_d + jv'_q\right) \tag{2.43}$$

From the above equation, the transient equivalent circuit of the induction motor is obtained as the following Figure 2.4:

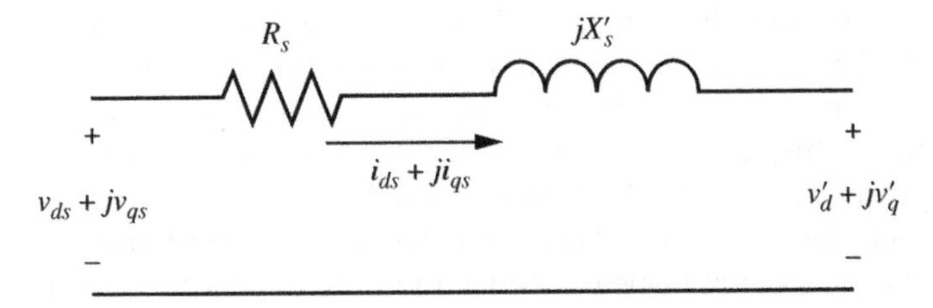

Figure 2.4 Transient equivalent circuit of the induction motor.

wherein R_s is the stator ohmic resistor and i_{ds} and i_{qs} and v_{ds} and v_{qs} are, respectively, the currents and the voltages of the d and q axes of the stator. In the equations of this model, s is the slip, H is the inertia constant, T'_0 is the open circuit time constant of the rotor, and X_s is the synchronous reactance. ω_s and ω_r are, respectively, the synchronous and rotor speeds, both in electric radians per second.

T_e is the electromagnetic torque that can be calculated as follows:

$$T_e = v'_d i_{ds} + v'_q i_{qs} \tag{2.44}$$

T_m is the motor load torque, which may be constant or a function of ω_r, as the following typical forms:

$$T_m = T_L \left(\frac{\omega_r}{\omega_s}\right)^m \tag{2.45}$$

or,

$$T_m = T_L \left(A\left(\frac{\omega_r}{\omega_s}\right)^2 + B\left(\frac{\omega_r}{\omega_s}\right) + C\right) \tag{2.46}$$

In (2.37)–(2.46), all variables except ω_r, ω_s, $T'_{0,}$ and H are in per unit. H and T'_0 are in seconds. The Eq. (2.36) shows that phasor representation is based on the peak value of current (I_m). Similar equation is available for voltage. Hence, the base values of voltages and currents are chosen as the peak values of the rated voltage and current.

It can be seen that there are the seven unknowns v'_d, v'_q, v_{ds}, v_{qs}, i_{ds}, i_{qs}, and ω_r in Eqs. (2.37)–(2.41). Therefore, two other equations are needed. These two equations are the power balance equations of the induction motor with the grid. The complex power received by the motor can be calculated as follows:

$$S = (v_{ds} + jv_{qs})(i_{ds} - ji_{qs}) = (v_{ds}i_{ds} + v_{qs}i_{qs}) + j(v_{qs}i_{ds} - v_{ds}i_{qs}) \tag{2.47}$$

The magnitude and angle of the motor terminal voltage are calculated according to Eqs. (2.48) and (2.49), as

$$V_{\mathrm{m}} = \sqrt{v_{\mathrm{ds}}^2 + v_{\mathrm{qs}}^2} \tag{2.48}$$

$$\theta = tg^{-1}\frac{v_{\mathrm{qs}}}{v_{\mathrm{ds}}} \tag{2.49}$$

Assuming that the induction motor is located at the bus i, the power balance equations at this bus can be written as follows:

$$\sum_{k=1}^{n} V_i V_k Y_{ik} \cos\left(\theta_i - \theta_k - \alpha_{ik}\right) = -\left(v_{\mathrm{ds}i} i_{\mathrm{ds}i} + v_{\mathrm{qs}i} i_{\mathrm{qs}i}\right) \tag{2.50}$$

$$\sum_{k=1}^{n} V_i V_k Y_{ik} \sin\left(\theta_i - \theta_k - \alpha_{ik}\right) = -\left(v_{\mathrm{qs}i} i_{\mathrm{ds}i} - v_{\mathrm{ds}i} i_{\mathrm{qs}i}\right) \tag{2.51}$$

The left sides of Eqs. (2.50) and (2.51) are the well-known equations of the injected active and reactive powers at the bus i used in the power flow calculations. θ_i and V_i are the voltage angle and magnitude at the bus i. Y_{ik} and α_{ik} are, respectively, the magnitude and angle of the array (i, k) of the admittance matrix. θ_i and V_i are related to $v_{\mathrm{ds}i}$ and $v_{\mathrm{qs}i}$ as Eqs. (2.48) and (2.49). Now, adding Eqs. (2.50) and (2.51) to the previous obtained Eqs. (2.37)–(2.42), the required seven equations are obtained.

2.4 The LTC Transformers

The LTC transformers are one of the most important devices to recover voltage. Although these devices are used to maintain the desired voltage, if the voltage becomes unstable, the operation of the LTCs will lead to voltage collapse. As stated in Chapter 1, the cause of this instability can be the impossibility of delivering the power that is required to pass through the transformer to regulate the voltage by its LTC.

2.4.1 The LTC Performance

Tap changing in EHV/HV and HV/MV transformers is usually done automatically and under load, which is called LTC. In most cases, the variable turn is placed on the windings of the side with higher voltage. This is to facilitate commutation current between the rings of the windings. Another reason is the higher number of turns on the higher voltage side, which make tap changing smoother [7]. Voltage control is performed in the side with lower short circuit level, so the LTCs usually control the voltage of the lower voltage side.

LTCs are divided into different types. In its mechanical type, the tap is changed by mechanical switches. One of the problems of this type is the required long time for each tap change [24]. Switching arc is another problem of the mechanical LTCs [25]. In another type of the LTCs, the mechanical switches are replaced by solid-state switches (electronic switches) such as GTOs. This replacement can include a part of the mechanical switches (electronically assisted tap changer) or include all mechanical switches (fully electronic tap changer) [26]. The electronic switches alleviate the problem of switching arc. Meanwhile, with these switches, the switching time for each tap change will decrease from a few seconds to a few milliseconds. Main disadvantages of the electronic switches are their high cost and large Ohmic losses.

Usually, in addition to the time delay caused by the operating time of the LTC devices, an intentional delay is added to prevent unnecessary operation of LTCs in the situations of transient voltage variations. The LTCs have two modes of operation: sequential mode and nonsequential mode [7]. In the sequential mode, the time delay of the first tap change is greater than the time delay of the next tap changes. The delay can be constant or dependent on the level of voltage change. In the nonsequential mode, there is no difference between the first tap change and other tap changes. In Chapter 4, it will be illustrated that in order to reduce the number of tap changes and reduce the probability of voltage instability occurrence, the time delay of tap changes in EHV/HV transformers should be less than those in HV/MV transformers. A dead band is defined for the LTC, which is the range in which if the voltage magnitude is in that, the LTC reaction stops. In order to prevent tap oscillation, each step of the tap change is considered smaller than the dead band.

2.4.2 The LTC Modeling

Accurate assessment of voltage stability needs proper modeling of related devices. Equations (2.52)–(2.59) make it possible to model the performance of different modes of LTCs in dynamic analyzing of voltage stability and to obtain the transient response of voltages. Wherein, db is the half of dead band, α is the tap value, and $\Delta\alpha$ is the step size of tap change, which is usually considered the half of the dead band [7]. α_{max} and α_{min} are, respectively, the upper and lower bounds of tap change. Δt is the simulation step and ΔV is equal to $|V_i| - |V_0|$, that is, i denotes the bus whose voltage is controlled by LTC. V_0 is the rated voltage. The Eq. (2.57) shows how the tap changes at the instant t. $f(t)$ equals to 1/−1 leads to one-step increase/decrease in the tap. e and c, respectively, show the voltage violation of the dead band and the time duration staying beyond the dead band.

The tap change needs voltage out of the dead band and elapsing the necessary time delays. Eq. (2.52) shows that at the instant t, when the voltage magnitude of the controlled bus becomes more than the upper limit of the dead band (less than the lower limit of the dead band) and the tap value is less than its upper limit (more than its lower limit), Δt will be added to the elapsed time to satisfy the necessary delay. Of course, this is provided in the previous step, and the voltage was not in the other side of the band dead, which is checked in Eq. (2.53). In the sequential mode, it is necessary to determine the first tap change. If the time interval between the stabilization of the voltage in the dead band and the re-out of this band is more than a certain limit, the resulted tap change is considered as the first change of the tap. This limit is shown in the following equations as T_{ns}. The distance from the last voltage stabilization is determined by the variable c'. According to Eq. (2.56), along with each tap change, the c' value decreases to zero and again increases only when either the voltage is in the dead band or the tap encounters one of its limits [Eq. (2.54)]. In the nonsequential mode, the variable c' and the T_{ns} parameter are removed. τ_{in} is the time delay required for the first tap change in the sequential mode and τ_s is the time delay for the next tap changes. These time delays can be in the form of Eqs. (2.58) and (2.59). In which, T_{inc} and T_{sc} are constant time delays. T_{in} and T_s are also coefficients that determine the amount of voltage-dependent delay. It is observed that the higher the voltage change, the lower the delay. By setting any of the above parameters to zero, constant or variable delays can be created. At the beginning of the simulation, c is set to zero and c' is equal to T_{ns}.

$$e(\Delta V(t), \alpha(t-\Delta t)) = \begin{cases} 1 & \text{if } \Delta V(t) > \text{db and } \alpha(t-\Delta t) < \alpha_{max} \\ -1 & \text{if } \Delta V(t) < -\text{db and } \alpha(t-\Delta t) > \alpha_{min} \\ 0 & \text{others} \end{cases} \tag{2.52}$$

$$c(e(t), c(t-\Delta t)) = \begin{cases} c(t-\Delta t) + \Delta t & \text{if } c(t-\Delta t) \geq 0 \text{ and } e(t) = 1 \\ c(t-\Delta t) - \Delta t & \text{if } c(t-\Delta t) \geq 0 \text{ and } e(t) = -1 \\ 0 & \text{others} \end{cases} \tag{2.53}$$

$$c'(e(t), c'(t-\Delta t), \alpha(t-\Delta t)) = \begin{cases} c'(t-\Delta t) + \Delta t & \text{if } \alpha(t-\Delta t) = \alpha_{max} \text{ or } \alpha(t-\Delta t) = \alpha_{min} \text{ or } e(t) = 0 \\ c'(t-\Delta t) & \text{if } e(t) = -1 \text{ or } e(t) = 1 \end{cases} \tag{2.54}$$

$$f(e(t), c(t), c'(t)) = \begin{cases} 1 & \text{if } c(t) \geq \tau_s(t) \text{ and } c'(t) \leq T_{ns} \text{ and } e(t) = 1 \\ 1 & \text{if } c(t) \geq \tau_{in}(t) \text{ and } c'(t) > T_{ns} \text{ and } e(t) = 1 \\ -1 & \text{if } c(t) \leq -\tau_s(t) \text{ and } c'(t) \leq T_{ns} \text{ and } e(t) = -1 \\ -1 & \text{if } c(t) \leq -\tau_{in}(t) \text{ and } c'(t) > T_{ns} \text{ and } e(t) = -1 \\ 0 & \text{others} \end{cases} \tag{2.55}$$

$$c'(f(t)) = \begin{cases} 0 & \text{if } f(t) = -1 \text{ or } f(t) = 1 \\ c'(t) & \text{others} \end{cases} \tag{2.56}$$

$$\alpha(t) = \alpha(t - \Delta t) + f(t) \times \Delta\alpha \tag{2.57}$$

$$\tau_{in}(t) = T_{in} \frac{db}{|\Delta V|} + T_{inc} \tag{2.58}$$

$$\tau_s(t) = T_s \frac{db}{|\Delta V|} + T_{sc} \tag{2.59}$$

2.4.3 The LTC Transformer Model

In the previous section, modeling of the LTC behavior when voltage changes was presented. To simulate the effect of tap changes on voltage, a suitable model for LTC transformers is required. Ignoring core losses and magnetic reactance, the transformer model is reduced to an ideal transformer connected to a series impedance. In the literature, depending on which side of the ideal transformer the series impedance is located, two per unit equivalent circuits as Figures 2.5 and 2.6 are available [27]. In these figures, α is the tap value that is the ratio of the number of off-nominal turns to the number of nominal turns in the tap-changing side. If the number of turns is nominal ($\alpha = 1$), the ideal transformer does not change the voltage and current in the per unit system and can be neglected in the equivalent circuit.

There are some important points about the common transformer models. The first point is that the use of each of the models shown in Figures 2.5 and 2.6 leads to different values of the loadability limit. To verify this, according to Figures 2.7 and 2.8, two simple systems are considered. In these figures, the transformers deliver the powers flowing through the transmission lines to the loads. It is assumed that the voltage magnitude at the bus 1 is 1 pu, and the impedances Z_L and Z_T are both equal to 0.05 pu. For both systems, the same impedance load

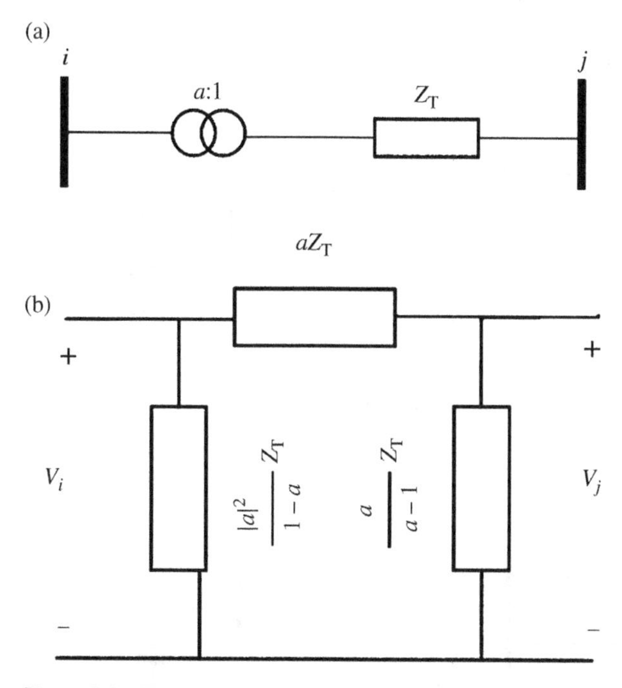

Figure 2.5 The common transformer model with impedance transferred to the constant turn side. (a) Ideal transformer with series impedance. (b) Equivalent circuit π.

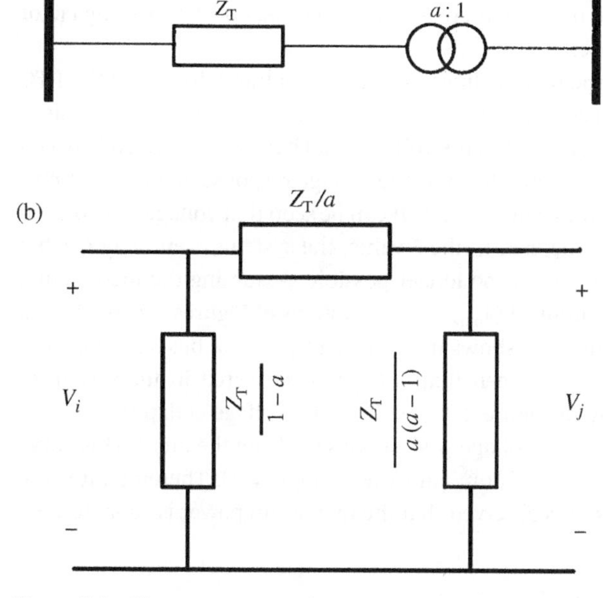

Figure 2.6 The common transformer model with impedance transferred to the variable turn side. (a) Ideal transformer with series impedance. (b) Equivalent circuit π.

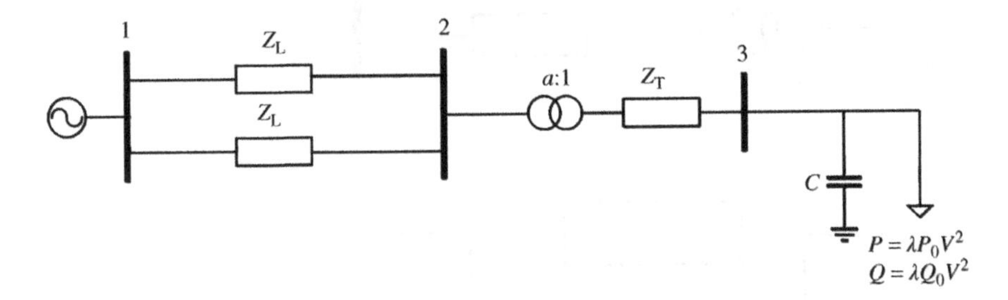

Figure 2.7 A simple system based on the model of Figure 2.5.

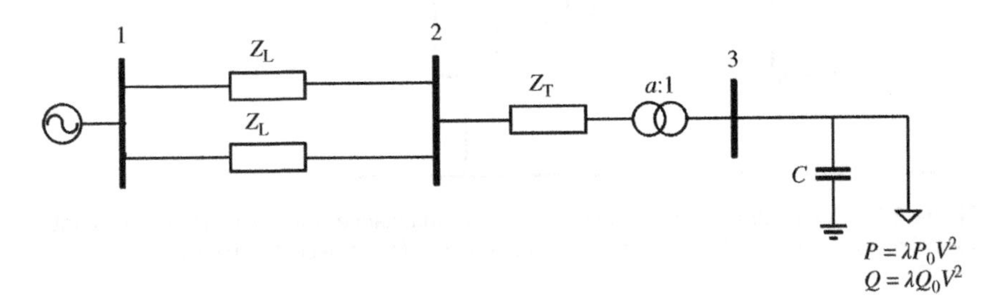

Figure 2.8 A simple system based on the model of Figure 2.6.

is considered with P_0 and Q_0 equal to 0.5 and 0.3 pu, respectively. To obtain better simulation results, next to the load, there is a parallel capacitor. The loading factor λ is to model load increase.

After outage of one of the parallel lines, the voltage of bus 3 drops and the LTC reacts to recover the voltage. The tap value is assumed to change in 0.01 pu steps with a time delay of 10 seconds. Also, the voltage dead band is considered between 0.99 and 1.01 pu. Figures 2.9 and 2.10 show the voltage response of bus 3 after outage of one of the parallel lines when $\lambda = 2.5$. It can be seen that voltage is recovered in the both systems. With increase in the λ value, the systems eventually reach a point at which voltage recovery is no longer possible. Assuming the incremental step of λ is 0.1 pu, the maximum λ (λ_{max}) for the systems of Figures 2.7 and 2.8 is 3 and 2.7, respectively. Figure 2.11 shows the voltage response of bus 3 for both systems when $\lambda = 2.85$. It can be seen that voltage is recovered in the system of Figure 2.7, but the system of Figure 2.8 experienced a voltage collapse.

Figure 2.12 shows the power voltage (PV) curve seen from the bus 3. This curve is plotted with the assumption of nominal turns ratio ($\alpha = 1$). Therefore, it is the same for the both models. It is observed that the maximum power is equal to 1.33

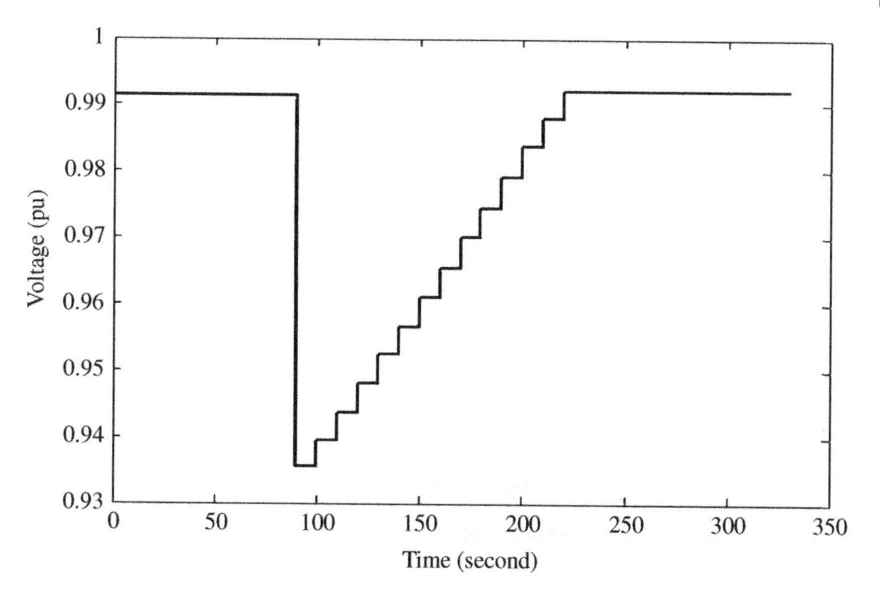

Figure 2.9 Voltage response of bus 3 in Figure 2.8 when $\lambda = 2.5$.

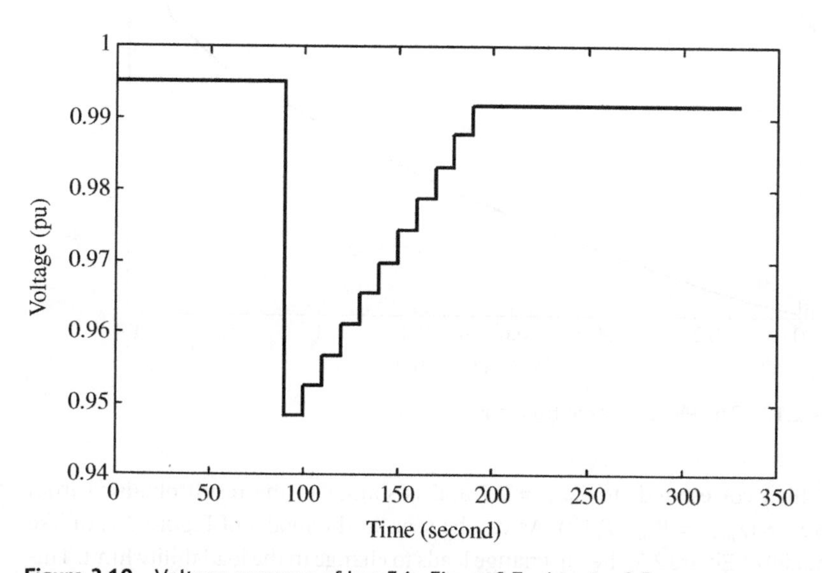

Figure 2.10 Voltage response of bus 3 in Figure 2.7 when $\lambda = 2.5$.

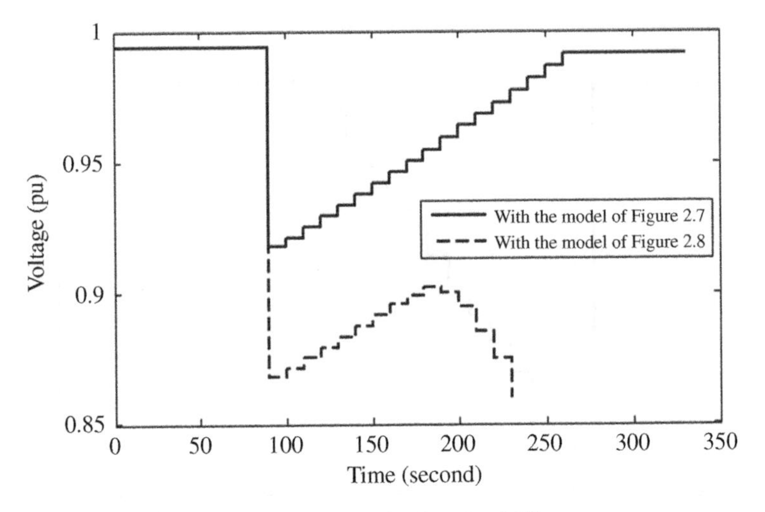

Figure 2.11 Voltage response of bus 3 when $\lambda = 2.85$.

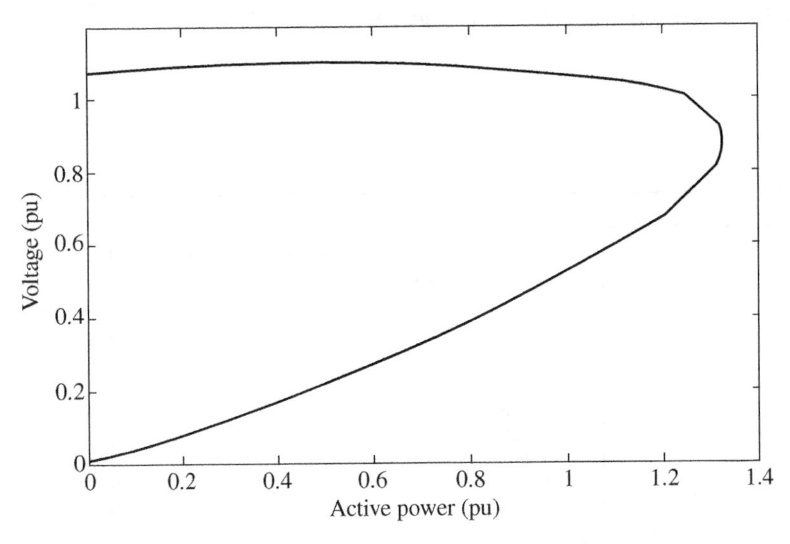

Figure 2.12 The PV curve seen from bus 3 with $\alpha = 1$.

pu, which corresponds to $\lambda_{\max} = 2.7$ and is similar to the result obtained from Figure 2.8 ($\lambda_{\max} = P_{\max}/P_0V^2$). As can be seen in the model of Figure 2.7, unlike the model of Figure 2.8, the tap change leads to change in the loadability limit. This can be justified in different words. For instance, it can be seen that the series impedance of the system, to which the loadability limit is proportional, is equal to $Z_L + Z_T$ in Figure 2.8 and $Z_L + \alpha^2 Z_T$ in Figure 2.7.

The question now is which model is closer to the behavior of a real LTC transformer. The point is that in each of the above models, the impedance of the transformer is the sum of the impedance of one side and the transferred impedance from the other side. If the number of turns on the primary and secondary windings is N_p and N_s and the nontransmitted impedances on each side are Z_p and Z_s, the relation between the impedances in ohms and the number of turns is as given in Eq. (2.60) [28]:

$$Z_p = Z_s(\Omega)\left(\frac{N_p}{N_s}\right)^2 \tag{2.60}$$

It is worthy to note that it is not possible to calculate $Z_p(\Omega)$ and $Z_s(\Omega)$ by the transformer tests. Meanwhile, Eq. (2.60) is an approximate relation. Now if, for example, the primary-side impedance is transferred to the secondary,

$$Z_p'(\Omega) = Z_p(\Omega)\left(\frac{N_s}{N_p}\right)^2 = Z_s(\Omega)\left(\frac{N_p}{N_s}\right)^2\left(\frac{N_s}{N_p}\right)^2 = Z_s(\Omega) \tag{2.61}$$

Therefore,

$$Z_p'(\Omega) = Z_s(\Omega) = \frac{Z_T(\Omega)}{2} \tag{2.62}$$

where Z_T is the total impedance of the transformer at the secondary side. To make these impedances per unit, the basic values of them are selected as follows:

$$Z_{base,s}(\Omega) = \frac{(V_{base,s})^2}{S_{base}} \tag{2.63}$$

$$Z_{base,p}(\Omega) = \frac{(V_{base,p})^2}{S_{base}} = \frac{(V_{base,s})^2}{S_{base}}\left(\frac{N_p}{N_s}\right)^2 \tag{2.64}$$

where $V_{base,\,s}$ and $V_{base,\,p}$ are the base voltages in the secondary and primary sides, respectively. Considering the relation between the impedances of the two sides and their base values,

$$Z_p(pu) = Z_s(pu) = \frac{Z_T(pu)}{2} \tag{2.65}$$

So, the transformer can be modeled as follows [29] (Figure 2.13):

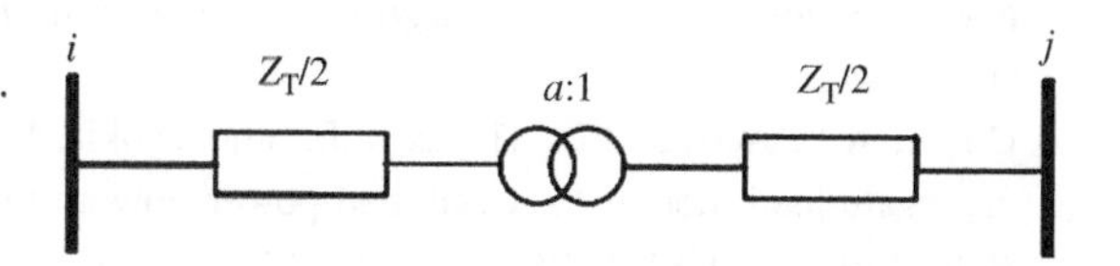

Figure 2.13 The improved model of LTC transformer [29].

With this model, after outage of one of the parallel lines, the series impedance of the simple system of Figures 2.7 and 2.8 becomes $Z_L + Z_T/2 + \alpha^2 Z_T/2$. In this situation, the loadability limit occurs in $\lambda_{max} = 2.87$. Another point about the above models is that, the transformer impedance change due to the change in its tap has not considered.

2.5 Summary

In this chapter, different models of loads and LTC transformers were presented. Loads are divided into two categories, static and dynamic. In static loads, when voltage changes, the power consumption by them reaches its final value instantly (or very quickly). But in dynamic loads, there is a transient period before reaching steady state. Electronic devices and induction motors are, respectively, examples of static and dynamic loads. Different static and dynamic models were introduced. Discussions were also held on buses which load modeling is done from view of them. It was pointed out that when the load modeling is done at a bus with lower voltage level, the model and consequently voltage stability assessment are more accurate. Of course, more computational burden is required.

Also, different modes of LTC performance and how to model them were described. Also, different methods to LTC transformers and their differences were investigated. Finally, an improved model was introduced, and its performance was compared with the previous ones.

References

1 Gaikwad, A., Markham, P., and Pourbeik, P. (2016). Implementation of the WECC Composite Load Model for utilities using the component-based modeling approach. *Proc. IEEE Power Eng. Soc. Transm. Distrib. Conf.* 2016: 1–5.

2 Khodabakhchian, B. and Vuong, G.T. (1997). Modeling a mixed residential-commercial load for simulations involving large disturbances. *IEEE Power Eng. Rev.* 17 (5): 68–69.

3 Yamashita, K., Martínez Villanueva, S., and Milanović, J.V. (2011). Initial results of international survey on industrial practice on power system load modelling conducted by CIGRE WG C4.605. *CIGRE 2011 Bologna Symposium – The Electric Power System of the Future: Integrating Supergrids and Microgrids,* Bologna (1 July 2011).

4 Collin, A.J., Acosta, J.L., Hayes, B.P., and Djokic, S.Z. (2010). Component-based aggregate load models for combined power flow and harmonic analysis. *IET Conf. Publ.* 2010 (572 CP): 1–10.

5 Li, Z., Guo, Q., Sun, H. et al. (2018). A distributed transmission-distribution-coupled static voltage stability assessment method considering distributed generation. *IEEE Trans. Power Syst.* 33 (3): 2621–2632.

6 Karbalaei, F., Abasi, S., Abedinzade, A., and Kaviani, M. (2015). A new method for considering distribution systems in voltage stability studies abstract. *Iran. J. Electr. Electron. Eng.* 12: 1–8.

7 Van Cutsem, T. and Vournas, C. (1988). *Voltage Stability of Electric Power Systems*. Norwell, MA: Kluwer Academic Publishers.

8 Milanović, J.V., Yamashita, K., Martínez Villanueva, S. et al. (2013). International industry practice on power system load modeling. *IEEE Trans. Power Syst.* 28 (3): 3038–3046.

9 Lu, N., Xie, Y., Huang, Z. et al. (2008). Load component database of household appliances and small office equipment. *2008 IEEE Power and Energy Society General Meeting – Conversion and Delivery of Electrical Energy in the 21st Century, PES*, Pittsburgh, PA, USA (20–24 July 2008), pp. 1–5.

10 Korunovic, L.M., Milanovic, J.V., Djokic, S.Z. et al. (2018). Recommended parameter values and ranges of most frequently used static load models. *IEEE Trans. Power Syst.* 33 (6): 5923–5934.

11 Korunovic, L.M., Sterpu, S., Djokic, S. et al. (2012). Processing of load parameters based on existing load models. *IEEE PES Innovative Smart Grid Technologies Conference Europe*, Berlin, Germany (14–17 October 2012), pp. 1–6.

12 Leinakse, M. and Kilter, J. (2018). Conversion error of second order polynomial ZIP to exponential load model conversion. *IET Conf. Publ.* 2018 (CP759): 1–5.

13 Kopczynski, L., Huppertz, P., Schallenburger, M., and Zeise, R. (2018). Optimal tap-operations of a regulated distribution transformer considering conservation voltage reduction effects in active low-voltage grids. *20th Power Systems Computation Conference PSCC 2018*, Dublin, Ireland (11–15 June 2018), pp. 1–7.

14 Hill, D.J. (1993). Nonlinear dynamic load models with recovery for voltage stability studies. *IEEE Trans. Power Syst.* 8 (1): 166–176.

15 Karlsson, D. and Hill, D.J. (1994). Modelung and identification of nonlinear dynamic loads in power systems. *IEEE Trans. Power Syst.* 9 (1): 157–166.

16 Xu, W., Mansour, Y., and Hydro, B.C. (1994). Voltage stability analysis using generic dynamic load models. *IEEE Trans. Power Syst.* 9 (1): 479–493.

17 Handschin, E., Ju, P., and Rehtanz, C. (1996). A comparative study on the nonlinear dynamic load models. *IEEE Power Tech Conference Proceedings*.

18 Pereira, L., Kosterev, D., Mackin, P. et al. (2002). An interim dynamic induction motor model for stability studies in the WSCC. *IEEE Trans. Power Syst.* 17 (4): 1108–1115.

19 Renmu, H., Jin, M., and Hill, D.J. (2006). Composite load modeling via measurement approach. *IEEE Trans. Power Syst.* 21 (2): 663–672.

20 Sauer, P.W. and Pai, M.A. (1998). *Power System Dynamics and Stability*, vol. 101. Upper Saddle River, NJ: Prentice Hall.

21 Kundur, P., Neal, J.B., and Mark, G.L. (1994). *Power System Stability and Control*, vol. 7. New York: McGraw-Hill.

22 Prasad, G.D. and Al-Mulhim, M.A. (1997). Performance evaluation of dynamic load models for voltage stability analysis. *Int. J. Electr. Power Energy Syst.* 19 (8): 533–540. https://doi.org/10.1016/s0142-0615(97)00025-2.

23 Borghetti, A., Caldon, R., Mari, A., and Nucci, C.A. (1997). On dynamic load models for voltage stability studies. *IEEE Power Eng. Rev.* 17 (2): 62.

24 Faiz, J. and Javidnia, H. (2000). Fast response solid-state on load transformers tap-changer. *IEE Conf. Publ.* 475: 355–359.

25 Cooke, G.H. and Williams, K.T. (1990). Thyristor assisted on-load tap changers for transformers. *Fourth International Conference on Power Electronics and Variable-Speed Drives (Conf. Publ. No. 324)*, IET, London, UK (17–19 July 1990) pp. 127–131.

26 Jiang, H., Shuttleworth, R., Al Zahawi, B.A.T. et al. (2001). Fast response GTO assisted novel tap changer. *IEEE Trans. Power Deliv.* 16 (1): 111–115.

27 Grainger, J.J. and Stevenson, W.D. (1994). *Power System Analysis*. New York: McGraw-Hill.

28 Choi, J.H. and Moon, S.I. (2009). The dead band control of LTC transformer at distribution substation. *IEEE Trans. Power Syst.* 24 (1): 319–326.

29 Ferreira, C.A. and Prada, R.B. (2013). Improved model for tap-changing transformer. *IET Gener. Transm. Distrib.* 7 (11): 1289–1295.

3

Generator Modeling

3.1 Introduction

Appropriate modeling of generators is very important in accurate analysis of voltage stability. Voltage stability analysis can be done in both static and dynamic methods. Static analysis is based on the calculation of the steady-state operating point, but in dynamic analysis, calculation of variables' time response is persuaded to evaluate the effect of amount and time of implementation of control measures. Obtaining the variables' time response requires the use of dynamic models of equipment. Therefore, generators modeling as one of the effective equipment should include both steady-state (static) and dynamic models. The limitations of generators, which can play an important role in the occurrence of voltage instability, should also be considered in the modeling. In this chapter, models of synchronous generators and two types of wind generators are presented. The accuracy of models is based on the accuracy required for voltage stability studies.

3.2 Synchronous Generator Modeling

A significant portion of electrical power in power systems is produced by synchronous generators [1]. These generators are the main sources of power generation in conventional power plants. Synchronous generators play an important role in controlling the network voltage. Adjusting their output voltage is one of the control measures to prevent voltage instability. One of the important causes of occurrence of voltage instability and collapse is reaching the synchronous generators to their reactive power limits and consequently losing their voltage control.

The dynamic reactive power capability of these generators, which allows overproduction of reactive power for a short period of time, can play an important role in maintaining voltage stability. For synchronous generators, various modelings

Voltage Stability in Electrical Power Systems: Concepts, Assessment, and Methods for Improvement, First Edition. Farid Karbalaei, Shahriar Abbasi, and Hamid Reza Shabani.

Published 2023 by John Wiley & Sons, Inc.

are presented in terms of the used simplifications, that in this section, some of them that are used in voltage stability studies are introduced.

3.2.1 Synchronous Machine Structure

The synchronous machine has three windings on the stator and one winding on the rotor, which is called the excitation winding. It is assumed that in addition to the excitation winding, there are also three damping windings on the rotor. The rotor windings are located along the d and q axes. The d-axis is in the direction of the excitation winding axis, and the q-axis is 90° ahead of the d-axis. Figure 3.1 shows a view of the different windings of synchronous machine. In this figure, the excitation winding with the label fd and the damping windings with the labels 1d, 1q, and 2q are shown. Windings 1q and 2q are located along the q-axis. The damping windings are short circuit.

3.2.2 Dynamic Equations

The equations of stator and rotor voltages in terms of currents and fluxes are as follows:

$$v_a = i_a r_s + \frac{d\lambda_a}{dt} \tag{3.1}$$

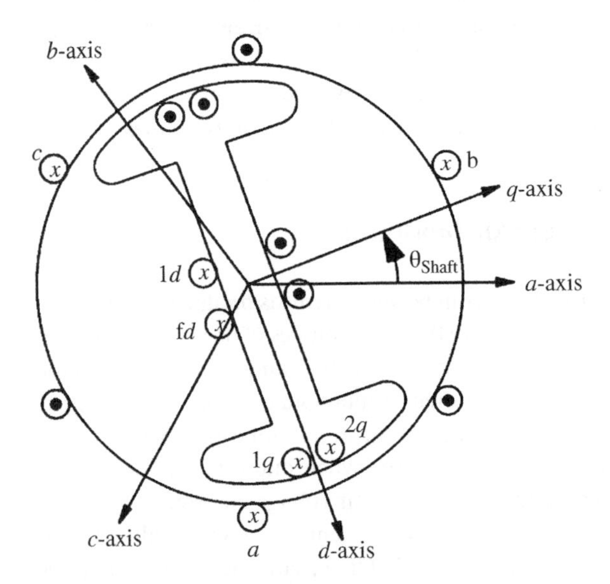

Figure 3.1 A view of a synchronous machine.

$$v_{\mathrm{b}} = i_{\mathrm{b}} r_{\mathrm{s}} + \frac{\mathrm{d}\lambda_{\mathrm{b}}}{\mathrm{d}t} \tag{3.2}$$

$$v_{\mathrm{c}} = i_{\mathrm{c}} r_{\mathrm{s}} + \frac{\mathrm{d}\lambda_{\mathrm{c}}}{\mathrm{d}t} \tag{3.3}$$

$$v_{\mathrm{fd}} = i_{\mathrm{fd}} r_{\mathrm{fd}} + \frac{\mathrm{d}\lambda_{\mathrm{fd}}}{\mathrm{d}t} \tag{3.4}$$

$$0 = i_{\mathrm{1d}} r_{\mathrm{1d}} + \frac{\mathrm{d}\lambda_{\mathrm{1d}}}{\mathrm{d}t} \tag{3.5}$$

$$0 = i_{\mathrm{1q}} r_{\mathrm{1q}} + \frac{\mathrm{d}\lambda_{\mathrm{1q}}}{\mathrm{d}t} \tag{3.6}$$

$$0 = i_{\mathrm{2q}} r_{\mathrm{2q}} + \frac{\mathrm{d}\lambda_{\mathrm{2q}}}{\mathrm{d}t} \tag{3.7}$$

where λ is the linkage flux and r is the ohmic resistance. In the above equations, for all windings, motor current rotation is used. If θ_{shaft} is the angle between the q-axis and the phase "a" axis, the motion equations are as follows:

$$\frac{\mathrm{d}\theta_{\mathrm{shaft}}}{\mathrm{d}t} = \frac{2}{P}\omega \tag{3.8}$$

$$J\frac{2}{P}\frac{\mathrm{d}\omega}{\mathrm{d}t} = T_{\mathrm{m}} - T_{\mathrm{e}} - T_{\mathrm{fw}} \tag{3.9}$$

where ω is the rotor speed in electric radians per second, P is the number of poles in each phase, J is the inertial torque, T_{m} is the mechanical torque, T_{e} is the electromagnetic torque, and T_{fw} is the torque due to air friction or the damping torque. Using the Park's transformation and the per unit system, Eqs. (3.1)–(3.9) are converted to Eqs. (3.10)–(3.18), in terms of d, q, and o variables.

$$\frac{1}{\omega_{\mathrm{s}}}\frac{\mathrm{d}\psi_{\mathrm{d}}}{\mathrm{d}t} = R_{\mathrm{s}} I_{\mathrm{d}} + \frac{\omega}{\omega_{\mathrm{s}}}\psi_{\mathrm{q}} + V_{\mathrm{d}} \tag{3.10}$$

$$\frac{1}{\omega_{\mathrm{s}}}\frac{\mathrm{d}\psi_{\mathrm{q}}}{\mathrm{d}t} = R_{\mathrm{s}} I_{\mathrm{q}} - \frac{\omega}{\omega_{\mathrm{s}}}\psi_{\mathrm{d}} + V_{\mathrm{q}} \tag{3.11}$$

$$\frac{1}{\omega_{\mathrm{s}}}\frac{\mathrm{d}\psi_{\mathrm{o}}}{\mathrm{d}t} = R_{\mathrm{s}} I_{\mathrm{o}} + V_{\mathrm{o}} \tag{3.12}$$

$$\frac{1}{\omega_{\mathrm{s}}}\frac{\mathrm{d}\psi_{\mathrm{fd}}}{\mathrm{d}t} = -R_{\mathrm{fd}} I_{\mathrm{fd}} + V_{\mathrm{fd}} \tag{3.13}$$

$$\frac{1}{\omega_{\mathrm{s}}}\frac{\mathrm{d}\psi_{\mathrm{1d}}}{\mathrm{d}t} = -R_{\mathrm{1d}} I_{\mathrm{1d}} \tag{3.14}$$

$$\frac{1}{\omega_s}\frac{d\psi_{1q}}{dt} = -R_{1q}I_{1q} \tag{3.15}$$

$$\frac{1}{\omega_s}\frac{d\psi_{2q}}{dt} = -R_{2q}I_{2q} \tag{3.16}$$

$$\frac{d\delta}{dt} = \omega - \omega_s \tag{3.17}$$

$$\frac{2H}{\omega_s}\frac{d\omega}{dt} = T_M - \left(\psi_d I_q - \psi_q I_d\right) - T_{FW} \tag{3.18}$$

where V_d, V_q, V_o, ψ_d, ψ_q, ψ_o and I_d, I_q, and I_o are the voltages, flux linkages, and stator currents on the axes d, q, and o. ω_s is the synchronous speed in electric radians per second. In the above equations, all variables except ω, δ, and H are in per unit. δ is the rotor angle with respect to a rotating axis, which rotates at a synchronous speed and is in the direction of the phase "a" axis at time zero.

$$\delta \triangleq \frac{P}{2}\theta_{shaft} - \omega_s t \tag{3.19}$$

H is the inertial constant in second. Another important point to note is that in making per unit, the conventional direction of the currents converts to generator rotation. That is, the positive conventional directions of I_d, I_q, and I_o is outward. The base value of stator voltages in abc variables is rated RMS line to neutral stator voltage, and in dq0 variables is rated peak line to neutral stator voltage. The base value of the three-phase power is the same throughout the system. The base values of the rotor currents are selected so that the off-diagonal entries in the inductance matrices on the axes d and q be equal as much as possible. (A complete discussion about selecting the base values is presented in Ref. [1].)

The relationships of flux linkages in terms of currents, assuming a linear relationship between them are given as Eqs. (3.20)–(3.25).

$$\Psi_d = X_d(-I_d) + X_{md}I_{fd} + X_{md}I_{1d} \tag{3.20}$$

$$\Psi_{fd} = X_{md}(-I_d) + X_{fd}I_{fd} + X_{md}I_{1d} \tag{3.21}$$

$$\Psi_{1d} = X_{md}(-I_d) + X_{md}I_{fd} + X_{1d}I_{1d} \tag{3.22}$$

$$\Psi_q = X_q(-I_q) + X_{mq}I_{1q} + X_{mq}I_{2q} \tag{3.23}$$

$$\Psi_{1q} = X_{mq}(-I_q) + X_{1q}I_{1q} + X_{mq}I_{2q} \tag{3.24}$$

$$\Psi_{2q} = X_{mq}(-I_q) + X_{mq}I_{1q} + X_{2q}I_{2q} \tag{3.25}$$

where X_d is the self-reactance of the axis d. X_{fd} is the self-reactance of the excitation winding and X_{1d} is the leakage reactance of the damping winding of the axis d. X_{md} is the mutual-reactance among the windings on the axis d. The equality of mutual reactances among the windings is due to the proper choosing of the base currents.

There are similar definitions for windings on the axis q. In modeling of synchronous machines, it is usual to define the following parameters:

$$X'_{\mathrm{d}} \triangleq X_{\mathrm{ls}} + \frac{1}{\frac{1}{X_{\mathrm{md}}} + \frac{1}{X_{\mathrm{lfd}}}} = X_{\mathrm{ls}} + \frac{X_{\mathrm{md}}X_{\mathrm{lfd}}}{X_{\mathrm{fd}}} = X_{\mathrm{d}} - \frac{X_{\mathrm{md}}^2}{X_{\mathrm{fd}}} \tag{3.26}$$

$$X'_{\mathrm{q}} \triangleq X_{\mathrm{ls}} + \frac{1}{\frac{1}{X_{\mathrm{mq}}} + \frac{1}{X_{\mathrm{l1q}}}} = X_{\mathrm{ls}} + \frac{X_{\mathrm{mq}}X_{\mathrm{l1q}}}{X_{\mathrm{1q}}} = X_{\mathrm{q}} - \frac{X_{\mathrm{mq}}^2}{X_{\mathrm{1q}}} \tag{3.27}$$

$$X''_{\mathrm{d}} \triangleq X_{\mathrm{ls}} + \frac{1}{\frac{1}{X_{\mathrm{md}}} + \frac{1}{X_{\mathrm{lfd}}} + \frac{1}{X_{\mathrm{l1d}}}} \tag{3.28}$$

$$X''_{\mathrm{q}} \triangleq X_{\mathrm{ls}} + \frac{1}{\frac{1}{X_{\mathrm{mq}}} + \frac{1}{X_{\mathrm{l1q}}} + \frac{1}{X_{\mathrm{l2q}}}} \tag{3.29}$$

$$T'_{\mathrm{d0}} \triangleq \frac{X_{\mathrm{fd}}}{\omega_{\mathrm{s}}R_{\mathrm{fd}}} \tag{3.30}$$

$$T'_{\mathrm{q0}} \triangleq \frac{X_{\mathrm{1q}}}{\omega_{\mathrm{s}}R_{\mathrm{1q}}} \tag{3.31}$$

$$T''_{\mathrm{d0}} \triangleq \frac{1}{\omega_{\mathrm{s}}R_{\mathrm{1d}}}\left(X_{\mathrm{l1d}} + \frac{1}{\frac{1}{X_{\mathrm{md}}} + \frac{1}{X_{\mathrm{lfd}}}}\right) \tag{3.32}$$

$$T''_{\mathrm{q0}} \triangleq \frac{1}{\omega_{\mathrm{s}}R_{\mathrm{2q}}}\left(X_{\mathrm{l2q}} + \frac{1}{\frac{1}{X_{\mathrm{mq}}} + \frac{1}{X_{\mathrm{l1q}}}}\right) \tag{3.33}$$

$$E'_{\mathrm{q}} \triangleq \frac{X_{\mathrm{md}}}{X_{\mathrm{fd}}}\Psi_{\mathrm{fd}} \tag{3.34}$$

$$E_{\mathrm{fd}} \triangleq \frac{X_{\mathrm{md}}}{R_{\mathrm{fd}}}V_{\mathrm{fd}} \tag{3.35}$$

$$E'_{\mathrm{d}} \triangleq -\frac{X_{\mathrm{mq}}}{X_{\mathrm{1q}}}\Psi_{\mathrm{1q}} \tag{3.36}$$

Using new variables and deleting the currents I_{fd}, I_{1d}, I_{1q}, and I_{2q}, the differential Eqs. (3.10)–(3.18) and the algebraic Eqs. (3.20)–(3.25) become as Eqs. (3.37)–(3.48).

$$\frac{1}{\omega_{\mathrm{s}}}\frac{d\Psi_{\mathrm{d}}}{dt} = R_{\mathrm{s}}I_{\mathrm{d}} + \frac{\omega}{\omega_{\mathrm{s}}}\Psi_{\mathrm{q}} + V_{\mathrm{d}} \tag{3.37}$$

$$\frac{1}{\omega_{\mathrm{s}}}\frac{d\Psi_{\mathrm{q}}}{dt} = R_{\mathrm{s}}I_{\mathrm{q}} - \frac{\omega}{\omega_{\mathrm{s}}}\Psi_{\mathrm{d}} + V_{\mathrm{q}} \tag{3.38}$$

$$\frac{1}{\omega_s}\frac{d\Psi_0}{dt} = R_s I_0 + V_0 \tag{3.39}$$

$$T'_{d0}\frac{dE'_q}{dt} = -E'_q - (X_d - X'_d)\left[I_d - \frac{X'_d - X''_d}{(X''_d - X_{ls})^2}\left(\Psi_{1d} + (X'_d - X_{ls})I_d - E'_q\right)\right] + E_{fd} \tag{3.40}$$

$$T''_{d0}\frac{d\Psi_{1d}}{dt} = -\Psi_{1d} + E'_q - (X'_d - X_{ls})I_d \tag{3.41}$$

$$T'_{q0}\frac{dE'_d}{dt} = -E'_d - \left(X_q - X'_q\right)\left[I_q - \frac{X'_q - X''_q}{\left(X'_q - X_{ls}\right)^2}\left(\Psi_{2q} + \left(X'_q - X_{ls}\right)I_q - E'_d\right)\right] \tag{3.42}$$

$$T''_{q0}\frac{d\Psi_{2q}}{dt} = -\Psi_{2q} - E'_d - \left(X'_q - X_{ls}\right)I_q \tag{3.43}$$

$$\frac{d\delta}{dt} = \omega - \omega_s \tag{3.44}$$

$$\frac{2H}{\omega_s}\frac{d\omega}{dt} = T_M - \left(\Psi_d I_q - \Psi_q I_d\right) - T_{FW} \tag{3.45}$$

$$\Psi_d = -X''_d I_d + \frac{(X''_d - X_{ls})}{(X'_d - X_{ls})}E'_q + \frac{(X'_d - X''_d)}{(X'_d - X_{ls})}\Psi_{1d} \tag{3.46}$$

$$\Psi_q = -X''_q I_q - \frac{\left(X''_q - X_{ls}\right)}{\left(X'_q - X_{ls}\right)}E'_d + \frac{\left(X'_q - X''_q\right)}{\left(X'_q - X_{ls}\right)}\Psi_{2q} \tag{3.47}$$

$$\Psi_0 = -X_{ls} I_0 \tag{3.48}$$

3.2.3 Voltage and Current Phasors

With the assumption that sinusoidal voltages and currents flow through the stator windings, their per unit values are as follows:

$$V_a = \sqrt{2}V_s \cos(\omega_s t + \theta_s) \tag{3.49}$$

$$V_b = \sqrt{2}V_s \cos\left(\omega_s t + \theta_s - \frac{2\pi}{3}\right) \tag{3.50}$$

$$V_c = \sqrt{2}V_s \cos\left(\omega_s t + \theta_s + \frac{2\pi}{3}\right) \tag{3.51}$$

$$I_a = \sqrt{2}I_s \cos(\omega_s t + \phi_s) \tag{3.52}$$

$$I_b = \sqrt{2}I_s \cos\left(\omega_s t + \phi_s - \frac{2\pi}{3}\right) \tag{3.53}$$

$$I_c = \sqrt{2}I_s \cos\left(\omega_s t + \phi_s + \frac{2\pi}{3}\right) \tag{3.54}$$

Assuming that the BABC and BDQ subtitles denotes the base value in the abc and dq0 variables, the per unit voltages and currents in the dq0 variables are as

$$V_d = \left(\frac{\sqrt{2}V_s V_{BABC}}{V_{BDQ}}\right) \sin\left(\frac{P}{2}\theta_{shaft} - \omega_s t - \theta_s\right) \tag{3.55}$$

$$V_q = \left(\frac{\sqrt{2}V_s V_{BABC}}{V_{BDQ}}\right) \cos\left(\frac{P}{2}\theta_{shaft} - \omega_s t - \theta_s\right) \tag{3.56}$$

$$V_0 = 0 \tag{3.57}$$

$$I_d = \left(\frac{\sqrt{2}I_s I_{BABC}}{I_{BDQ}}\right) \sin\left(\frac{P}{2}\theta_{shaft} - \omega_s t - \phi_s\right) \tag{3.58}$$

$$I_q = \left(\frac{\sqrt{2}I_s I_{BABC}}{I_{BDQ}}\right) \cos\left(\frac{P}{2}\theta_{shaft} - \omega_s t - \phi_s\right) \tag{3.59}$$

$$I_0 = 0 \tag{3.60}$$

Given the base values, the following equations can be written as

$$\frac{\sqrt{2}V_s V_{BABC}}{V_{BDQ}} = V_s \quad \frac{\sqrt{2}I_s I_{BABC}}{I_{BDQ}} = I_s \tag{3.61}$$

Also, according to the definition of δ,

$$V_d = V_s \sin(\delta - \theta_s) \tag{3.62}$$

$$V_q = V_s \cos(\delta - \theta_s) \tag{3.63}$$

$$I_d = I_s \sin(\delta - \phi_s) \tag{3.64}$$

$$I_q = I_s \cos(\delta - \phi_s) \tag{3.65}$$

From Eqs. (3.62)–(3.65), the bellow phasor equations can be obtained:

$$\left(V_d + jV_q\right)e^{j\left(\delta - \frac{\pi}{2}\right)} = V_s e^{j\theta_s} \tag{3.66}$$

$$\left(I_d + jI_q\right)e^{j\left(\delta - \frac{\pi}{2}\right)} = I_s e^{j\phi_s} \tag{3.67}$$

The right sides of Eqs. (3.66) and (3.67) are phasors based on a rotating reference frame that rotates at synchronous speed and in the time zero is in the direction of the phase "*a*" axis. $V_d + jV_q$ and $I_d + jI_q$ are the phasors of these variables in the dq rotating frame of generator. $e^{j\left(\delta - \frac{\pi}{2}\right)}$ states the relation between the obtained phasors from the synchronously rotating reference frame and the generator *dq* rotating frame. As shown in Figure 3.2, there is the angle $\frac{\pi}{2} - \delta$ between these rotating

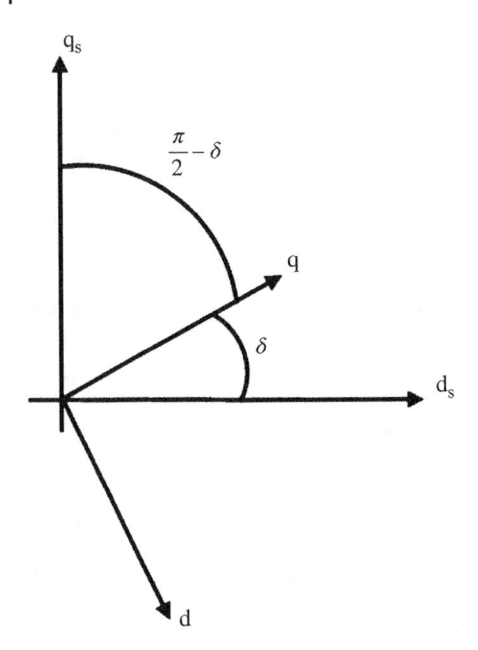

Figure 3.2 The synchronously rotating reference and dq rotating frames of generator.

frames. In this figure, d_s and q_s are the axes of the synchronously rotating reference frame. The real and imaginary parts of a variable phasor are in respect equal to the value of this variable in the phase "*a*" when the axes d and q pass through the axis of this phase.

3.2.4 Steady-State Equations

By equating the derivatives of Eqs. (3.37)–(3.45) to zero, the steady-state equations are obtained. In the steady-state conditions, the current of the damping windings is zero. Also, assuming the system is a balanced three-phase type, the zero component variables are equal to zero. Accordingly, the steady-state equations become:

$$V_d = -R_s I_d - \Psi_q \tag{3.68}$$

$$V_q = -R_s I_q + \Psi_d \tag{3.69}$$

$$0 = -E'_q - (X_d - X'_d) I_d + E_{fd} \tag{3.70}$$

$$0 = -\Psi_{1d} + E'_q - (X'_d - X_{ls}) I_d \tag{3.71}$$

$$0 = -E'_d + \left(X_q - X'_q\right) I_q \tag{3.72}$$

$$0 = -\Psi_{2q} - E'_d - \left(X'_q - X_{ls}\right) I_q \tag{3.73}$$

$$0 = T_{\mathrm{M}} - \left(\Psi_{\mathrm{d}} I_{\mathrm{q}} - \Psi_{\mathrm{q}} I_{\mathrm{d}}\right) - T_{\mathrm{FW}} \tag{3.74}$$

$$\Psi_{\mathrm{d}} = E'_{\mathrm{q}} - X'_{\mathrm{d}} I_{\mathrm{d}} \tag{3.75}$$

$$\Psi_{\mathrm{q}} = -E'_{\mathrm{d}} - X'_{\mathrm{q}} I_{\mathrm{q}} \tag{3.76}$$

By substituting Ψ_{d} and Ψ_{q} in (3.68) and (3.69),

$$V_{\mathrm{d}} = -R_{\mathrm{s}} I_{\mathrm{d}} + E'_{\mathrm{d}} + X'_{\mathrm{q}} I_{\mathrm{q}} \tag{3.77}$$

$$V_{\mathrm{q}} = -R_{\mathrm{s}} I_{\mathrm{q}} + E'_{\mathrm{q}} - X'_{\mathrm{d}} I_{\mathrm{d}} \tag{3.78}$$

These two algebraic equations can be written in the form of a complex equation,

$$\left(V_{\mathrm{d}} + jV_{\mathrm{q}}\right) e^{j(\delta - \pi/2)} = -\left(R_{\mathrm{s}} + jX_{\mathrm{q}}\right)\left(I_{\mathrm{d}} + jI_{\mathrm{q}}\right) e^{j(\delta - \pi/2)} + \overline{E} \tag{3.79}$$

where

$$\begin{aligned} \overline{E} &= \left[\left(E'_{\mathrm{d}} - \left(X_{\mathrm{q}} - X'_{\mathrm{q}}\right) I_{\mathrm{q}}\right) + j\left(E'_{\mathrm{q}} + (X_{\mathrm{q}} - X'_{\mathrm{d}}) I_{\mathrm{d}}\right)\right] e^{j(\delta - \pi/2)} \\ &= j\left[(X_{\mathrm{q}} - X'_{\mathrm{d}}) I_{\mathrm{d}} + E'_{\mathrm{q}}\right] e^{j(\delta - \pi/2)} \end{aligned} \tag{3.80}$$

It is clear that depending on the variables in $\overline{E}$, different complex equations are obtained. Based on Eq. (3.79), the steady-state equivalent circuit of the synchronous machine is as shown in Figure 3.3.

The internal voltage $\overline{E}$ can be further simplified using Eq. (3.70).

$$\overline{E} = j\left[(X_{\mathrm{q}} - X_{\mathrm{d}}) I_{\mathrm{d}} + E_{\mathrm{fd}}\right] e^{j(\delta - \pi/2)} = \left[(X_{\mathrm{q}} - X_{\mathrm{d}}) I_{\mathrm{d}} + E_{\mathrm{fd}}\right] e^{j\delta} \tag{3.81}$$

The important point to be observed is that the angle of $\overline{E}$ is equal to δ. Therefore, having the equivalent circuit of Figure 3.3, the angle δ can be obtained. It can also be written from Eq. (3.35),

$$I_{\mathrm{fd}} = E_{\mathrm{fd}} / X_{\mathrm{md}} \tag{3.82}$$

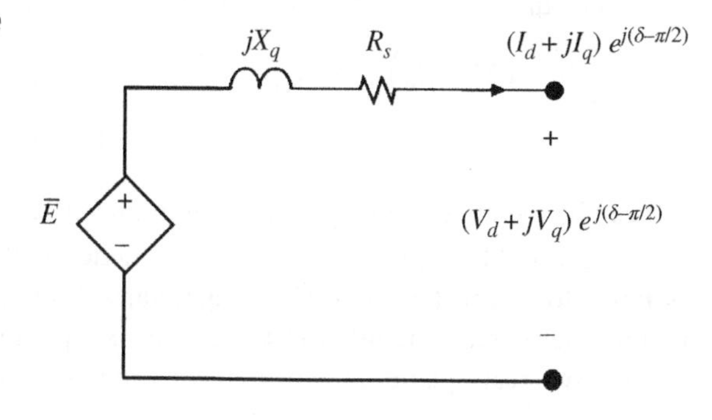

Figure 3.3 The steady-state equivalent circuit of the synchronous machine [1].

In the steady-state conditions, T_{FW} is zero and electrical torque is equal to mechanical torque,

$$T_{ELEC} = T_M = \Psi_d I_q - \Psi_q I_d = V_d I_d + V_q I_q + R_s\left(I_d^2 + I_q^2\right) \tag{3.83}$$

Which is exactly equal to the active power delivered by the internal source. Also, for $I = (I_d + jI_q)e^{j(\delta - \pi/2)}$,

$$T_{ELEC} = T_M = \text{Real}\left[\overline{EI^*}\right] \tag{3.84}$$

3.2.5 Simplification of Synchronous Machine Equations

Machine equations can be simplified by some assumptions; it is common to ignore the dynamics of the stator windings, i.e. the derivative terms in Eqs. (3.37)–(3.39) are assumed to be zero. By doing this, the stator equations become algebraic ones. The dynamics of transmission lines are also neglected. This makes it possible to use phasor equations for the network. Another simplification that is usually done is to set the time constants T''_{q0} and T''_{d0} to zero. This makes it possible to ignore flux linkages Ψ_{1d} and Ψ_{2q} in the equations. By these simplifications, the resulting model, known as the two-axis model, includes four differential equations and two algebraic equations (a complex equation) (of course, the excitation system, turbine, and governor models will be added if necessary.). These equations are

$$T'_{d0}\frac{dE'_q}{dt} = -E'_q - (X_d - X'_d)I_d + E_{fd} \tag{3.85}$$

$$T'_{q0}\frac{dE'_d}{dt} = -E'_d + \left(X_q - X'_q\right)I_q \tag{3.86}$$

$$\frac{d\delta}{dt} = \omega - \omega_s \tag{3.87}$$

$$\frac{2H}{\omega_s}\frac{d\omega}{dt} = T_M - E'_d I_d - E'_q I_q - \left(X'_q - X'_d\right)I_d I_q - T_{FW} \tag{3.88}$$

$$\left[E'_d + \left(X'_q - X'_d\right)I_q + jE'_q\right]e^{j\left(\delta - \frac{\pi}{2}\right)} = \left(R_s + jX'_d\right)\left(I_d + jI_q\right)e^{j\left(\delta - \frac{\pi}{2}\right)} + \left(V_d + jV_q\right)e^{j\left(\delta - \frac{\pi}{2}\right)} \tag{3.89}$$

It can be seen that there are four differential variables E'_q, E'_d, ω, and δ and four algebraic variables I_d, I_q, V_d, and V_q. Until modeling the excitation system, E_{fd} is assumed to be constant, too. These eight variables require eight equations to be found. Therefore, in addition to the above six equations (the complex equation is in fact two real equations), two other equations are needed that are the equations

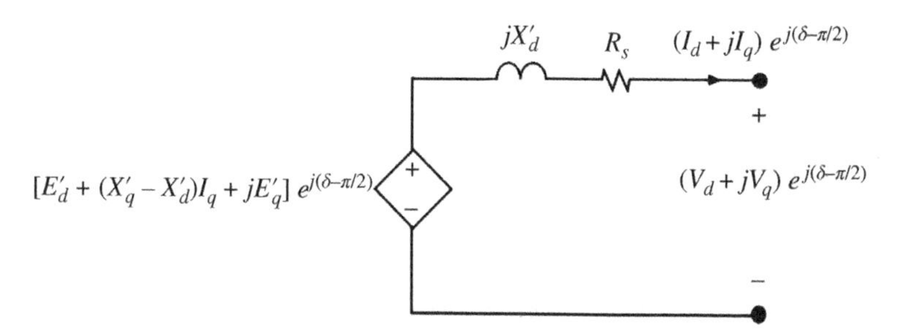

Figure 3.4 Two-axis dynamic model of synchronous machine [1].

related to the generator connection to the network. From the algebraic equation, the two-axis model of dynamic equivalent circuit is formed. This equivalent circuit is as shown in Figure 3.4.

A further simplification that is usually done is assuming that T'_{q0} is small, and this time constant is also set to zero. With this assumption, E'_d can be removed from the equations. The result of this simplification known as the one-axis model consists of three-differential equations and two algebraic equations as follows:

$$T'_{d0}\frac{dE'_q}{dt} = -E'_q - (X_d - X'_d)I_d + E_{fd} \tag{3.90}$$

$$\frac{d\delta}{dt} = \omega - \omega_s \tag{3.91}$$

$$\frac{2H}{\omega_s}\frac{d\omega}{dt} = T_M - E'_q I_q - (X_q - X'_d)I_d I_q - T_{FW} \tag{3.92}$$

$$\left[(X_q - X'_d)I_q + jE'_q\right]e^{j\left(\delta-\frac{\pi}{2}\right)} = (R_s + jX'_d)\left(I_d + jI_q\right)e^{j\left(\delta-\frac{\pi}{2}\right)} + \left(V_d + jV_q\right)e^{j\left(\delta-\frac{\pi}{2}\right)} \tag{3.93}$$

From the complex equation, the dynamic equivalent circuit of the one-axis model is shown in Figure 3.5.

3.2.6 Saturation Modeling

A control measure to increase the voltage stability margin is to increase the voltage of synchronous generators. Increasing the voltage is done by increasing the excitation current and consequently the flux. This increase in flux can lead to the machine saturation, in which case the self- and mutual-inductances (and consequently the self- and mutual-reactances in Eqs. (3.20)–(3.25)) are no longer fixed

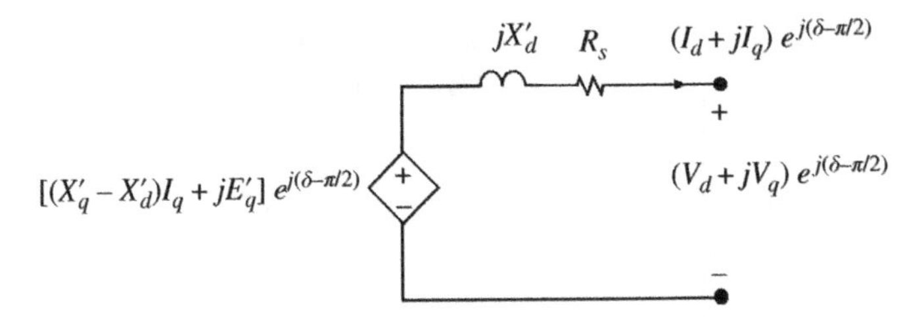

Figure 3.5 One-axis dynamic model of synchronous machine [1].

values and are a function of the machine operation point. For saturation modeling, the self-reactances of the windings are first decomposed as follows:

$$X_{\mathrm{d}} = X_{\mathrm{ls}} + X_{\mathrm{md}} \tag{3.94}$$

$$X_{\mathrm{q}} = X_{\mathrm{ls}} + X_{\mathrm{mq}} \tag{3.95}$$

$$X_{\mathrm{fd}} = X_{\mathrm{lfd}} + X_{\mathrm{md}} \tag{3.96}$$

$$X_{\mathrm{1d}} = X_{\mathrm{l1d}} + X_{\mathrm{md}} \tag{3.97}$$

$$X_{\mathrm{1q}} = X_{\mathrm{l1q}} + X_{\mathrm{mq}} \tag{3.98}$$

$$X_{\mathrm{2q}} = X_{\mathrm{l2q}} + X_{\mathrm{mq}} \tag{3.99}$$

where X_{ls}, X_{lfd}, X_{l1d}, X_{l1q}, and X_{l2q} are leakage reactances that are independent from saturation. Therefore, the values of the reactances in the saturation conditions denoted by the superscript "s" are expressed as Eqs. (3.100)–(3.105).

$$X_{\mathrm{d}}^{\mathrm{s}} = X_{\mathrm{ls}} + X_{\mathrm{md}}^{\mathrm{s}} \tag{3.100}$$

$$X_{\mathrm{q}}^{\mathrm{s}} = X_{\mathrm{ls}} + X_{\mathrm{mq}}^{\mathrm{s}} \tag{3.101}$$

$$X_{\mathrm{fd}}^{\mathrm{s}} = X_{\mathrm{lfd}} + X_{\mathrm{md}}^{\mathrm{s}} \tag{3.102}$$

$$X_{\mathrm{1d}}^{\mathrm{s}} = X_{\mathrm{l1d}} + X_{\mathrm{md}}^{\mathrm{s}} \tag{3.103}$$

$$X_{\mathrm{1q}}^{\mathrm{s}} = X_{\mathrm{l1q}} + X_{\mathrm{mq}}^{\mathrm{s}} \tag{3.104}$$

$$X_{\mathrm{2q}}^{\mathrm{s}} = X_{\mathrm{l2q}} + X_{\mathrm{mq}}^{\mathrm{s}} \tag{3.105}$$

Saturation modeling is done by defining the saturation coefficients K_{d} and K_{q}, as follows:

$$K_{\mathrm{d}} \triangleq \frac{X_{\mathrm{md}}^{\mathrm{s}}}{X_{\mathrm{md}}} \tag{3.106}$$

$$K_{\mathrm{q}} \triangleq \frac{X_{\mathrm{mq}}^{\mathrm{s}}}{X_{\mathrm{mq}}} \tag{3.107}$$

where X_{md} and X_{mq} are the reactances values with the assumption of unsaturation. In salient pole machines, only saturation on the *d*-axis is usually considered, but in nonsalient pole machines, saturation modeling is done on both axes. In nonsalient pole machines, the values of K_d and K_q are different but close to each other, which are assumed to be equal due to the fact that in most cases the saturation characteristics in the *d*-axis are not available [2].

Saturation coefficients when entering the saturation area have values close to 1 and along with increasing saturation, their value decreases. The saturation degree depends on the air-gap flux linkage. The air-gap flux linkage in the direction of axes d and q is obtained using Eqs. (3.108) and (3.109) [3].

$$\Psi_{ad} = \Psi_d + X_{ls}I_d = V_q + R_sI_q + X_{ls}I_d \tag{3.108}$$

$$\Psi_{aq} = \Psi_q + X_{ls}I_q = -V_d - R_sI_d + X_{ls}I_q \tag{3.109}$$

Now, the air-gap flux linkage becomes:

$$\Psi_{ag} = \sqrt{\Psi_{ad}^2 + \Psi_{aq}^2} = \sqrt{(V_q + R_sI_q + X_{ls}I_d)^2 + (-V_d - R_sI_d + X_{ls}I_q)^2} = V_l \tag{3.110}$$

where V_l is the magnitude of a voltage called voltage behind leakage reactance, which is obtained from the following equation:

$$V_l = V_s e^{j\theta_s} + (R_s + jX_{ls})I_s e^{j\varphi_s} \tag{3.111}$$

Therefore, having the machine's output voltage and current, the air-gap flux can be calculated from Eq. (3.111). Various functions have been proposed to express the relationship between the air-gap flux and saturation coefficient, an example of which is shown as Eq. (3.112) [2].

$$K_q = K_d = \frac{1}{1 + mV_l^n} \tag{3.112}$$

where *m* and *n* are positive real numbers.

3.2.7 Synchronous Generator Capability Curve

The synchronous generator capability curve expresses the limits of active and reactive power generation. This curve shows that the reactive power generation limit is not a fixed number and depends on the generated active power. The same is true about active power generation. The limits of active and reactive power generations at each output voltage value are specified by the maximum allowable values of stator and rotor (excitation) currents. Of course, turbine capability can also be a limiting factor in active power generation.

- Limitation due to the stator maximum current

The relationship between the complex power in per unit at a certain voltage magnitude V_s with the stator current is as (3.113).

$$\overline{S} = (V_s e^{j\theta_s}) \times \left(I_s e^{j\emptyset_s}\right)^* \tag{3.113}$$

Assuming that the stator current is at its maximum value, the complex power change curve on the P–Q plane is a circle with the center of the coordinate origin and the radius $V_s I_{s\ max}$, which $I_{s\ max}$ is the maximum value of stator current. Due to violation of the stator maximum allowable current, it is not possible to operate the generator continuously beyond this circle.

- Limitation due to the excitation maximum current

Ignoring the salience of poles and the stator resistance from the steady-state equivalent circuit, the following equation can be written:

$$X_{md} I_{fd} e^{j\delta} = V_s e^{j\theta_s} + jX_s I_s e^{j\emptyset_s} \tag{3.114}$$

where $X_s = X_d = X_q$. By multiplying both sides of Eq. (3.114) by $e^{-j\theta_s}$:

$$X_{md} I_{fd} e^{j(\delta-\theta_s)} = V_s + jX_s I_s e^{j(\emptyset_s-\theta_s)} \tag{3.115}$$

By equating the real and the imaginary parts,

$$X_{md} I_{fd} \cos(\delta-\theta_s) = V_s - X_s I_s \sin(\emptyset_s-\theta_s) \tag{3.116}$$

$$X_{md} I_{fd} \sin(\delta-\theta_s) = X_s I_s \cos(\emptyset_s-\theta_s) \tag{3.117}$$

So, the per unit values of active and reactive powers are

$$P = V_s I_s \cos(\theta_s-\emptyset_s) = \frac{X_{md}}{X_s} V_s I_{fd} \sin(\delta-\theta_s) \tag{3.118}$$

$$Q = V_s I_s \sin(\theta_s-\emptyset_s) = \frac{X_{md}}{X_s} V_s I_{fd} \cos(\delta-\theta_s) - \frac{V_s^2}{X_s} \tag{3.119}$$

Assuming that the current I_{fd} is fixed at its maximum value, at a certain voltage V_s, the change curve of P and Q is a circle with the center of $-\frac{V_s^2}{X_s}$ and radius of $\frac{X_{md}}{X_s} V_s I_{fd\ max}$ that $I_{fd\ max}$ is the maximum value of the excitation current. Figure 3.6 shows the synchronous generator capability curve due to the limitation of stator and excitation currents. It is observed that except for large amounts of active power, the maximum amount of reactive power generation is limited by the excitation current limit. In this figure, the limitation imposed by the turbine in generating active power is shown, too.

Figure 3.6 The synchronous generator capability curve

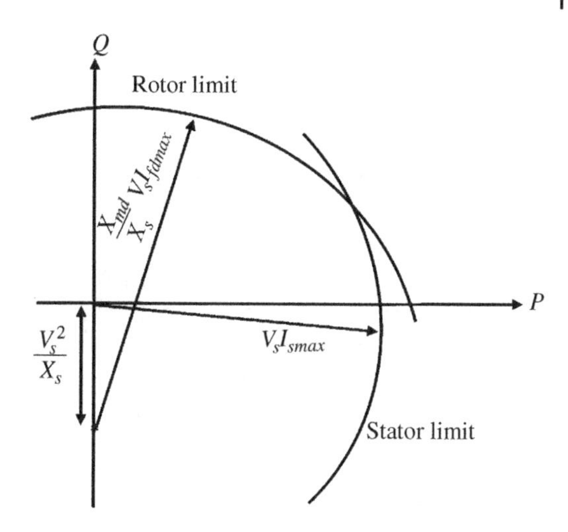

The limits specified in Figure 3.6 change as the output voltage changes. The stator limit changes further as the voltage changes because the center of the circle related to this limit is fixed and just its radius changes. But both of the center and the radius of the circle related to the rotor limit depend on the output voltage, and this dependence is such that it reduces the effect of the output voltage change on the change of rotor limit. For example, although increasing the voltage increases the radius, it also causes the center of the circle to move in the negative direction of the Q axis. Even the saturation effect may reduce the rotor limit, when the voltage increases. Because increasing the voltage causes decrease of X_s and more displacement of the circle center. The stator current and turbine limits do not change due to machine saturation. Figure 3.7 shows the effect of voltage change on the stator and rotor limits.

If the salience of the poles are considered, the equations of the active and reactive powers are given as (3.120) and (3.121). The salience of the poles has no effect on the stator limit but changes the rotor limit. At each certain value of V_s, it is possible to calculate this limit by substituting I_{fd} with $I_{fd\ max}$ and numerically solving Eqs. (3.120) and (3.121) for different values of I_s, θ_s, and $\emptyset_s$.

$$P = \frac{V_s}{X_q}\left[X_{md}I_{fd} + \left(X_q - X_d\right)I_d\right]\sin\left(\delta - \theta_s\right) \tag{3.120}$$

$$Q = \frac{V_s}{X_q}\left[X_{md}I_{fd} + \left(X_q - X_d\right)I_d\right]\cos\left(\delta - \theta_s\right) - \frac{V_s^2}{X_q} \tag{3.121}$$

According to Eq. (3.119), reducing the excitation current leads to the reactive power generation decrement. Under low load conditions, large reduction in the

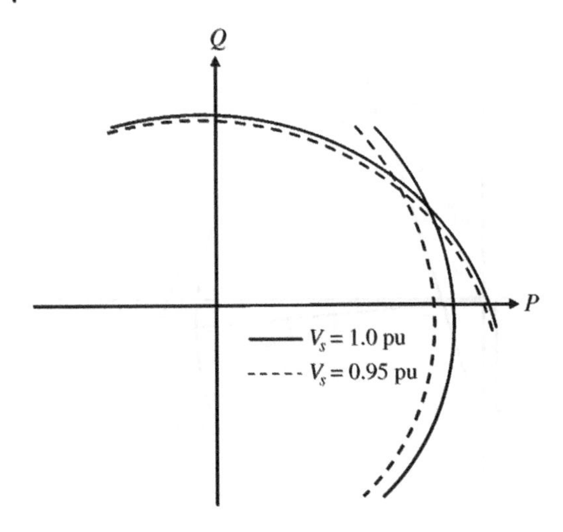

Figure 3.7 The effect of voltage changes on generator capability curve.

excitation current can cause absorption of reactive power by the synchronous generator, which is called the under-excitation state. Further reduction of the excitation current causes the end region of stator core to heat up due to eddy current loss, which limits the absorption of reactive power [2]. It is obvious that in the voltage instability occurrence conditions that the maximum reactive power capacity of generators is used to maintain voltage, the under-excitation state does not occur.

3.2.8 Excitation System Modeling

The excitation system, also known as the automatic voltage regulator (AVR), is responsible for supplying the DC voltage to the excitation winding to control the generator output voltage. There are different types of excitation systems. Older types use DC and AC generators to supply the required power to the excitation winding. But in newer types, the required power is supplied from the output of a synchronous generator through a transformer. All components of this excitation system are static, and for this reason they are called static excitation system. A complete description of the excitation systems types is given in the Ref. [3]. The block diagram of Figure 3.8 is usually used to model the excitation system,

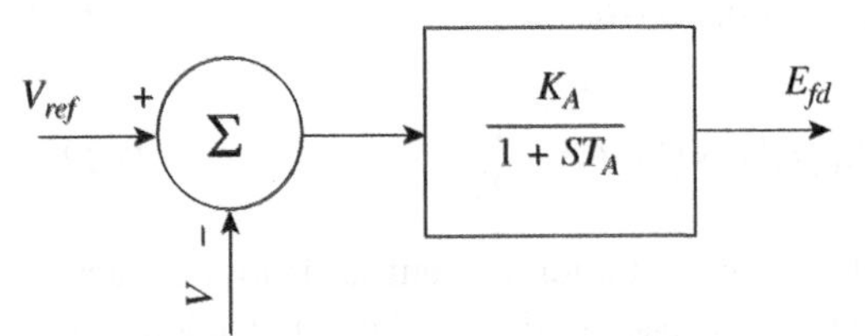

Figure 3.8 The block diagram of static excitation system.

which is a simple first-order model of a static exciter [1]. In this block diagram, V_{ref} is the excitation system reference voltage and V is the magnitude of the measured voltage, which can be the same as the generator output voltage or the voltage of another point inside or outside the generator. K_{A} and T_{A} are the gain and time constant of the excitation system.

The differential equation of this model is also as Eq. (3.122).

$$T_{\mathrm{A}}\frac{\mathrm{d}E_{\mathrm{fd}}}{\mathrm{d}t} + E_{\mathrm{fd}} = K_{\mathrm{A}}(V_{\mathrm{ref}} - V) \tag{3.122}$$

In order to improve the voltage stability, instead of the generator output voltage, voltage of another point outside the generator that has shorter distance from load centers may be adjusted. In this case, V is determined from Eq. (3.123) that $\overline{V}_{\mathrm{s}}$ and $\overline{I}_{\mathrm{s}}$ are the generator output phasor voltage and current and $R_{\mathrm{C}} + jX_{\mathrm{C}}$ is the impedance between the generator and the desired point. This impedance is usually considered to be between 50 and 90% of the step-up transformer impedance. This technique is called load or line drop compensation [2, 3]. The important point to consider is that the point at which the voltage is regulated should not be common among two or more generators. Because, in these case, a small difference between their excitation systems, which is inevitable, will cause one of the generators to try to regulate the desired voltage to a higher value than the other generators. In this situation, this generator uses its maximum reactive power generation capacity, and the other generators absorb the reactive power as much as their excitation limit allows [2].

$$V = \left|\overline{V}_{\mathrm{s}} - (R_{\mathrm{C}} + jX_{\mathrm{C}})\overline{I}_{\mathrm{s}}\right| \tag{3.123}$$

Several generators may have a common output bus connected to the network via a step-up transformer. For example, in hydroelectric power plants, several small generators use a common step-up transformer. In this case, the negative sign in Eq. (3.123) is changed to positive sign as Eq. (3.124), to control the voltage of a hypothetical point inside the generators instead of the output bus voltage or voltage at a common point outside the generators. This weakens the network voltage regulation, but instead it distributes reactive power evenly among generators.

$$V = \left|\overline{V}_{\mathrm{s}} + (R_{\mathrm{C}} + jX_{\mathrm{C}})\overline{I}_{\mathrm{s}}\right| \tag{3.124}$$

3.2.9 Governor Modeling

By controlling the input power to the turbine, the governor is responsible for regulating the frequency and output active power of the generators. The governor steady-state characteristic is shown in Figure 3.9. In this figure, the generator speed or frequency changes when its output active power changes are shown. It is observed that the governor's performance is such that increasing the output

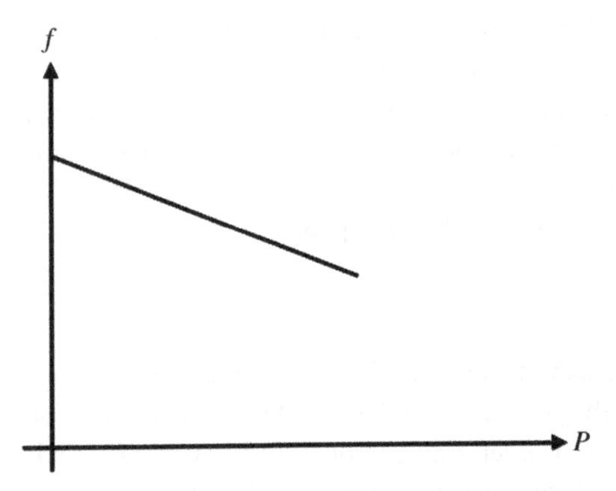

Figure 3.9 Steady-state characteristic of governor with speed droop.

power causes the generator speed to decrease, which is called droop characteristic. Due to the fact that, at the steady state, the frequency is the same all over the system, the droop characteristic allows more than one generator to be connected to the system. In the absence of a droop characteristic, the generators try to adjust the system frequency to different values, which is not possible. During changing the system active power consumption (for example, due to a change in load or a generator outage), the slope of the governor's droop characteristic determines the amount of active power change of each generator.

When the governor characteristic is constant, increasing the output active power caused by the consumption increment will cause the system frequency to drop. In order to recover frequency to the nominal value, it is possible to create parallel characteristics in governors according to Figure 3.10. To do this, a device called speed changer is used.

Eqs. (3.125) and (3.126) are usually used to model the governor [1]. In these equations, ω_r is the generator speed, ω_s is the synchronous speed, and R_D is the

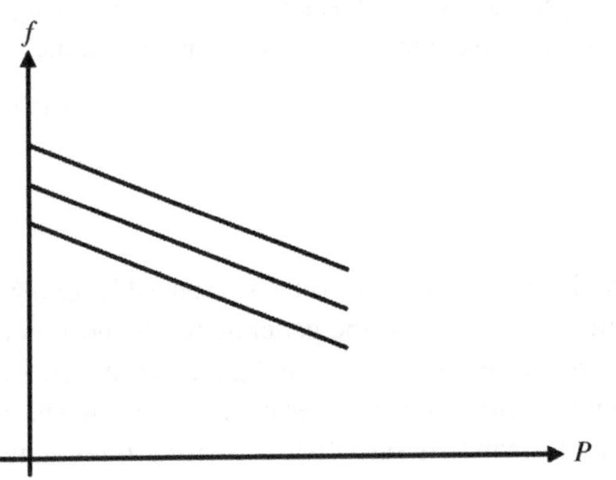

Figure 3.10 Parallel characteristics in governors.

speed droop. Also, P_{SV} is the valve position and P_C is power change setting adjusted by speed changer. T_{SV} and T_{CH} are the time constants of valve and steam chest.

$$T_{SV}\frac{dP_{SV}}{dt} = -P_{SV} + P_C - \frac{1}{R_D}\left(\frac{\omega_r}{\omega_s} - 1\right) \quad (3.125)$$

$$T_{CH}\frac{dT_M}{dt} = -T_M + P_{SV} \quad (3.126)$$

Using these equations, it is possible to calculate the frequency change and the generators output active power change for each change in the system. Due to the dependence of the system loadability limit on the generators output active power, accurate voltage stability assessment may need the governors modeling.

3.2.10 Overexcitation Limiter (OXL) Modeling

In the synchronous generator, there are various limiters such as overexcitation limiter, underexcitation limiter, voltage/Hz limiter, and armature current limiter, which just the overexcitation limiter is effective in the occurrence of voltage instability. Hence, here only modeling of this limiter is discussed.

The overexcitation limiter prevents over current of excitation winding. Reduction of excitation current can be done immediately, but usually to improve the network stability, a temporary over current is allowed [2]. The delay in the operation of limiter depends on the value of over current, which is determined by means of the inverse time characteristic of limiter. This characteristic does not increase the steady-state loadability limit, but can prevent the instability occurred due to leaving the attraction region. Leaving the attraction region is the main cause of the occurrence of first-swing angle instability and as mentioned in the Chapter 1 it can be a reason of voltage collapse. Figure 3.11 shows an example of the inverse time characteristic of overexcitation limiters [3].

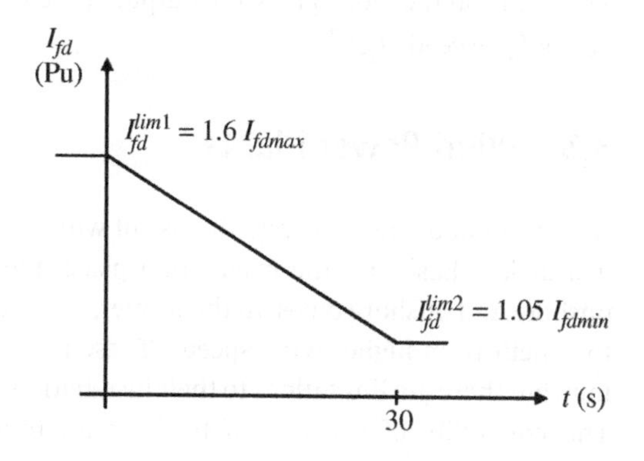

Figure 3.11 The inverse time characteristic of overexcitation limiters (OXL) [3].

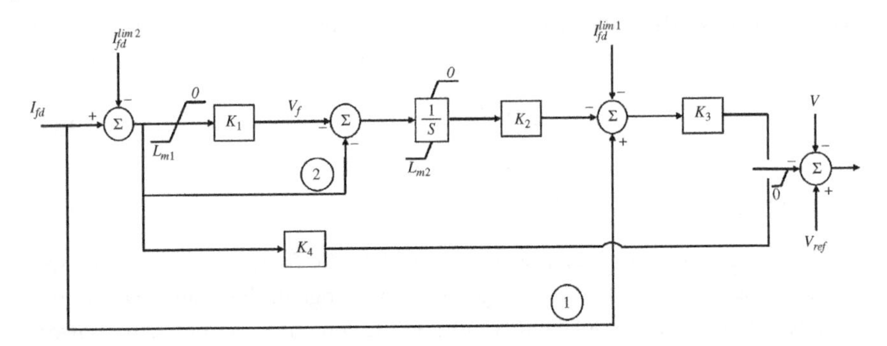

Figure 3.12 The diagram block of overexcitation limiter [3].

It is observed that $I_{fd}^{lim\,2}$ is chosen 5% more than $I_{fd\,max}$. If the excitation current exceeds $I_{fd}^{lim\,1}$, the limitation process is almost instantaneous. The block diagram of Figure 3.12 can be used to implement the characteristic of Figure 3.11 in time simulations [3]. The output of the block diagram enters the input collector of the excitation system (Figure 3.8) and is considered as a part of the measured voltage, from the excitation system view point. Therefore, if it is necessary to reduce the excitation current, a positive signal inserts to the collector, which is considered as overvoltage from the excitation system point of view. A zero value of this signal means that the excitation current is not reduced. When I_{fd} is less than $I_{fd\,max}$, the switch changes to the low position, causing zero output signal. If the current is greater than $I_{fd}^{lim\,1}$, the limitation is almost instantaneous using the control loop 1. If the current is less than $I_{fd}^{lim\,1}$, the limiting operation is performed by control loop 2. The gain K_2 as well as the value of current I_{fd} determine the limitation time delay. The signal V_F only gets a value when I_{fd} becomes less than $I_{fd}^{lim\,2}$. In this state, using the large gain of K_1, the output of the integrator quickly becomes zero. As long as the current I_{fd} is less than $I_{fd}^{lim\,2}$, the output of the integrator remains zero. The limits on the block allow the output of the integral change to a negative number as I_{fd} exceeds $I_{fd}^{lim\,2}$.

3.3 Wind Power Plants

As a clean and cheap energy, the use of wind to generate electricity is increasing worldwide. Based on the installation place, the wind turbines are divided into onshore and offshore types. In the former, turbines must be installed in high areas to benefit from higher wind speeds. These turbines cost less to install and maintain, but there are limitations to their installation due to noise and visual problems. The possibility of collision of blades with birds is another problem of these turbines. The above restrictions are not for offshore turbines installed at sea. Therefore, it is possible to build larger wind farms with these turbines. However, connecting offshore turbines to transmission network usually requires longer cables.

Wind turbines deliver the wind power through a gearbox to an electric generator. The output voltage of the generator is about 700 V, which is changed to a suitable value by a transformer. The available wind power, in addition to the wind speed, also depends on the swept area of the turbine blades as follows [4].

$$P_{\text{air}} = \frac{1}{2}\rho A \nu^3 \tag{3.127}$$

where

ρ= Air density, kg/m
A= Swept area of the blades, m^2
ν= Wind speed, m/s

Only part of the available wind power is transferred to the turbine rotor, determined by the power factor C_P [4].

$$P_{\text{wind-turbine}} = C_P P_{\text{air}} = C_P \times \frac{1}{2}\rho A \nu^3 \tag{3.128}$$

C_P depends on the pitch angle of turbine blade and tip-speed ratio. The largest value of C_P is obtained with the blade pitch angle equal to zero and by increasing it, C_P will decrease [5]. The tip-speed ratio is shown as Eq. (3.129) [4].

$$\lambda = \frac{\omega R}{\nu} \tag{3.129}$$

where

ω = Rotational speed of rotor
R = Radius to tip of rotor
ν = Wind speed, m/s

An example of the curve of C_P changes in terms of λ at zero pitch angle is as shown in Figure 3.13.

It can be seen that, with this pitch angle, the maximum value of C_P is when λ is about 8. In variable speed turbines, it is possible to adjust λ to a desired value at different wind speeds. But in a constant speed turbine, the value of λ changes with the change of wind speed.

The output power of a wind turbine at different wind speeds is expressed by its power curve. Figure 3.14 shows an example of power curve for a 4.4 MW wind turbine [5]. This curve is obtained by assuming that at all wind speeds, C_P is set to its

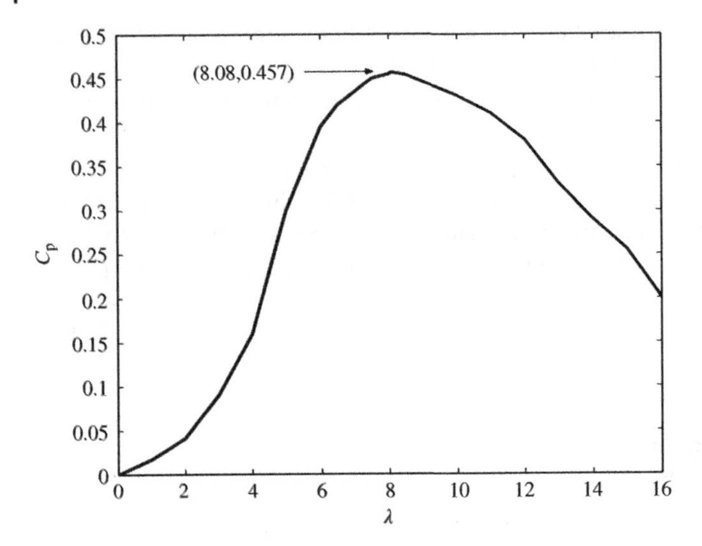

Figure 3.13 The C_p changes in terms of λ at zero pitch angle.

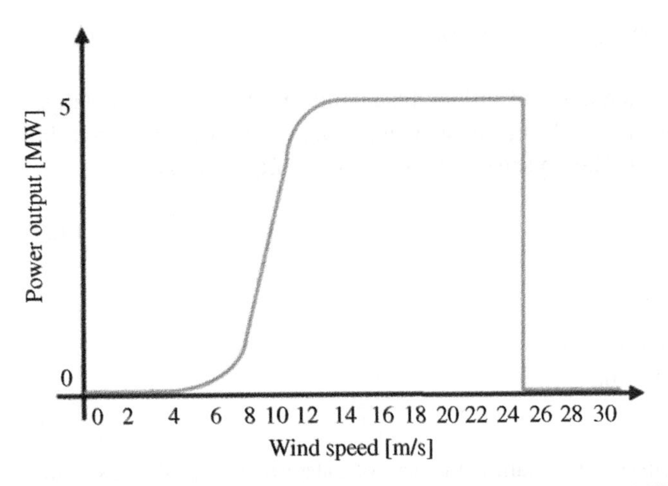

Figure 3.14 Changes of generator power in terms of wind speed [5].

maximum value. The cut-in speed is the wind speed at which the generator begins to generate power. In this turbine, this speed is equal to 4 m/s, which the power of 0.1 MW is produced at it. After that, the generated power increases with increasing wind speed. At the wind speed of 14 m/s, the nominal power of 4.4 MW is reached. Rated power is usually determined using the maximum output power of the generator [4]. Until this wind speed, the value of the pitch angle is set to zero in order to receive the maximum possible wind power. But at wind speeds above 14 m/s, this angle is increased to limit the received power. Until the wind speed reaches the cut-out speed (Here, 24 m/s), the output power is kept constant at its maximum value by increasing the pitch angle. At wind speeds more than the cut-out speed, the turbine is shut down to prevent probable damages.

3.3.1 Fixed-Speed Induction Generator (FSIG)-based Wind Turbine

3.3.1.1 Physical Description

Although most new wind farms include variable speed types, fixed speed wind farms are still used that the main reasons are low cost, brushless, maintenance free, and operational simplicity [6]. Figure 3.15 shows the main components of an fixed speed induction generator (FSIG)-based wind turbine, which includes the turbine rotor, gearbox, induction generator (Asynchronous), soft starter, capacitor bank, and turbine transformer.

Squirrel cage induction generators are commonly used in FSIG-based wind turbines. The speed (Slip) of induction generators changes with the change of power, but due to the fact that the change of speed in the normal operating range of the generators are very small, these turbines are called constant speed wind turbines. A capacitive bank is used to supply the reactive power consumed by the induction generator. Unlike synchronous generators, these generators always consume reactive power unless, like doubly fed induction generators (DFIGs), a controllable voltage is injected into the rotor windings. A soft-starter unit is used to apply the voltage slowly. This minimizes transient currents during starting.

3.3.1.2 Induction Machine Steady-State Model

The induction machine model is presented in Section 2.3.2 and is briefly mentioned in this section and the next section, too. Figure 3.16 shows the steady-state single-phase equivalent circuit of the induction machine. In this figure, x_s and x_m

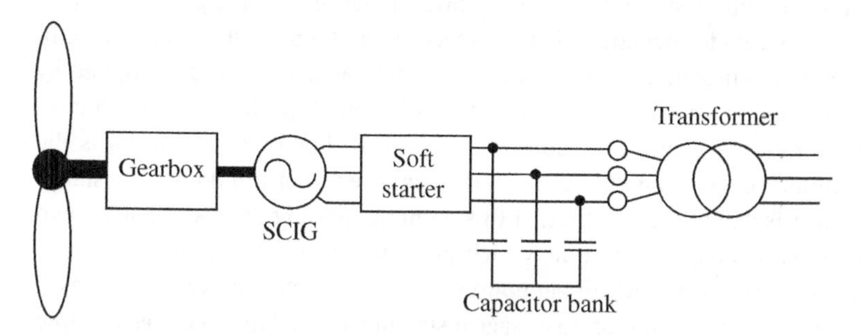

Figure 3.15 Components of FSIG-based wind turbine.

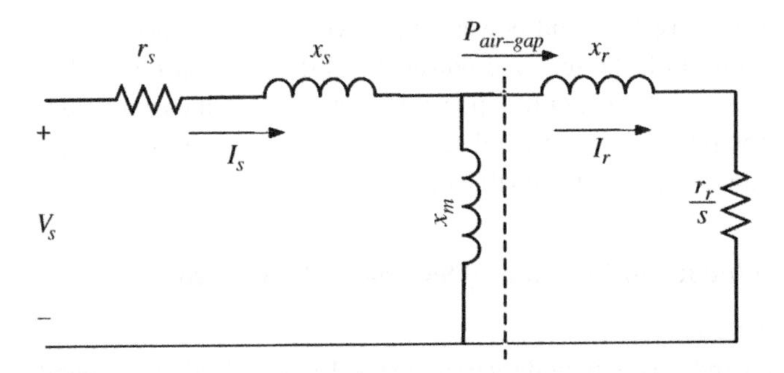

Figure 3.16 Steady-state single-phase equivalent circuit of induction machine.

are the stator leakage reactance and the magnetizing reactance, respectively. r_s is the ohmic resistance of the stator. x_r and r_r are in respect the leakage reactance and the rotor resistance that are transmitted to the stator. S is slip that defined according to Eq. (3.130). In this equation, ω_s is the synchronous speed and ω_r is the rotor speed in radians per second.

$$S = \frac{\omega_s - \omega_r}{\omega_s} \tag{3.130}$$

$P_{\text{air-gap}}$ is the power passes through the air gap that is obtained from the following equation:

$$P_{\text{air-gap}} = \frac{r_r}{S} I_r^2 \tag{3.131}$$

From this power, the electromagnetic torque is calculated as Eq. (3.132).

$$T_e = 3 \frac{P_f}{2} \frac{r_r}{S\omega_s} I_r^2 \tag{3.132}$$

where P_f is the number of machine poles. By substituting I_r with V_s and the impedances in Figure 3.15, the changes in electromagnetic torque with respect to slip for a certain amount of stator voltage are shown in Figure 3.17. The generator operation mode appears for negative slips, i.e. speeds more than synchronous speed. In this figure, the approximate variation curve of the mechanical torque applied by the turbine rotor is also shown. The point of collision of mechanical and electromagnetic torques indicates the equilibrium point. In Figure 3.16, point A is the stable equilibrium point and point B is the unstable equilibrium point. The stability of point A is due to the fact that, at this point, increasing the speed leads to an increase in electromagnetic torque, which prevents further acceleration.

Due to the fact that induction generators consume reactive power, FSIG-based wind turbines are always subject to voltage instability and collapse. Voltage collapse happens when there is no equilibrium point for the wind turbine. Under these

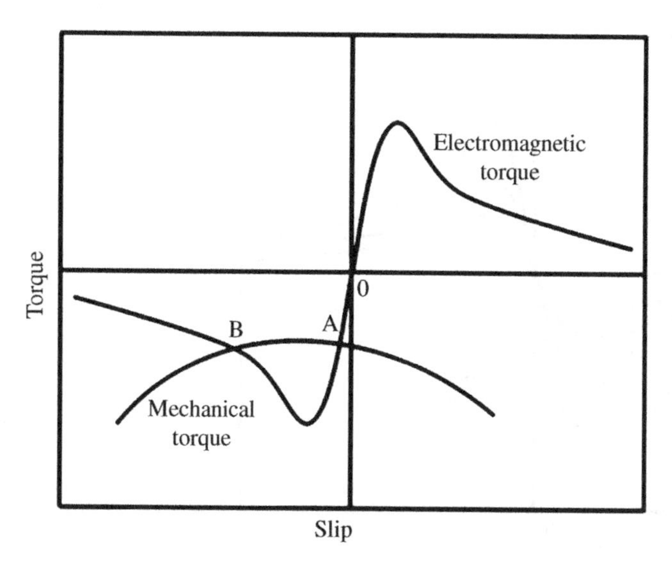

Figure 3.17 Induction machine torque–slip changes.

conditions, the speed of the generator rotor and consequently the amount of slip increases, which causes high absorption of reactive power and a sharp decrease in voltage. Two factors cause the lack of equilibrium point: (i) increasing the input mechanical torque due to increasing wind speed and (ii) decreasing the generator output voltage, which is caused by various reasons such as increasing the system load and occurrence of contingencies [6]. Figure 3.18 shows the effect of decreasing voltage and increasing wind speed on electromagnetic and mechanical torques, respectively. It is observed that each of these two factors can lead to a mode of equilibrium point lack.

Another point that can be seen in Figure 3.18 is that increasing the input mechanical torque itself leads to increase in the slip value and absorption more reactive power. Consequently, it is possible that increasing the mechanical torque, due to the output active power increment, initially increases the output voltage. However, along with more increase of this torque, reactive power absorption may dominate active power generation and causing the occurance of voltage drop. Impact of power generation and consumption on voltage increment and reduction can be illustrated using the simple system of Figure 3.19.

In this system, a generator is connected to the infinite bus via a transmission line. The relationship between the generator voltage and its output power can be approximately described using Eq. (3.133) [7]. In this equation, a positive sign of powers denotes production of them.

$$V_2 = V_1 + rP + xQ \tag{3.133}$$

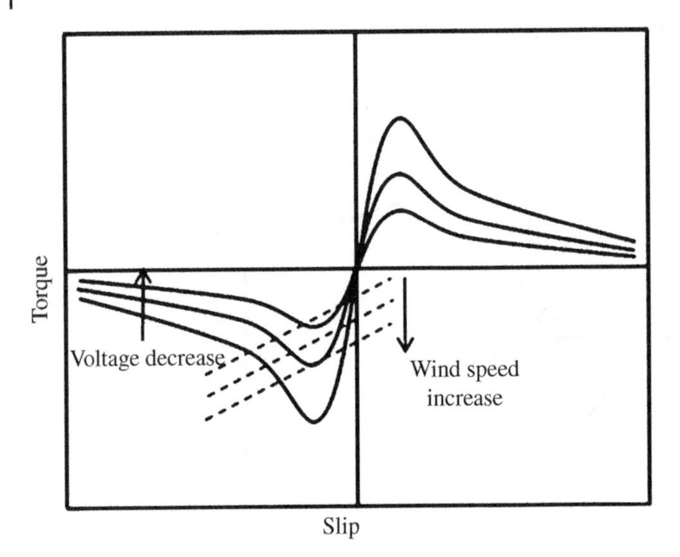

Figure 3.18 The effect of decreasing voltage and increasing wind speed.

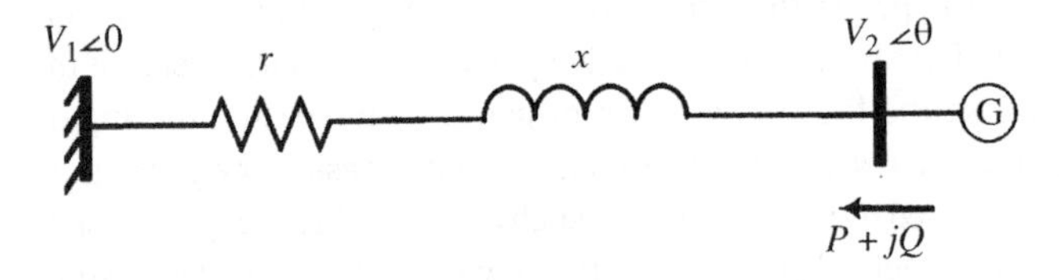

Figure 3.19 Simple two-bus system.

It is observed that the generation of each of the active and reactive powers increases the voltage and their consumption reduces the voltage. This equation also shows that the relative effect of each power on the voltage change depends on the values of r and x.

The steady-state equivalent circuit is used to determine the equilibrium point as well as to diagnosis of existence or nonexistence of an equilibrium point [6]. Of course, the model parameters must be carefully found to correctly predict the machine behavior. In Ref. [8], it is shown that in some cases, it may be necessary to use the double-cage model (Figure 3.20) instead of the single-cage mode, as shown in Figure 3.16. This is much more necessary, especially for machines with high starting torque.

3.3.1.3 Induction Generator Dynamic Model

The induction motor dynamic model was described in detail in Section 2.3.2. Here, the model equations are briefly repeated for the generator convention with

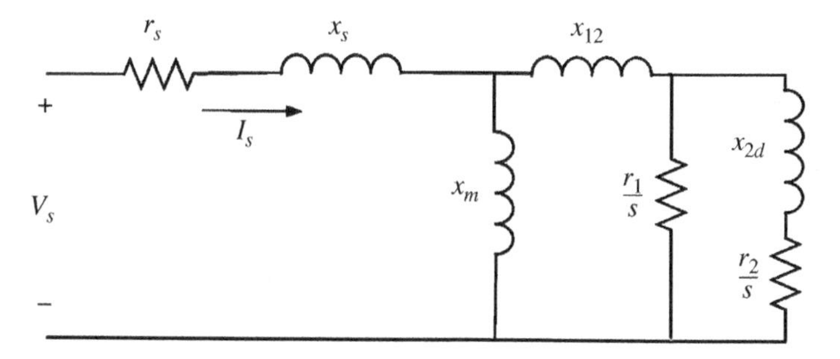

Figure 3.20 A double cage for steady-state equivalent circuit of induction machine [8].

considering the rotor supply voltage. These equations can be used for both FSIGs and DFIGs. The only difference is in the rotor supply voltage, which is zero in FSIG.

The equations of stator and rotor voltages on the d and q axes in terms of currents and flux linkages in per unit system are as Eqs. (3.134)–(3.137). In these equations, ω_b is the base speed in electrical radians per second.

$$V_{ds} = -R_s I_{ds} - \frac{\omega_s}{\omega_b}\Psi_{qs} + \frac{1}{\omega_b}\frac{d}{dt}\Psi_{ds} \tag{3.134}$$

$$V_{qs} = -R_s I_{qs} + \frac{\omega_s}{\omega_b}\Psi_{ds} + \frac{1}{\omega_b}\frac{d}{dt}\Psi_{qs} \tag{3.135}$$

$$V_{dr} = R_r I_{dr} - S\frac{\omega_s}{\omega_b}\Psi_{qr} + \frac{1}{\omega_b}\frac{d}{dt}\Psi_{dr} \tag{3.136}$$

$$V_{qr} = R_r I_{qr} - S\frac{\omega_s}{\omega_b}\Psi_{dr} + \frac{1}{\omega_b}\frac{d}{dt}\Psi_{qr} \tag{3.137}$$

Also, the equations of the stator and rotor flux linkage in terms of currents in per unit system are as (3.138)–(3.141).

$$\Psi_{ds} = -L_{ss} I_{ds} + L_m I_{dr} \tag{3.138}$$

$$\Psi_{qs} = -L_{ss} I_{qs} + L_m I_{qr} \tag{3.139}$$

$$\Psi_{dr} = L_{rr} I_{dr} - L_m I_{ds} \tag{3.140}$$

$$\Psi_{qr} = L_{rr} I_{qr} - L_m I_{qs} \tag{3.141}$$

Obtaining equations for I_{dr} and I_{qr} from (3.140) and (3.141) and inserting them in (3.136) and (3.137), the following equation is obtained [4]:

$$\frac{dV'_d}{dt} = -\frac{1}{T'_0}\left[V'_d - (X_s - X'_s)I_{qs}\right] + S\omega_s V'_d - \omega_s \frac{L_m}{L_{rr}} V_{qr} \tag{3.142}$$

$$\frac{dV'_q}{dt} = -\frac{1}{T'_0}\left[V'_q + (X_s - X'_s)I_{ds}\right] - S\omega_s V'_d + \omega_s \frac{L_m}{L_{rr}} V_{dr} \tag{3.143}$$

where

$$V'_d = -\frac{\omega_s L_m}{\omega_b L_{rr}} \Psi_{qr} \tag{3.144}$$

$$V'_q = \frac{\omega_s L_m}{\omega_b L_{rr}} \Psi_{dr} \tag{3.145}$$

$$T'_0 = -\frac{L_{rr}}{\omega_b R_r} \tag{3.146}$$

$$X_s = \frac{\omega_s}{\omega_b} L_{ss} \tag{3.147}$$

And,

$$X'_s = \frac{\omega_s}{\omega_b} \left(L_{ss} - \frac{L_m^2}{L_{rr}} \right) \tag{3.148}$$

For FSIGs, $V_{dr} = V_{qr} = 0$.
Similarly, there are following equations for the stator voltages:

$$V_{ds} = -R_s I_{ds} + X'_s I_{qs} + V'_d - \frac{X'_s}{\omega_s} \frac{d}{dt} I_{ds} + \frac{1}{\omega_s} \frac{d}{dt} V'_q \tag{3.149}$$

$$V_{qs} = -R_s I_{qs} - X'_s I_{ds} + V'_q - \frac{X'_s}{\omega_s} \frac{d}{dt} I_{qs} + \frac{1}{\omega_s} \frac{d}{dt} V'_d \tag{3.150}$$

In the stability studies, it is common to ignore the stator transitions. Therefore, the derivative terms in Eqs. (3.149)–(3.150) are omitted. This ensures that the stator equations, which are usually assumed to be in sinusoidal steady state, be compatible with the network equations. Accordingly, Eqs. (3.149) and (3.150) change as follows:

$$V_{ds} = -R_s I_{ds} + X'_s I_{qs} + V'_d \tag{3.151}$$

$$V_{qs} = -R_s I_{qs} - X'_s I_{ds} + V'_q \tag{3.152}$$

The next differential equation is the swing equation, defined as follows:

$$\frac{d}{dt} \omega_r = \frac{\omega_b}{2H} (T_m - T_e) \tag{3.153}$$

H is the inertia constant in seconds and T_e is electromagnetic torque, obtained as Eq. (3.154).

$$T_e = \frac{V'_d I_{ds} + V'_q I_{qs}}{\omega_s / \omega_b} \tag{3.154}$$

In the above equations, except ω_r, ω_s, ω_b, T'_0, and H, other variables and parameters are per unit. V'_d, V'_q, and ω_r are state variables and V_{ds}, V_{qs}, I_{ds}, and I_{qs} are algebraic variables. The relationship between V_{ds} and V_{qs} with I_{ds} and I_{qs} is determined by the network equations.

3.3.2 Doubly Fed Induction Generator (DFIG)-based Wind Turbine

3.3.2.1 Physical Description

Figure 3.21 shows the typical structure of a DFIG-based wind turbine. In this type of wind turbine, the stator of induction generator is connected directly to the network, and its wound rotor is powered by a back-to-back power converter that provides adjustable voltage and frequency. Supplying rotor with variable frequency causes the mechanical speed of the rotor to be independent of the network frequency, which allows the turbine to operate at variable speeds. The over current crowbar is used to protect the converters. Unlike the FSIG-based wind turbine, which is only capable of producing active power at speeds higher than synchronous, the DFIG-based wind turbine has the ability that by adjusting the rotor supply voltage, the active power be delivered to the network through the stator at both lower and higher speeds than the synchronous speed.

When the generator speed is more than the synchronous speed, the active power is also delivered to the network through the rotor, but when the rotor speed is less than the synchronous speed, the rotor absorbs the active power from the network through the converters.

The rotor-side converter (RSC) is responsible for controlling the rotor voltage and current. As will be shown in the following sections, by controlling the rotor current, it is possible to control the stator active and reactive powers. The main duty of the grid-side converter (GSC) is to regulate the DC link voltage regardless of the amount of active power passing through it [9]. In addition to the stator, reactive power can also be injected into the network through this converter if the capacity of the GSC converter allows. Of course, the priority is to use the GSC capacity to pass active power. Passing active power through the converters is necessary for keeping the DC link voltage constant.

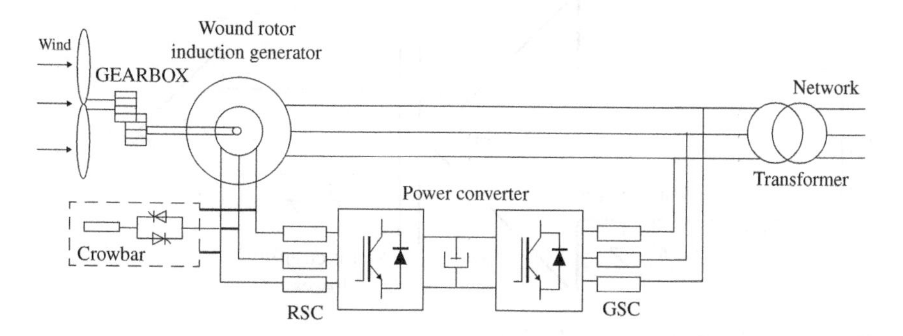

Figure 3.21 The typical structure of a DFIG-based wind turbine.

3.3.2.2 DFIG Steady-State Characteristic

Figure 3.22 shows the DFIG steady-state single-phase equivalent circuit using motor convention [4]. The difference between this figure and the conventional model of induction machine is the presence of voltage source V_r/S, which is produced due to injection of voltage into the rotor.

The electromagnetic torque, which in the per unit system is equal to the power passing through the air gap, is obtained from Eq. (3.155) [4].

$$T_e = \left(I_r^2 \frac{R_r}{S} \right) + \frac{P_r}{S} \tag{3.155}$$

where P_r is the rotor input/output active power received from/sent to the back-to-back converter, which is calculated as follows:

$$P_r = \mathrm{Re}\left(\frac{\overline{V}_r}{S} \overline{I}_r^* \right) \tag{3.156}$$

Figure 3.23 illustrates the steady-state relationships between the mechanical power and the rotor and stator active powers in a DFIG [4]. In this figure, P_m is

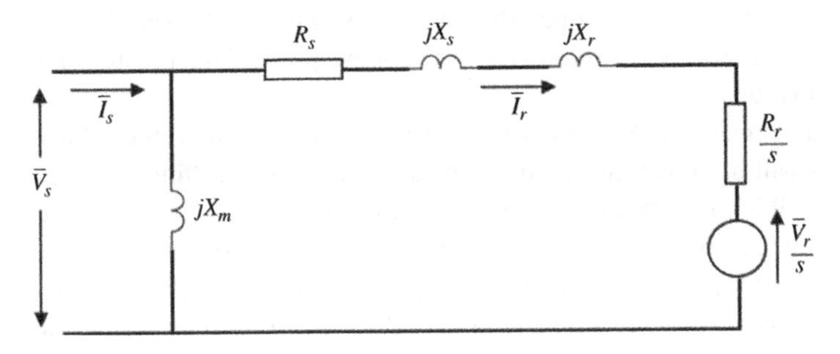

Figure 3.22 The DFIG steady-state single-phase equivalent circuit [4].

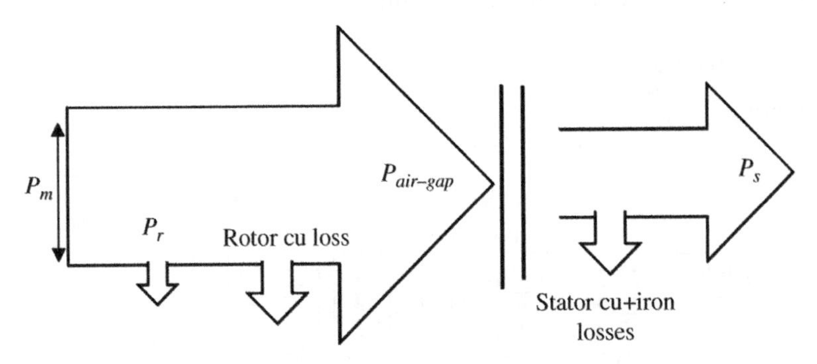

Figure 3.23 Power equations in DFIG [4].

the mechanical power generated by the turbine, P_r is the active power delivered by the rotor to the RSC converter, $P_{air-gap}$ is the active power passing through the air gap, and P_s is the stator output power. This figure also shows the rotor and stator ohmic losses and the iron losses. Ignoring the stator and rotor losses, there are the following equations:

$$P_{\text{air-gap}} = P_s \tag{3.157}$$

$$P_{\text{air-gap}} = P_m - P_r \tag{3.158}$$

Consequently,

$$P_s = P_m - P_r \tag{3.159}$$

At steady state, the mechanical and electromagnetic torques are equal. Hence, the Eq. (3.159) can be written as follows:

$$T\omega_s = T\omega_r - P_r \tag{3.160}$$

Therefore,

$$P_r = -T(\omega_s - \omega_r) \tag{3.161}$$

Using Eq. (3.130),

$$P_r = -ST\omega_s = -sP_s \tag{3.162}$$

And from Eq. (3.159),

$$P_m = (1 - s)P_s \tag{3.163}$$

Equation (3.162) states that in the negative slips (speeds more than synchronous), P_r is positive, which is based on Figure 3.23, which means the active power delivery to the network through the rotor. Similarly, it is obvious that at speeds below synchronous, active power is transmitted from the network to the rotor. The total generated active power by generator is

$$P_g = P_s + P_r \tag{3.164}$$

3.3.2.3 Optimum Wind Power Extraction [5]

As stated in Eq. (3.128), only a part of the wind power can be transferred to the generator, the value of which is determined by the coefficient C_P. As mentioned, the value of C_P in addition to the blade pitch angle is also related to the ratio of the turbine rotation speed (consequently the generator speed) to wind speed, and the maximum value of C_P at each wind speed is obtained for a given generator speed. Figure 3.24 shows an example of generator power changes in terms of rotor speed at different wind speeds [4]. The maximum point of each curve is obtained for the

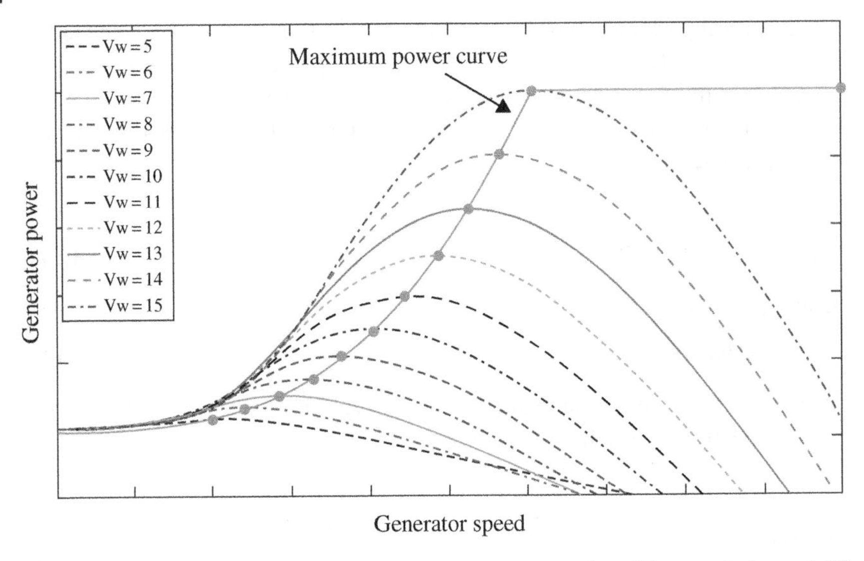

Figure 3.24 Generator power changes in terms of rotor speed at different wind speeds [4].

speed at which the maximum value C_P is obtained. In this figure, the maximum power curve P_{Opt} is also drawn, which is the result of connecting the maximum points of curves. The goal is to set the generating speed on this curve at any wind speed, provided that various constraints are not violated.

3.3.2.4 Torque Control

Stable operating point is achieved when the input torque is equal to the electromagnetic torque. At each wind speed, the maximum input torque is obtained at a certain speed of generator. According to the swing equation (Eq. (3.153)), by controlling the electromagnetic torque, the generator speed can be changed and reached the desired speed to extract the maximum possible power and torque from the wind. The important thing is that you do not need to know the wind speed to do this. For this purpose, the torque–speed curve of Figure 3.25 is used. This figure shows the maximum torque that can be extracted from the wind at each generator speed. Different points of this curve are obtained at different wind speeds. Therefore, at each generator speed, the operating point is on this curve only when the wind speed be also at the value corresponding to that point. In other words, at any wind speed, only one generator speed can be placed on this curve. The purpose of electromagnetic torque control is to achieve this generator speed. To illustrate this, suppose the current speed of generator is ω_1, but the optimum generator speed to place the operating point on the torque-speed curve be ω_2 (Figure 3.25).

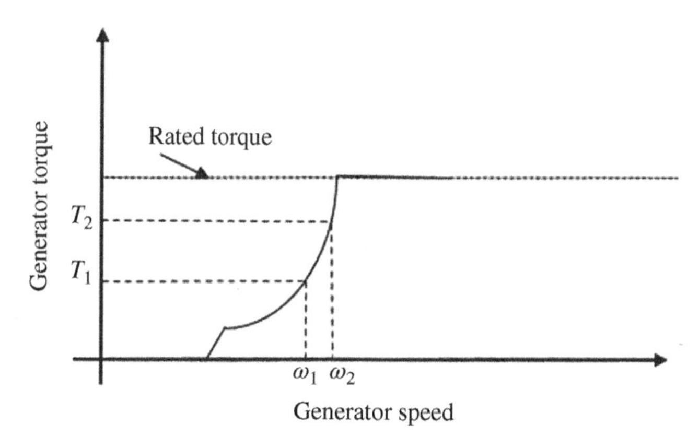

Figure 3.25 The DFIG speed–torque characteristic [4].

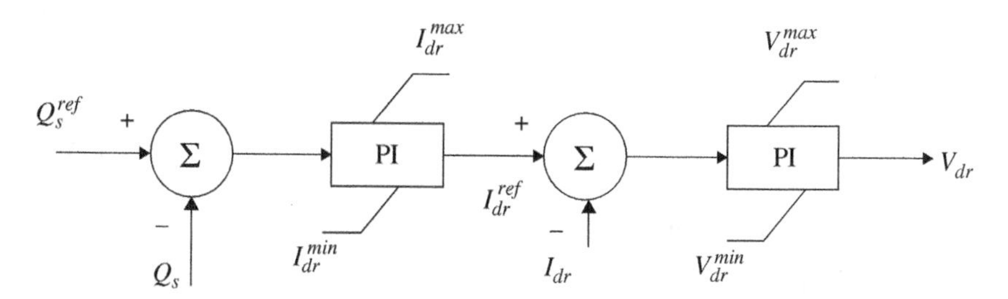

Figure 3.26 The block diagram of the stator reactive power control in DFIG.

Since $\omega_1 < \omega_2$, the wind generated torque at the generator speed ω_1 is greater than T_1 (Figure 3.24). Therefore, if the electromagnetic torque is set to T_1, according to the swing equation, the generator speed will increase. When the generator speed reaches ω_2, if the electromagnetic torque of the generator is set to T_2, the speed change will stop because the wind generated torque also is equal to T_2. Therefore, if at each generator speed, the corresponding torque on the torque–speed curve of Figure 3.26 is selected as the reference value for the electromagnetic torque, the generator speed is adjusted so that the maximum possible power and torque is extracted from the wind.

Electromagnetic torque setting is possible by controlling the rotor current. Using V'_{d} and V'_{q} [Relations (3.144) and (3.145)] as well as rotor flux equations [Relations (3.140) and (3.141)], the electromagnetic torque is expressed as follows [4]:

$$T_{\mathrm{e}} = L_{\mathrm{m}}\left(I_{\mathrm{dr}}I_{\mathrm{qs}} - I_{\mathrm{qr}}I_{\mathrm{ds}}\right) \tag{3.165}$$

Also, neglecting the stator resistance and transients and substituting Ψ_{qs},

$$V_{ds} = \frac{-\omega_s}{\omega_b}\left(L_{ss}I_{qs} + L_m I_{qr}\right) \tag{3.166}$$

From this equation,

$$I_{qs} = \frac{\omega_b}{\omega_s L_{ss}} V_{ds} + \frac{L_m}{L_{ss}} I_{qr} \tag{3.167}$$

By selecting the stator flux oriented reference frame, V_{ds} becomes zero. Therefore, Eq. (3.167) becomes:

$$I_{qs} = \frac{L_m}{L_{ss}} I_{qr} \tag{3.168}$$

Now, from Relation (3.135), ignoring the stator resistance and stator transients as well as replacing Ψ_{ds} from (3.138):

$$I_{qs} = \frac{\omega_s}{\omega_b}(-L_{ss}I_{ds} + L_m I_{dr}) \tag{3.169}$$

From (3.169), I_{ds} becomes:

$$I_{ds} = -\frac{\omega_b}{\omega_s L_{ss}} V_{qs} + \frac{L_m}{L_{ss}} I_{dr} \tag{3.170}$$

By replacing I_{ds} and I_{qs} in (3.165):

$$T_e = L_m\left[I_{dr}\left(\frac{L_m}{L_{ss}} I_{qr}\right) - I_{qr}\left(-\frac{\omega_b}{\omega_s L_{ss}} V_{qs} + \frac{L_m}{L_{ss}} I_{dr}\right)\right] = I_{qr}\frac{\omega_b L_m V_{qs}}{\omega_s L_{ss}} \tag{3.171}$$

Equation (3.171) indicates the relationship between T_e and I_{qr}. Of course, the amount of generator output voltage is also affected by T_e. However, due to the limited range of voltage changes, it is expected that the electromagnetic torque be regulated at the desired value by changing the *q*-axis component of the rotor current.

3.3.2.5 Voltage Control

In the previous section, it was shown that the electromagnetic torque regulation is possible using controlling the component q of the rotor current. The following shows that the component d of the rotor current can be used to control the stator injection reactive power.

The stator generation reactive power can be expressed as follows:

$$Q_s = \text{Im}\left(V_s I_s^*\right) = V_{qs}I_{ds} - V_{ds}I_{qs} \tag{3.172}$$

where $\overline{V}_s$ and $\overline{I}_s$ are the stator voltage and current phasors.

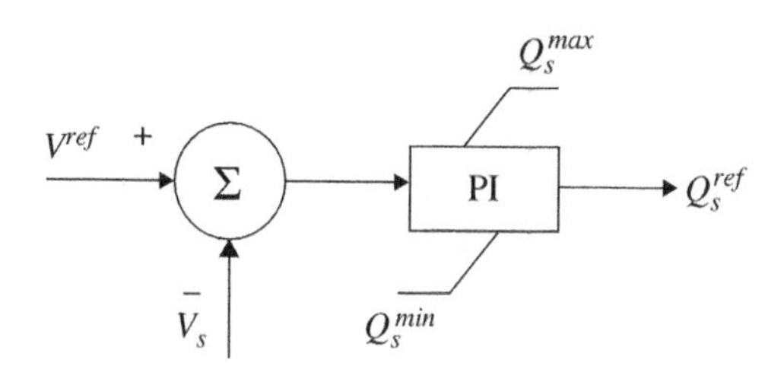

Figure 3.27 The block diagram of Q_s^{ref} determination in control voltage mode [9].

Using Eq. (3.138), the current I_{ds} becomes:

$$I_{ds} = -\frac{1}{L_{ss}}\Psi_{ds} + \frac{L_m}{L_{ss}}I_{dr} \tag{3.173}$$

Given that V_{ds} is zero,

$$Q_s = V_{qs}\left(-\frac{1}{L_{ss}}\Psi_{ds} + \frac{L_m}{L_{ss}}I_{dr}\right) \tag{3.174}$$

Also, ignoring the stator resistance and transient is,

$$\Psi_{ds} = \frac{\omega_b V_{qs}}{\omega_s} \tag{3.175}$$

By replacing Ψ_{ds} from (3.175) into (3.174)

$$Q_s = -\frac{\omega_b V_{qs}^2}{\omega_s L_{ss}} + \frac{L_m V_{qs}}{L_{ss}}I_{dr} \tag{3.176}$$

It is observed that by changing the component *d* of the rotor current, it is possible to regulate the stator reactor power. Figure 3.26 shows the block diagram of the stator reactive power control. The reference value I_{dr} (I_{dr}^{ref}) is determined based on the desired reactive power of stator. I_{dr}^{ref} is also produced by changing V_{dr}.

In the voltage control mode, Q_s^{ref} can be determined from the desired voltage value (V_{ref}) using the block diagram of Figure 3.27. In this figure, V_s is the stator voltage magnitude.

3.4 Summary

This chapter was devoted to synchronous generators modeling as well as modeling of FSIG-based and DFIG-based wind turbines. In synchronous generator modeling, controllers and limiters effective in voltage stability were considered. How to generate power in wind farms and the advantages of variable speed wind power plants in extracting maximum power from wind were described. Also, equations related to power and voltage control in the DFIG-based wind turbine were presented.

References

1 Sauer, P.W. and Pai, M.A. (1998). *Power System Dynamics and Stability*, vol. 101. Upper Saddle River, NJ: Prentice Hall.

2 Van Cutsem, T. and Vournas, C. (1988). *Voltage Stability of Electric Power Systems*. Norwell, MA: Kluwer.

3 Kundur, P., Neal, J.B., and Mark, G.L. (1994). *Power System Stability and Control*, vol. 7. New York: McGraw-Hill.

4 Anaya-Lara, O., Jenkins, N., Ekanayake, J.B. et al. (2011). *Wind Energy Generation: Modelling and Control*. Wiley.

5 Christ, F. and Abeykoon, C. (2015). Modelling of a wind power turbine. *2015 Internet Technol. Appl. ITA 2015 - Proc. 6th Int. Conf.*, (September), Wrexham, UK (8–11 September 2015), pp. 207–212. https://doi.org/10.1109/ITechA.2015.7317396.

6 Liu, J.H. and Chu, C.C. (2014). Long-term voltage instability detections of multiple fixed-speed induction generators in distribution networks using synchrophasors. *IEEE Trans. Smart Grid* 6 (4): 2069–2079.

7 Vovos, P.N., Kiprakis, A.E., Wallace, A.R., and Harrison, G.P. (2007). Centralized and distributed voltage control: impact on distributed generation penetration. *IEEE Trans. Power Syst.* 22 (1): 476–483. https://doi.org/10.1109/TPWRS.2006.888982.

8 Pedra, J., Córcoles, F., Monjo, L. et al. (2012). On fixed-speed WT generator modeling for rotor speed stability studies. *IEEE Trans. Power Syst.* 27 (1): 397–406. https://doi.org/10.1109/TPWRS.2011.2161779.

9 Kayikçi, M. and Milanović, J.V. (2007). Reactive power control strategies for DFIG-based plants. *IEEE Trans. Energy Convers.* 22 (2): 389–396. https://doi.org/10.1109/TEC.2006.874215.

4

Impact of Distributed Generation and Transmission–Distribution Interactions on Voltage Stability

4.1 Introduction

Many voltage stability studies, which include stability assessment and determining proper control actions to prevent voltage instability, use the power system model. The power system model, which includes distribution and transmission networks, consists of a large number of differential and algebraic equations. Exact determination of system response requires solving all these equations at the same time, which may not be possible for real-time studies and when a large number of probable contingencies are considered. Hence, simplifications are usually done in the model used. One of these simplifications is the separate analysis of voltage stability in distribution and transmission networks; in studying the transmission network voltage stability, the distribution system is replaced with a load model, and in analyzing the distribution network voltage stability, the transmission network is replaced by a voltage source. This chapter examines this issue that how much these simplifications can lead to error in voltage stability assessments. Another topic covered in this chapter is the effect of distributed generation (DG) units on voltage stability. For this purpose, doubly fed induction generator (DFIG)-based wind turbine, which is one of the most widely used sources, is used. DIgSILENT PowerFactory software is used to simulate this chapter.

4.2 Interactions of Transmission and Distribution Networks

4.2.1 The Studied System

In this chapter, the simple system of Figure 4.1 is used for simulation [1]. In this figure, a simple transmission network consisting of an ideal voltage source with 1

Voltage Stability in Electrical Power Systems: Concepts, Assessment, and Methods for Improvement, First Edition. Farid Karbalaei, Shahriar Abbasi, and Hamid Reza Shabani.

Published 2023 by John Wiley & Sons, Inc.

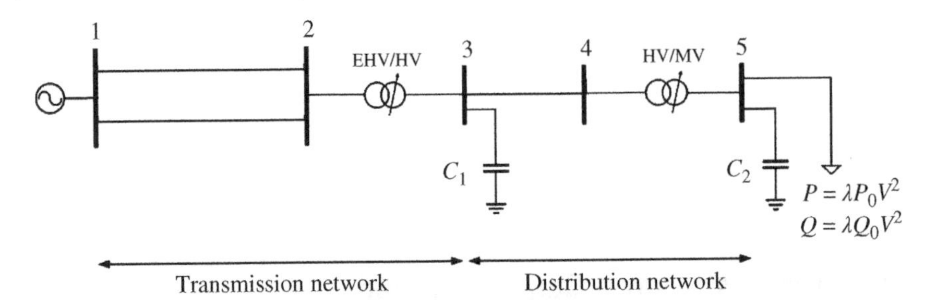

Figure 4.1 The studied system.

pu voltage magnitude, two parallel lines, and an under load tap changer (ULTC) transformer is connected to a distribution network with one line, one ULTC transformer, and one constant impedance load. Similar to the assumption used in Ref. [1], in simulations, the variable tap is placed on the low voltage side of the transformers. Two fixed capacitors are also used to compensate reactive power. The purpose of the capacitors is to provide lead conditions in order to increase the probability of voltage instability occurrence. Otherwise, the nose power voltage (PV) curve voltage (voltage at collapse point) will be very low. This causes that the distance of system operating point (which normally should be in a limited range around the rated voltage) from the PV curve nose be so large that voltage instability be impossible. For this reason, it is said that voltage instability occurs mainly in systems with high reactive power compensation. Of course, it is clear that reactive power compensation has many advantages, including increase in the loadability limit, but the loadability margin must always be maintained at an acceptable value.

Load tap changer (LTC) of transmission and distribution transformers controls voltage of buses 3 and 5, respectively. The dead band of voltage control is assumed to be from 0.99 to 1.01 pu. Also, the time delays of each tap movement for transmission and distribution transformers are in respect 5 and 10 seconds. An extended range is considered to change the taps (from 0.85 to 1.15 pu or from 0.8 to 1.2 pu). P_0 and Q_0 are selected 4.53 and 0.2 pu, respectively. The susceptances of capacitors C_1 and C_2 are in respect 2.25 and 2.1 pu. Voltage instability occurs when the tap changer of one of two transmission and distribution transformers fails to regulate the bus voltage under its control. This failure first causes the voltage of that bus to drop and then leads to voltage reduction at other buses.

As described in Chapter 1, the limitation of power transfer to the controlled bus makes it impossible to voltage recovery at that bus. This limitation depends on the impedance of the upstream grid, the power factor of power received from the bus, and also the Thevenin voltage seen from the bus. Therefore, in the system of Figure 4.1, the power transfer limit to bus 3 is determined by the source voltage magnitude, the impedance of the transmission transformer, the impedance of the line between buses 1 and 2, as well as the power factor of the transformer output power. The power factor depends on the susceptance of the capacitors, the load power and the power factor at bus 5, and the distribution network impedance. The power transfer limit to bus 5 also depends on the voltage of bus 3, distribution line and transformer impedances, susceptance of capacitor C_2, and load power factor. In all simulations, the outage of one of the parallel lines between buses 1 and 2 is considered as contingency.

4.2.2 Stable Case (Case 1)

Figure 4.2 shows the voltage response at buses 3 and 5 for an example of a stable case. The black curve is the voltage of bus 3, and the gray curve represents the voltage of bus 5. Before line outage, which occurs at 200 seconds, the bus voltages are regulated by tap changers. The parameters of this mode are given in Table 4.1. Also

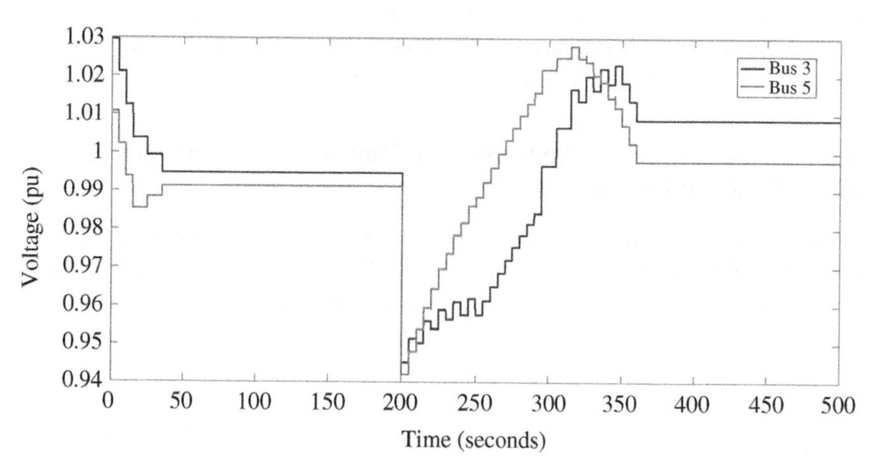

Figure 4.2 Voltage response at buses 3 and 5 in case 1.

Table 4.1 System parameters in case 1.

Transmission line impedance (pu)	Distribution line impedance (pu)	Transmission transformer impedance (pu)	Distribution transformer impedance (pu)	λ
0.025	0.02	0.03	0.04	1.3

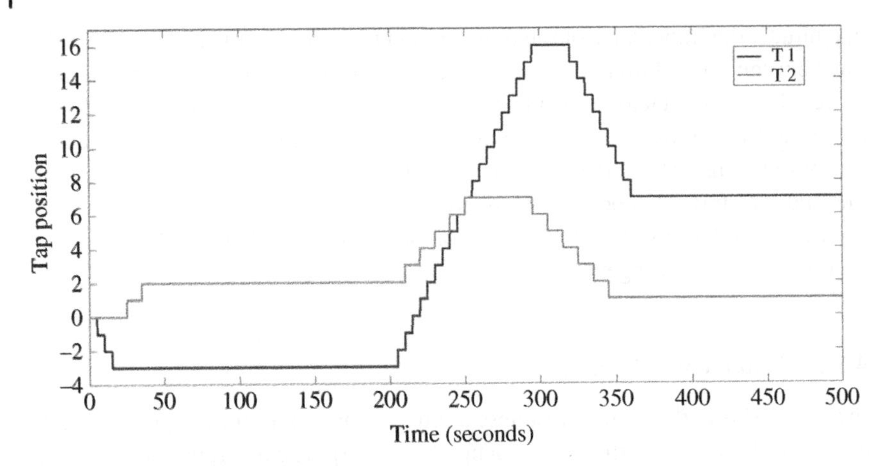

Figure 4.3 Tap changes of transformers in case 1.

in Figure 4.3 changes of transmission and distribution transformers taps are given. In this figure, the black curve corresponds to the transmission transformer tap.

As can be seen, due to interaction between tap changers, voltage changes are not monotonic, but eventually fall within the dead band range.

4.2.3 Instability Due to the Inability of Transmission Transformer's LTC to Regulate Voltage (Case 2)

Figure 4.4 shows the voltage response at buses 3 and 5 in an unstable case. In this case, instability occurs due to the inability of LTC of transmission transformer in regulating the voltage of bus 3. It can be seen that the voltage of bus 5 is recovered

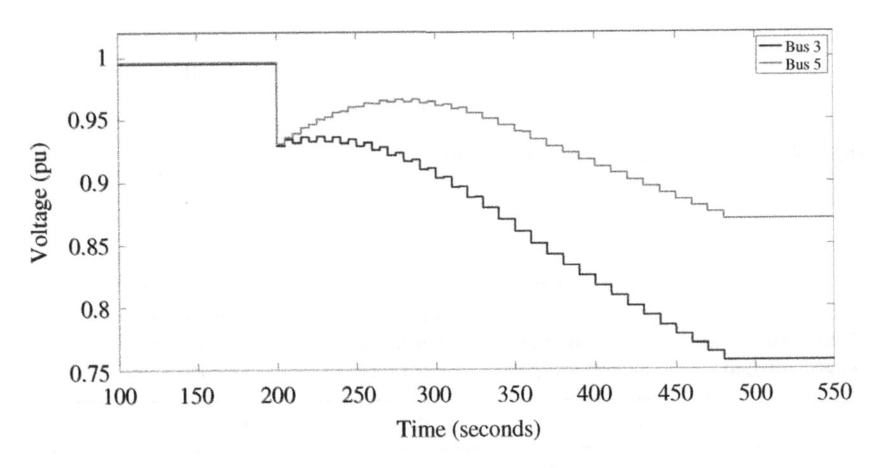

Figure 4.4 Voltage response at buses 3 and 5 in case 2.

Table 4.2 System parameters in case 2.

Transmission line impedance (pu)	Distribution line impedance (pu)	Transmission transformer impedance (pu)	Distribution transformer impedance (pu)	λ
0.025	0.02	0.03	0.04	1.3591

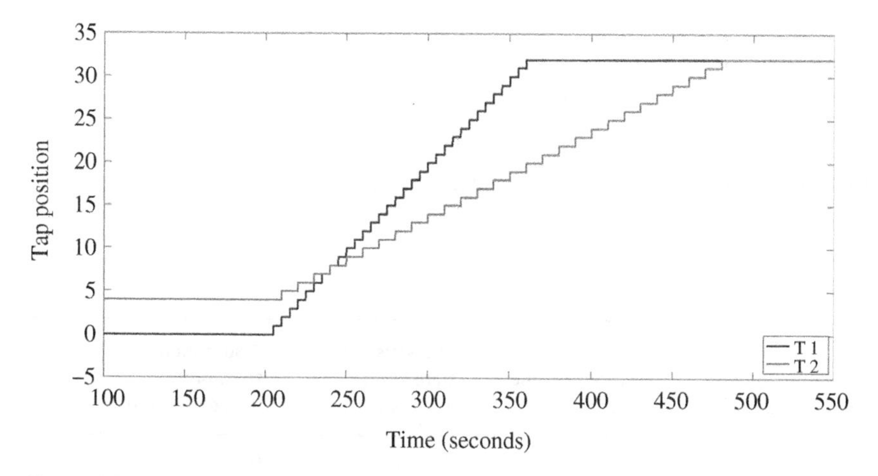

Figure 4.5 Tap changes of transformers in case 2.

by the LTC operation in the distribution transformer, but the voltage instability and drop at bus 3 reduces the transformable power to bus 5 and finally causes voltage instability and drop at this bus. Table 4.2 shows the system parameters in this case. The value of λ in the table is in fact the minimum value of the loading factor, which in this case causes instability. Figure 4.5 also shows the tap changes that have been increased to their maximum value to regulate voltage. In this case, since voltage instability is due to limitation on transformable power in transmission network, replacing the distribution network with a suitable load model does not cause much error in detecting proximity to voltage instability. But replacing the transmission network with a voltage source is not appropriate.

4.2.4 Instability Due to the Inability of Distribution Transformer's LTC to Regulate Voltage (Case 3)

Figure 4.6 shows the voltage response at buses 3 and 5 in an unstable case caused by the inability of LTC of distribution transformer in regulating the voltage of bus 5. It can be seen that the voltage of bus 3 is recovered at first, but after the voltage of

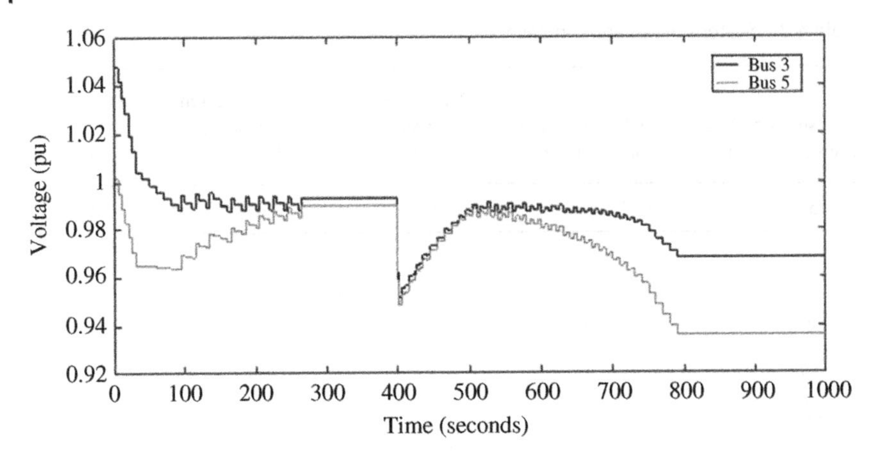

Figure 4.6 Voltage response at buses 3 and 5 in case 3.

Table 4.3 System parameters in case 3.

Transmission line impedance (pu)	Distribution line impedance (pu)	Transmission transformer impedance (pu)	Distribution transformer impedance (pu)	λ
0.02	0.07	0.0358	0.08	1.1

bus 5 decreases, regulation of the voltage of bus 3 becomes impossible and this bus also experiences voltage drop. Table 4.3 shows the system parameters in this case. It can be seen that high impedance of the distribution network has caused these conditions.

In this case, since voltage instability is due to limitation on transformable power in distribution network, modeling the transmission network alone and replacing the distribution network with a load model cannot provide an accurate assessment of voltage instability risk. If this modeling is used, a suitable loading margin must be remained to ensure voltage stability. In this case, the voltage stability state in the distribution network can be diagnosed by replacing the transmission network with a voltage source.

In the instabilities presented in cases 2 and 3, it is possible to predict voltage instability risk using a separate analysis in transmission or distribution network. But if the instability does not occur due to the transformable power limitation but due to exiting from the attraction region, certainly none of the separate analyses will be able to detect the occurrence of voltage instability and only using dynamic analysis with complete model of the system the instability can be detected. An example of voltage instability due to exiting from the attraction region is given in Section 1.2.5. Ref. [1] provides an example of this type of instability in the system of Figure 4.1.

4.3 Impact of Distribution Generation (DG) Units

Given that DG units are mainly located near load centers and supply a portion of power consumption of loads, it seems that they can reduce the power passing through the transmission lines and have a positive effect on voltage stability. DG units are usually operated in PQ mode with unit power factor. The reason for choosing unit power factor is to use the maximum capacity of the units to generate active power. According to Eq. (3.133), increasing the output power of units may cause overvoltage at them. In this situation, to control the voltage, the PQ mode is converted into PV mode [2]. In conditions of high active power generation, voltage control requires reactive power absorption, which has a negative effect on voltage stability. DG units are mainly connected to the distribution network at the medium voltage (MV) level, but it is possible to connect wind farms consisting of several wind turbines to the network at the high voltage (HV) level.

4.3.1 Connecting DG Units to MV Distribution Networks

MV distribution networks are fed by HV/MV transformers. Since automatic voltage control is not performed by LTC in MV/LV transformers, MV network loads usually have voltage-dependent characteristic. For this reason, in Figure 4.1, the constant impedance load is used for the load connected to bus 5. Figure 4.7 shows an example of an MV distribution network.

A method to reduce voltage instability risk is to decrease the power consumption of loads. To this aim, when loads have voltage-dependent characteristic, their power consumption can be reduced by deliberately reducing the loads voltage [3]. Voltage reduction is usually done by reducing the LTCs voltage set point, which reduces the supply bus voltage of distribution network. However, the reduction of voltage must be such as not to reduce voltage at buses connected to end of feeders to a value less than a minimum acceptable value. This limitation in voltage reduction makes it impossible to reduce voltage effectively at loads close to the supply bus of distribution networks. The presence of DG units can greatly help to solve this problem [4]. Proper regulation of voltage in DG units can cause almost uniform voltage along the feeder, which allows maximum voltage drop at more buses.

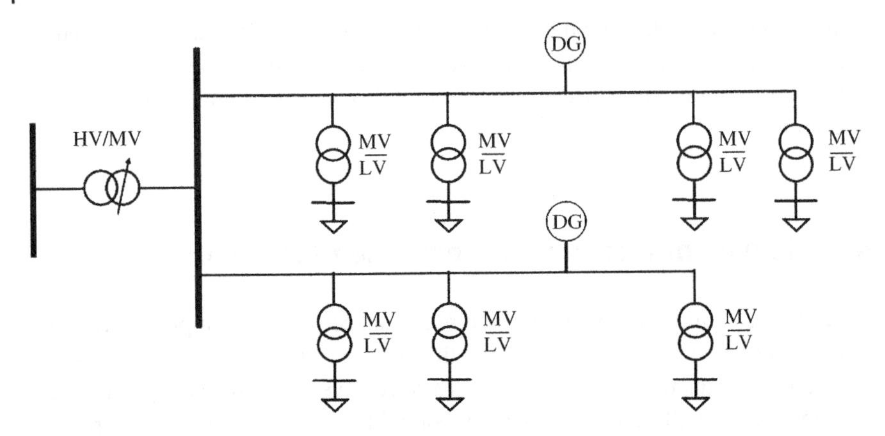

Figure 4.7 A simple MV distribution network.

4.3.2 Connecting DG Units to HV Distribution Networks

For this section, a DFIG-based wind power plant with two series transformers as shown in Figure 4.8a, b are connected to buses 3 and 4, respectively. Using two series transformers is to provide suitable conditions to connect the wind power plant to the HV network. The impedance of transformers is considered to be 0.06 pu, and the system parameters are chosen similar to Section 4.2.3 (Case 2).

Similar to the previous sections, by cutting out one of the two parallel lines, the minimum value of λ which causes voltage instability is obtained. This loading factor is determined by increasing λ in small steps and time simulation of system. The results are given in Tables 4.4–4.7. In Tables 4.4 and 4.5, the PQ mode for the wind power plants connected to buses 3 and 4 is considered. Tables 4.6 and 4.7 also show the results of connecting the wind power plant with PV mode to the network. In these tables, the output active power is increased from 2 to 4 pu. Also, the DFIG terminal voltage is presented. It is observed that when the wind power plant has PQ mode, increase in its active power generation causes increase of λ and in other words, increase of the stability margin. But, when the wind power plant is connected to bus 3 and is operated in the PV mode, increase of active power generation reduces the stability margin. The reason for this is that according to Table 4.4, the connection of wind power plant to bus 3 causes overvoltage, and voltage control system of the wind power plant is forced to absorb reactive power to regulate the voltage to 1 pu, this leads to reduction of stability margin.

It can be seen from Table 4.5 that connecting the wind power plant to bus 4 does not cause overvoltage even for the output active power of 4 pu. Therefore, connecting the wind power plant with PV mode to this bus leads to generate reactive power and increase stability margin (Table 4.7). Another result is that in both

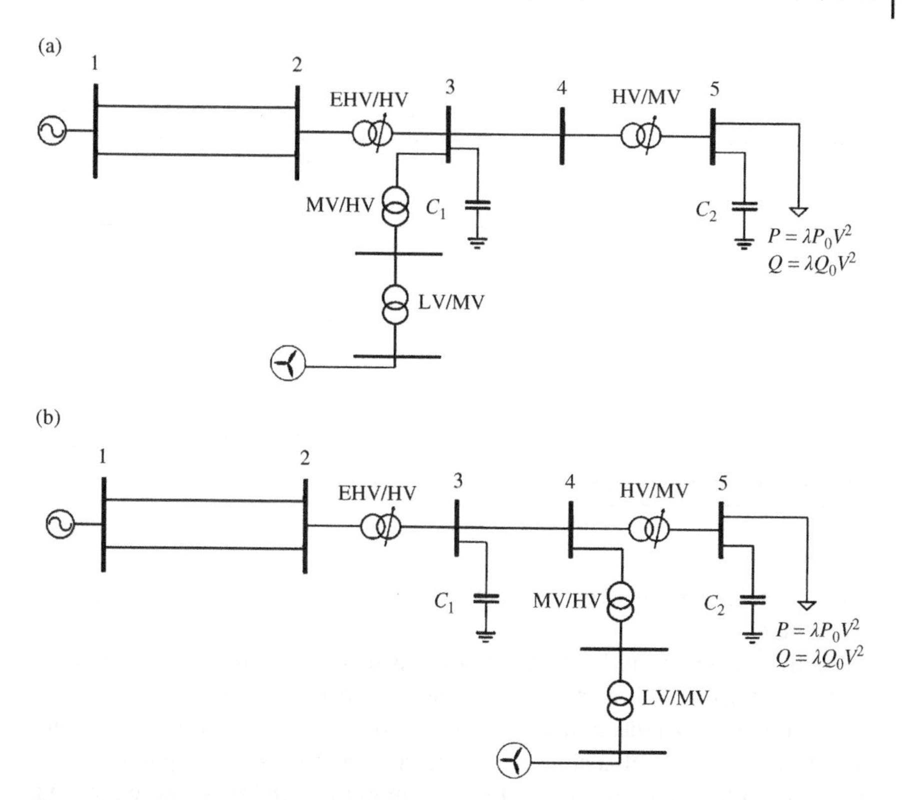

Figure 4.8 Connecting wind power plant to the system of Figure 4.1 (a) connecting plant to bus 3 (b) connecting plant to bus 4.

Table 4.4 Connecting wind power plant with PQ mode to bus 3.

	P **= 2 pu**	***P*** **= 2.5 pu**	***P*** **= 3 pu**	***P*** **= 3.5 pu**	***P*** **= 4 pu**
λ	1.4641	1.4701	1.4731	1.4761	1.4791
The DFIG terminal voltage (pu)	1.02	1.02	1.03	1.03	1.035

Table 4.5 Connecting wind power plant with PQ mode to bus 4.

	P **= 2 pu**	***P*** **= 2.5 pu**	***P*** **= 3 pu**	***P*** **= 3.5 pu**	***P*** **= 4 pu**
λ	1.5571	1.5571	1.5601	1.5631	1.5661
The DFIG terminal voltage (pu)	0.95	0.96	0.96	0.965	0.967

Table 4.6 Connecting wind power plant with PV mode to bus 3.

	$P = 2$ pu	$P = 2.5$ pu	$P = 3$ pu	$P = 3.5$ pu	$P = 4$ pu
λ	1.4551	1.4521	1.4521	1.4491	1.4461
The DFIG terminal voltage (pu)	1	1	1	1	1

Table 4.7 Connecting wind power plant with PV mode to bus 4.

	$P = 2$ pu	$P = 2.5$ pu	$P = 3$ pu	$P = 3.5$ pu	$P = 4$ pu
λ	1.6081	1.6201	1.6261	1.6351	1.6411
The DFIG terminal voltage (pu)	1	1	1	1	1

modes, installing the wind power plant closer to the load buses increases the loadability margin.

Reactive power absorption by wind power plants, when these plants are in PV mode, increases as their active power production increases. For two reasons, this can reduce the loadability limit and consequently reduce the voltage stability margin. The first reason is that absorption of reactive power occupies a portion of the transmission line capacity. Hence, the maximum transferable power to the load will be reduced. This is the reason of reduction of the maximum transferable power. Another reason is that increasing in the reactive power consumption of wind power plants will increase the reactive power generation of synchronous generators and consequently will increase the probability of reaching these generators to their reactive power limits and losing their voltage control. Figure 4.9 shows the same system under study in which a synchronous generator is connected to bus 2 via a transformer.

Figures 4.10 and 4.11 show the curves of synchronous generators reactive power output in term of the generated active power of the wind power plants when these power plants are connected to buses 3 and 4, respectively. It is observed that in both cases, the reactive power produced by the synchronous generator increases. Of course, this increase is less when the wind power plant is connected to bus 4. Because in this case, increasing the active power of the wind power plant causes a less increase in voltage than the case wherein the wind power plant is connected to bus 3. Less increase in the voltage of wind power plant reduces the need for reactive power absorption by this power plant.

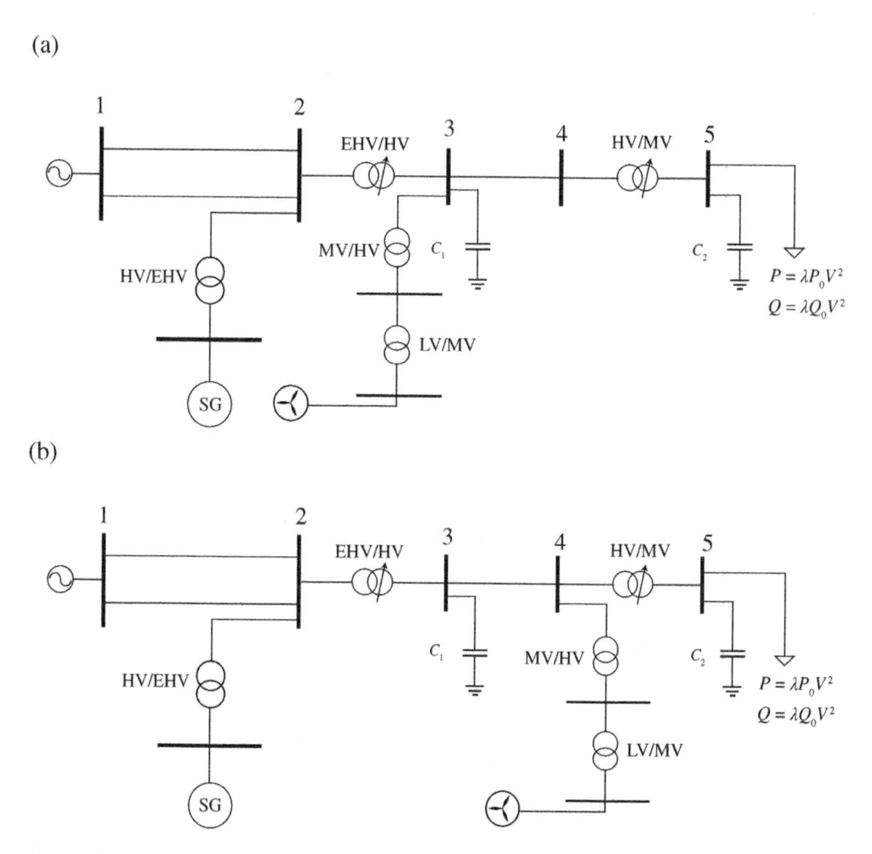

Figure 4.9 Connecting a synchronous generator to the system of Figure 4.2 (a) wind power plant connecting to bus 3 and (b) wind power plant connecting to bus 4.

Figure 4.12 shows the results of a sample of the voltage instability occurrence in this system that occurs after the synchronous generator reaches its reactive power generation limit. In this figure, the voltage variations at buses 3 and 5 are provided. Rapid voltage drop after voltage drop due to instability of LTCs is due to loss of synchronism of synchronous generator.

4.4 Summary

This chapter consists of two parts. In the first part, the interaction of transmission and distribution networks in appearing voltage instability in a simple system was investigated. It was observed that these interactions may cause that separate

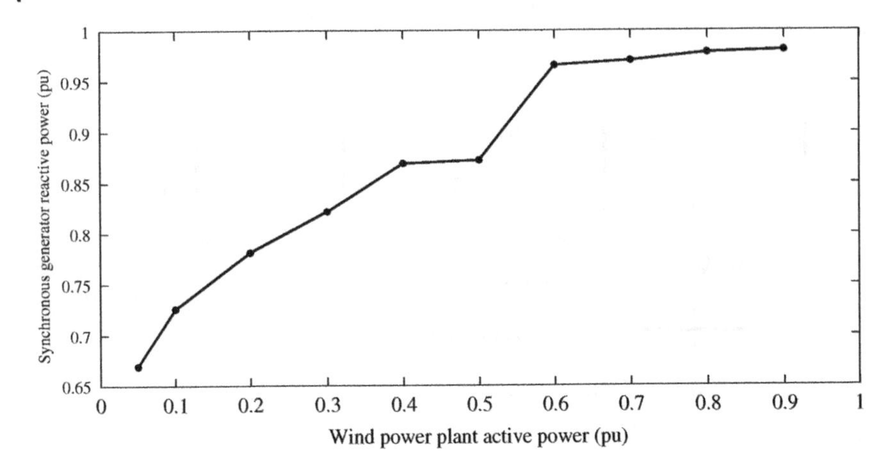

Figure 4.10 Increase of generated reactive power of the synchronous generator when the active power of the wind power plant connected to bus 3 increases.

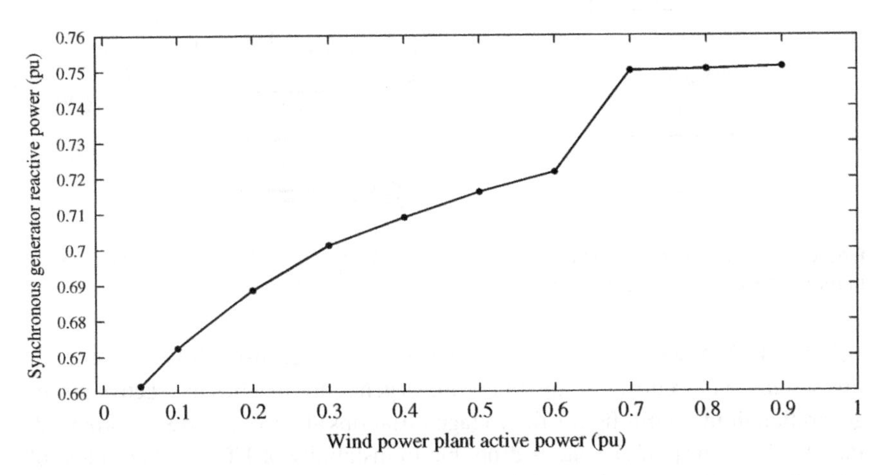

Figure 4.11 Increase of generated reactive power of the synchronous generator when the active power of the wind power plant connected to bus 4 increases.

analysis of voltage stability in distribution and transmission networks cannot properly indicate the risk of voltage instability. The second part was dedicated to the effect of DG units on voltage stability. These sources can reduce the flowing power in transmission lines and thus help maintain voltage stability. Of course,

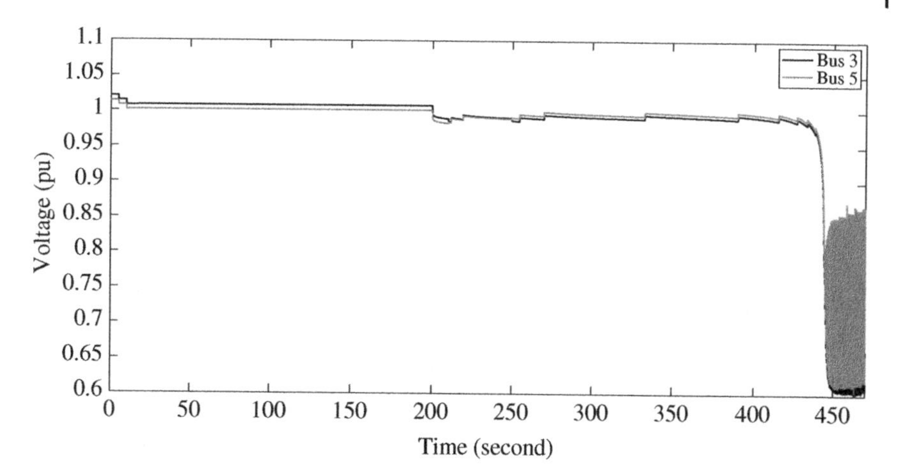

Figure 4.12 Voltage instability occurrence in the system of Figure 4.9a.

their effect can be different when used in different modes and connected to different parts of the network.

References

1 Abbasi, S.M., Karbalaei, F., and Badri, A. (2019). The effect of suitable network modeling in voltage stability assessment. *IEEE Trans. Power Syst.* 34 (2): 1650–1652. https://doi.org/10.1109/TPWRS.2019.2892598.

2 Vovos, P.N., Kiprakis, A.E., Wallace, A.R., and Harrison, G.P. (2007). Centralized and distributed voltage control: impact on distributed generation penetration. *IEEE Trans. Power Syst.* 22 (1): 476–483. https://doi.org/10.1109/TPWRS.2006.888982.

3 Zamani, V. and Baran, M.E. (2018). Meter placement for conservation voltage reduction in distribution systems. *IEEE Trans. Power Syst.* 33 (2): 2109–2116. https://doi.org/10.1109/TPWRS.2017.2737402.

4 Aristidou, P., Valverde, G., and Van Cutsem, T. (2017). Contribution of distribution network control to voltage stability: a case study. *IEEE Trans. Smart Grid* 8 (1): 106–116. https://doi.org/10.1109/TSG.2015.2474815.

Part II

Voltage Stability Assessment Methods

5

The Continuation Power Flow (CPF) Methods

5.1 Introduction

It is well known that the greater distance between the current operating point and the steady-state loadability limit leads to less risk of voltage instability. Therefore, determining the steady-state loadability limit is very important. The loadability limit of a power system consists of infinite points that depend on the direction of load and generation increment, and the system operating point approaches one of these points. Determining the steady-state loadability limit, which is usually done using power flow equations, is possible by both iterative and direct methods.

One of the iteration-based methods is the step-by-step increase of load and generation to push the operating point toward the loadability limit. Divergence of power flow calculations at each load level probably means exceeding the system's loadability limit. If the incremental steps of load and generation are chosen small enough, the last point of convergence can be considered as the loadability limit. However, due to the singularity of the Jacobian matrix at the loadability limit, the power flow calculations may diverge before reaching the loadability limit that this divergence is not due to exceeding the loadability limit. To relieve this problem, in a method known as continuation power flow (CPF), the original power flow equations are modified in such a way that the Jacobian matrix of the equations is changed so that it does not become singular at loadability limit and even after this limit [1]. This makes it possible to obtain the lower points of the PV curve as much as possible. In other one of the iteration-based methods, the quadratic approximation of the PV curve is used [2, 3]. Wherein, using one or two operating points, a quadratic function is initially obtained. Then, the maximum value of this function is introduced as the first approximation of the PV curve nose (loadability limit). Then, this approximated point is added to the previous two operating points and a new quadratic function is obtained based on these three points. Then, the maximum point of this function is considered as the second approximation of

Voltage Stability in Electrical Power Systems: Concepts, Assessment, and Methods for Improvement, First Edition. Farid Karbalaei, Shahriar Abbasi, and Hamid Reza Shabani.

Published 2023 by John Wiley & Sons, Inc.

the PV curve nose. The procedure continues until the distance between two consecutive approximated maximum points becomes less than a predefined threshold. In the direct method, determination of the loadability limit is formulated as an optimization problem with the objective function of the system loadability limit [4, 5].

This chapter describes the CPF methods, and the PV curve approximation methods will be discussed in Chapter 6. The direct method has many applications in determining preventive measures and is discussed in Chapter 10.

5.2 The CPF Elements

As mentioned, CPF is based on gradually increasing load and generation in a power system to obtain different points of PV curve. This curve can be drawn at any bus of the system. In this curve, the bus voltage magnitude is plotted in terms of the received active power, assuming that the power factor is constant. The horizontal axis can be the active power amount or the loading factor. According to Figure 5.1, the strategy is that at each operating point, the next point is predicted and then the predicted point is corrected to get the next actual point. Since the aim of the correction step is determining the actual operating point; therefore, the nonlinear equations of the system, i.e. the power flow equations must be used in this step. Of course, in order to avoid divergence of calculations, a change must

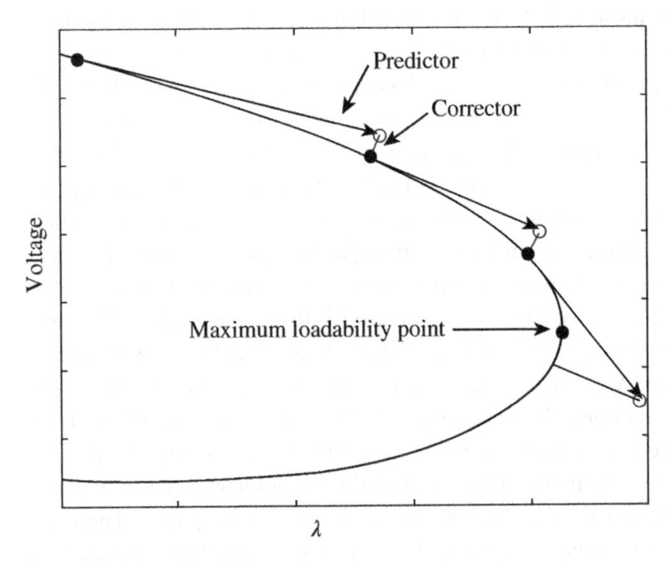

Figure 5.1 Prediction and correction steps in CPF.

be made in the power flow equations. This change is usually in the form of adding an equation to the power flow equations, which is called parameterization. The added equation determines the amount of movement on the PV curve, i.e. the distance from the current operating point to the next operating point.

Another important factor that is effective in the convergence of correction step calculations is the distance from the predicted point to the actual point. The shorter the distance, the less probability of calculations divergence and the less number of the required iterations. Since prediction is based on approximation of PV curve changes, shorter distance between the next predicted point and the current point leads to shorter distance between the predicted point and the next actual point. Of course, when this distance become shorter, movement on the PV curve slows down and consequently more calculation to determine the maximum loadability point is required. Therefore, the size of the prediction step must be carefully adjusted so that in addition to the convergence of the correction step calculations, it leads to acceptable speed in determining of the maximum loadability point. According to what stated earlier, four main components for CPF can be considered: prediction, parameterization, correction, and prediction step size determination.

5.3 Predictors

When prediction of the next operating point is more accurate, solving the correction step equations will require less iterations. Predictors are divided into linear and nonlinear.

5.3.1 Linear Predictors

5.3.1.1 Tangent Method

In this method, prediction of the next operating point is done using the tangent vector on the PV curve (Figure 5.1). In order to describe this method, it is assumed that the power flow equations are in the form of Eqs. (5.1) and (5.2) and are represented in the compact form of Eq. (5.3).

$$\lambda P_{\mathrm{Gi0}} - \lambda P_{\mathrm{Li0}} = \sum_{j=1}^{n} V_i V_j y_{ij} \cos\left(\delta_i - \delta_j - \theta_{ij}\right), \quad i\epsilon[\text{PV and PQ buses}] \tag{5.1}$$

$$-\lambda Q_{\mathrm{Li0}} = \sum_{j=1}^{n} V_i V_j y_{ij} \sin\left(\delta_i - \delta_j - \theta_{ij}\right), \quad i\epsilon[\text{PQ buses}] \tag{5.2}$$

$$\underline{f}(\underline{x}, \lambda) = 0 \tag{5.3}$$

where $\boldsymbol{x}$ is the vector of magnitude and angle of voltage of buses. P_{Li0} and P_{Gi0} are the consumed and generated active powers at bus i in the base case, and Q_{Li0} is the consumed reactive power at bus i in the base case. λ is loading factor used to increase load and generation. The value of λ increases from 1 at the base case to λ^* at the maximum loadability point.

The tangent vector is obtained by differentiation from Eq. (5.3):

$$\mathrm{d}\left[\underline{\boldsymbol{f}}(\underline{\boldsymbol{x}}, \lambda)\right] = \underline{\boldsymbol{f}}_{\underline{\boldsymbol{x}}}\mathbf{d}\underline{\mathbf{x}} + \underline{\boldsymbol{f}}_{\lambda}\mathrm{d}\lambda = 0 \tag{5.4}$$

where $\boldsymbol{f}_{\boldsymbol{x}}$ and $\boldsymbol{f}_{\lambda}$ are, respectively, the gradients of $\boldsymbol{f}$ with respect to $\boldsymbol{x}$ and λ, calculated at the current operating point. Eq. (5.4) can be re-written as

$$\begin{bmatrix} \underline{\boldsymbol{f}}_{\underline{\boldsymbol{x}}} & \underline{\boldsymbol{f}}_{\lambda} \end{bmatrix} \begin{bmatrix} \mathbf{d}\underline{\mathbf{x}} \\ d\lambda \end{bmatrix} = 0 \tag{5.5}$$

In Eq. (5.5), the left matrix is the original power flow Jacobian matrix that the column $\boldsymbol{f}_{\lambda}$ is added to it. Adding this column prevents singularity of the matrix at the loadability limit and even after this limit. The right vector is a tangent vector. It should be noted that in Eq. (5.5), λ is considered as a variable, and the number of variables is more than the number of equations. In this case, Eq. (5.5) has infinite solutions. To relieve this problem, one of the variables in the tangent vector is assumed to be known and its value is set to 1 [6]. In other words, if $\boldsymbol{t}$ represents the tangent vector,

$$\underline{\boldsymbol{t}} = \begin{bmatrix} \mathbf{d}\underline{\mathbf{x}} \\ \mathrm{d}\lambda \end{bmatrix} \quad t_k = \mp 1 \tag{5.6}$$

where k represents the number of the element of vector $\boldsymbol{t}$ whose value is set to $+1$ or -1. According to Eq. (5.6), the tangent vector is calculated as follows:

$$\begin{bmatrix} \underline{\boldsymbol{f}}_{\underline{\boldsymbol{x}}} & \underline{\boldsymbol{f}}_{\lambda} \\ \multicolumn{2}{c}{\underline{\boldsymbol{e}_k}} \end{bmatrix} [\underline{\boldsymbol{t}}] = \begin{bmatrix} \underline{\mathbf{0}} \\ \mp 1 \end{bmatrix} \tag{5.7}$$

where $\boldsymbol{e}_k$ is a vector whose all elements except kth element is equal to 1 are zero. Choosing the value of $+1$ or -1 depends on the direction of the variable change during the steps of the PV curve tracking. For example, if a load bus voltage magnitude is selected as a known variable, due to its reduction during moving toward the nose and the lower part of the PV curve, the value of t_k is set to -1. In the early steps of tracking the PV curve, λ can be selected as a known variable in Eqs. (5.6) and (5.7). But this choice must change as it approaches the nose of the PV curve. Because, if a large value is assigned to λ, the predicted point will be too far from the PV curve, making it difficult to determine the next actual operating point. Detection of proximity to the nose of the PV curve is possible by observing the values of $\Delta x/\Delta\lambda$. Near the nose of the PV curve, this ratio notably increases. In each step, it is better to consider the variable that has the maximum element of the tangent vector in the previous step as the known variable [6]. Once the tangent vector is determined, the next operating point is predicted as follows:

$$\begin{bmatrix} \hat{\underline{\boldsymbol{x}}}^{i+1} \\ \hat{\lambda}^{i+1} \end{bmatrix} = \begin{bmatrix} \underline{\boldsymbol{x}}^{i} \\ \lambda^{i} \end{bmatrix} + \sigma \begin{bmatrix} \mathbf{d}\underline{\boldsymbol{x}}^{i} \\ \mathrm{d}\lambda^{i} \end{bmatrix} \tag{5.8}$$

where the vector $(\boldsymbol{x}^i, \lambda^i)$ is the current point and the vector $\left(\hat{\boldsymbol{x}}^{i+1}, \hat{\lambda}^{i+1}\right)$ represents the predicted point. σ is the prediction step size and its value must be selected so that the predicted point be inside the convergence radius of the correction step.

5.3.1.2 Secant Method

In this method, Eq. (5.9) is used to predict the next operating point [7]. According to this equation, the prediction of the $i + 1$th operating point is done using the two previous operating points and is based on the line that passes through these two points. Figure 5.2 illustrates the prediction strategy in the secant method.

$$\begin{bmatrix} \hat{\underline{\boldsymbol{x}}}^{i+1} \\ \hat{\lambda}^{i+1} \end{bmatrix} = \begin{bmatrix} \underline{\boldsymbol{x}}^{i} \\ \lambda^{i} \end{bmatrix} + \sigma \begin{bmatrix} \underline{\boldsymbol{x}}^{i} - \underline{\boldsymbol{x}}^{i-1} \\ \lambda^{i} - \lambda^{i-1} \end{bmatrix} \tag{5.9}$$

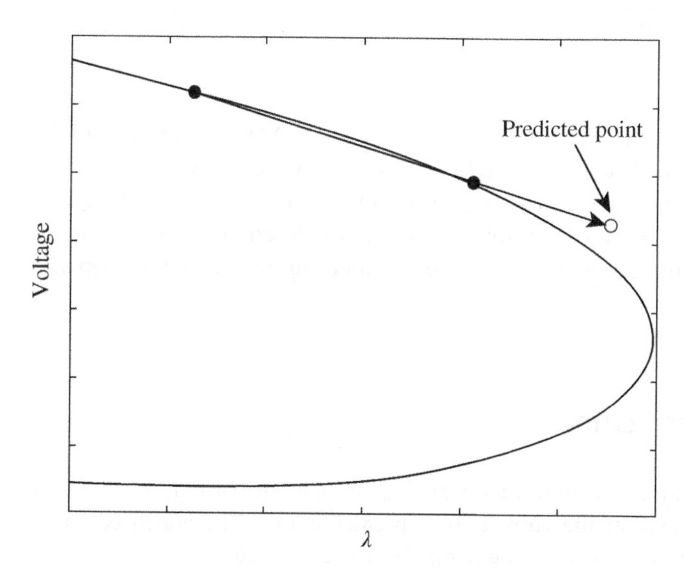

Figure 5.2 Next operating point prediction using secant method.

5.3.2 Nonlinear Predictors

In Refs. [8] and [9], the Lagrange polynomial interpolation equation is used to predict the next operating point. In this equation, in order to predict the new operating point, several previous operating points are used, as given in Eq. (5.9).

$$\underline{\hat{x}}(s_{j+1}) = \sum_{l=0}^{m} L_{j-l}(s)\underline{x}(s_{j-l}) \tag{5.10}$$

where $\hat{x}(s_{j+1})$ is the predicted point at the step $j+1$th, m denotes the order of predictor that shows the previous $m+1$ points are used for prediction, and $L_{j-l}(s)$ are called Lagrangian interpolation coefficients, which are calculated as

$$L_{j-l}(s_{j+1}) = \frac{\prod_{\substack{i=0 \\ i \neq l}}^{m} (s_{j-i} - s_{j+1})}{\prod_{i=0}^{m} (s_{j-i} - s_{j-l})} \tag{5.11}$$

For instance, when the aim is prediction of point s_4 and the previous three points are used, $m = 2$ and the Lagrange polynomial interpolation equation is given as

$$\underline{\hat{x}}(s_4) = L_3(s_4)\underline{x}(s_3) + L_2(s_4)\underline{x}(s_2) + L_1(s_4)\underline{x}(s_1) \Rightarrow \tag{5.12}$$

$$\underline{\hat{x}}(s_4) = \frac{(s_1 - s_4)(s_2 - s_4)}{(s_1 - s_3)(s_2 - s_3)}\underline{x}(s_3) + \frac{(s_3 - s_4)(s_1 - s_4)}{(s_3 - s_2)(s_1 - s_2)}\underline{x}(s_2) + \frac{(s_3 - s_4)(s_2 - s_4)}{(s_3 - s_1)(s_2 - s_1)}\underline{x}(s_1) \tag{5.13}$$

s_{j+1} is obtained from the equation $s_{j+1} = s_j + \Delta s$, where Δs is the arclength of the PV curve between the two points s_j and s_{j+1}. Proper selection of Δs has an important role in improving the prediction quality. Small values of Δs reduce the prediction error but increase the time required to determine the maximum loadability point. In Section 5.5, a method for choosing the prediction step size is presented.

5.4 Parameterization

As mentioned, parameterization is adding an equation to the power flow equations to prevent divergence of the correction step calculations. The added equation determines the amount of movement on the PV curve, i.e. the distance from the current operating point to the next actual operating point.

5.4.1 Local Parameterization

In this type of parameterization, a certain value is assigned to one of the variables and that variable becomes a parameter [6]. Accordingly, the added equation to the power flow equations is given in Eq. (5.14).

$$x_k = \eta \tag{5.14}$$

In the tangent prediction method, the selected variable is the variable that is given in Eq. (5.6), and its magnitude is set to 1.

5.4.2 Arclength Parameterization

In the arclength parameterization method [9], the selected parameter is the arclength of the PV curve between the current and next points. In order to determine the next point on the PV curve, a circle is drawn with the center of the previous operating point and the radius of the distance from the previous operating point to the predicted point (Figure 5.3). The intersection of this circle with the PV curve indicates the next operating point (corrected point). By doing this, the arclength is approximately equal to Δs. In this method, the parameterization equation is in the form of Eq. (5.15). In this equation, x_i and x_i^c are, respectively, the ith current and corrected components of x.

$$\sum_{i=1}^{n} \left(x_i^c - x_i\right)^2 + (\lambda^c - \lambda)^2 = (\Delta s)^2 \tag{5.15}$$

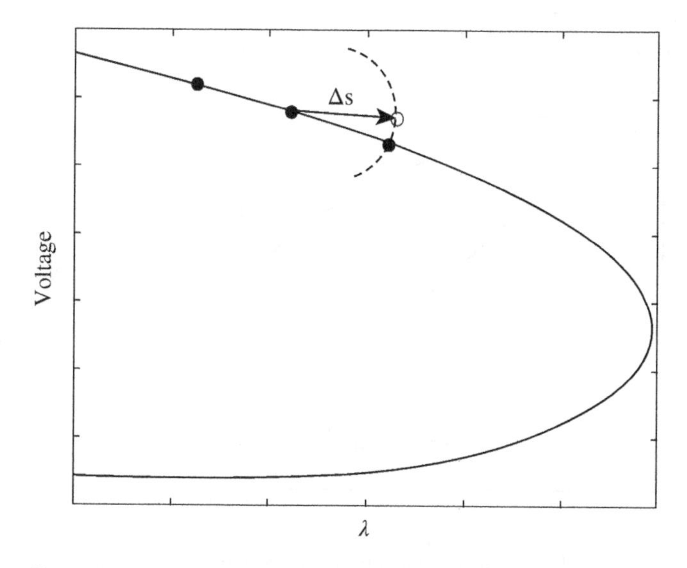

Figure 5.3 Arclength parameterization.

5.4.3 Local Geometric Parameterization

In [10], the parameterization equation is the equation of a straight line drawn from the first determined point on the PV curve (At the base load) with a negative slope. Given that the PV curve is initially almost horizontal, this line certainly intersects the PV curve. In this method, a key parameter called t_β^k is defined according to Eq. (5.16). In which, V_{pa} is the magnitude of the bus voltage chosen for parameterization. $\left(V_{\text{Pa}}^0, \lambda^0\right)$ corresponds to the base value of power flow, and $\left(V_{\text{Pa}}^k, \lambda^k\right)$ corresponds to the kth point of CPF. Figure 5.4 shows the angle β^k.

$$t_\beta^k = \tan \beta^k = \frac{\lambda^k - \lambda^0}{V_{\text{Pa}}^k - V_{\text{Pa}}^0} \tag{5.16}$$

To select a bus whose voltage magnitude is used for parameterization, Eq. (5.17) is used.

$$V_{\text{Pa}}^k : \left|\frac{\Delta V_{\text{Pa}}^k}{\Delta \lambda}\right| = \max\left\{\left|\frac{\Delta V_1}{\Delta \lambda}\right|, \left|\frac{\Delta V_2}{\Delta \lambda}\right|, \ldots, \left|\frac{\Delta V_n}{\Delta \lambda}\right|\right\} \tag{5.17}$$

The equation of the straight line that is used as the parameterization equation and added to Eq. (5.3) is given in Eq. (5.18).

$$t_\beta\left(V_{\text{pa}} - V_{\text{Pa}}^0\right) - \left(\lambda - \lambda^0\right) = 0 \tag{5.18}$$

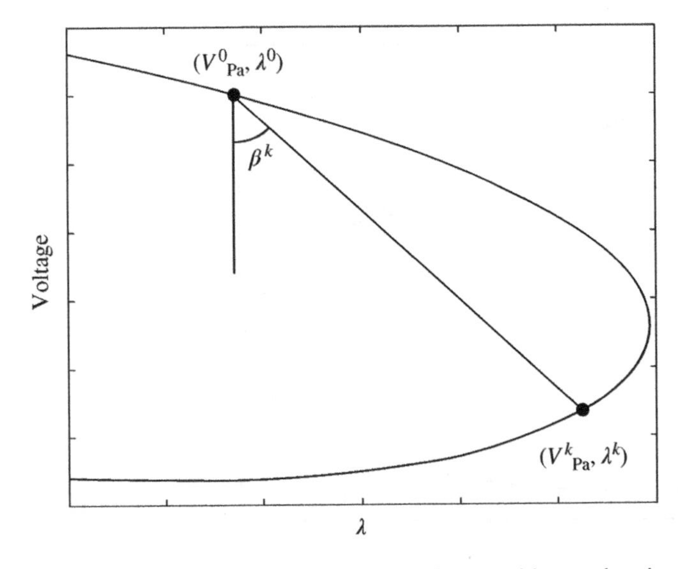

Figure 5.4 Parameterization using a straight line with negative slope.

t_β is chosen at the prediction step. For this aim, Δt_β is determined according to Eq. (5.19) and added to t_β obtained from the previous step.

$$\Delta t_\beta = \frac{\lambda^1 - \lambda^0}{N\left(V^0_{\text{Pa}} - V^1_{\text{Pa}}\right)}, \quad N = 20, \dots, 100 \tag{5.19}$$

$\left(V^1_{\text{Pa}}, \lambda^1\right)$ is the second point determined on the CPF that the slope of the line connecting this point and the point $\left(V^0_{\text{Pa}}, \lambda^0\right)$ with respect to the vertical axis is used as the initial t_β. N is between 20 and 100, that the larger it is selected, the more accurately the maximum loadability point is determined.

When one of the generators reaches its reactive power generation limits, the corresponding bus is converted from PV to PQ. In this state, the reactive power produced by the generator is set at the maximum limit, and its bus voltage changes from a constant value to a dependent variable. This causes the PV curve to change as shown in Figure 5.5. The point at which the generator converts from PV to PQ is called the constraint exchange point (CEP).

From the point of on CEP, the CPF must move on the PV curve with setting the generator reactive power at its maximum limit. Accordingly, the starting point from which the lines are drawn must change from $\left(V^0_{\text{Pa}}, \lambda^0\right)$ to $\left(V^0_{\text{Pa}-\text{new}}, \lambda^0\right)$. Determining a new starting point is possible by setting the base load to λ^0 and solving power flow assuming the generator bus is converted from PV to PQ mode.

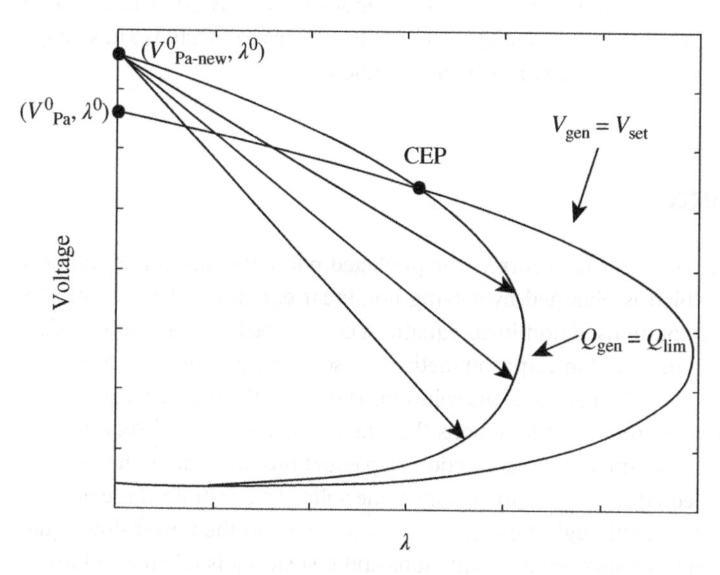

Figure 5.5 PV curve change when a generator reaches its reactive power generation limit.

5.4.4 Alternative Parameterization

In [8], another type of parameterization is presented in which the equation added to Eq. (5.3) is related to different functions such as the whole system active and reactive power losses, the slack generator active or reactive power, the generators reactive power, the transmission line losses, and the power passing through the transmission lines. All of these functions increase uniformly as the system load increases, so they can be used to track the system changes. The mentioned equations are in the form of Eq. (5.20).

$$W(\underline{x}, \lambda, \mu) = \mu W^0 - F(\underline{x}, \lambda) = 0 \tag{5.20}$$

where $\boldsymbol{F}$ is the considered function, $\boldsymbol{W}^0$ is the value of this function in the base case, and μ is a new parameter that by changing it the points on the PV curve are obtained. After linearizing Eqs. (5.3) and (5.20), we get

$$\begin{bmatrix} \underline{f}_{\underline{x}} & \underline{f}_{\lambda} \\ F_{\underline{x}} & F_{\lambda} \end{bmatrix} \begin{bmatrix} \mathrm{d}\underline{x} \\ \mathrm{d}\lambda \end{bmatrix} = \begin{bmatrix} 0 \\ \Delta k_W \end{bmatrix} \tag{5.21}$$

where $\Delta \boldsymbol{k_W} = \boldsymbol{W}^0 \Delta\mu$, and the next operating point is predicted by it. To do this, the tangent vector $\begin{bmatrix} \mathrm{d}x \\ \mathrm{d}\lambda \end{bmatrix}$ is calculated from Eq. (5.21), and the next operating point is predicted according to Eq. (5.8).

Ref. [5] proposes a method for selecting lines whose losses are used as a parameter. This method uses a line voltage stability index as well as other line-related variables such as losses and the voltage magnitude of buses connected to it. Voltage stability indices are introduced in Chapters 7 and 8.

5.5 Correctors

The aim of the corrector is to correct the predicted point and determine the next actual point, which is obtained by solving nonlinear equations. To achieve this aim, numerical solutions of nonlinear equations can be used, and the most widely used of which is the Newton-Raphson method. In solving equations, the predicted point is used as the initial guess of the solution. The closer the predicted point is to the actual solution, the less interactions that the numerical method require.

If λ is taken as parameter, the correction step equations are exactly the same as the power flow equations. But if, for example, the voltage magnitude of the bus K is taken as parameter, although the equations are the same as the power flow equations, λ is taken as an unknown in equations and instead V_K is taken as a known parameter. In these conditions, in the Jacobian matrix, the column of derivatives relative to V_K is replaced by derivatives relative to λ. In other parameterizations, i.e. arclength, alternative, and local geometric parameterizations, an equation is added to the power flow equations. In these conditions, λ is also added to the unknowns to equalize the number of equations and unknowns. Of course, a row and a column will be added to the power flow Jacobian matrix.

5.6 Determining the Prediction Step Size

As mentioned earlier, the size of prediction step has a great impact on the performance of the CPF method. The smaller step size leads to the lower prediction error. Low prediction error means that there is not much distance between the predicted point and the actual points obtained in the correction step. This increases the convergence probability of calculations in the corrector step and reduces the number of required iterations. On the other hand, the small prediction step increases the required calculations in determining the maximum loadability point. Given the shape of the PV curve, it is obvious that approaching the maximum loadability point, the PV curve changes become faster. This makes it more difficult to predict the next point. Therefore, some larger prediction step sizes can be selected at the flat part of PV curve, and approaching the maximum loadability point the step sizes should be reduced. Detection of approaching the maximum loadability point can be done in several ways. For example, increase in the number of required iterations in the correction step can be a sign of approaching this point. Increase in the sensitivity $\frac{dx}{d\lambda}$ can also be another sign. Also when equations such as the system active and reactive losses, the slack generator active or reactive powers and reactive power generated by the generators are used for parameterization, the Euclidean norm of the tangent vector [Eq. (5.21)] increases when approaching the maximum loadability point. Therefore, in Refs. [1, 4, 5], Eq. (5.22) is used to determine the magnitude of the prediction step, where $\|t\|_2$ is the Euclidean norm of the tangent vector and σ_0 is the initial prediction step.

$$\sigma = \frac{\sigma_0}{\|t\|_2} \tag{5.22}$$

It should be noted that if the voltage magnitude is used as parameter, when approaching the maximum loadability point, the Euclidean norm of the tangent vector decreases. Therefore, in these conditions, it may not be appropriate to use Eq. (5.22). Consequently, to determine prediction step in tangent and secant methods, Eq. (5.23) is proposed.

$$\sigma = \frac{\sigma_0}{\|Z\|_2}, \quad \sigma \leq \sigma_{\max} \tag{5.23}$$

where

$$Z = \begin{bmatrix} \frac{\Delta V_1}{\Delta \lambda} \\ \vdots \\ \frac{\Delta V_n}{\Delta \lambda} \end{bmatrix} / m_0 \tag{5.24}$$

$$m_0 = \max \left[\left| \frac{\Delta V_1}{\Delta \lambda} \right|, \ldots, \left| \frac{\Delta V_n}{\Delta \lambda} \right| \right] \tag{5.25}$$

In the nonlinear predictor (Section 5.3.2), the prediction step size determination means to determine the size Δs. For example, the second-order predictor uses the previous three points. Assume that s_1 corresponds to the base load and s_2 and s_3 correspond to the second and third points, respectively. It is also assumed that these points are predicted by the tangent method and corrected by the arclength parameterization. Given that these three points are on the flat part of the PV curve, the distance between the points (Δs) and the length of the tangent vector are probably very close. In Ref. [8] to select Δs to predict the new point (fourth point onward), the PV curve is divided into three parts according to Figure 5.6.

In the first part, which is the flat part of the curve, Δs s are considered same and equal to the length of the tangent vector corresponding to the first three points (h_0) provided that the length of the tangent vector is slightly different from Δs. At all

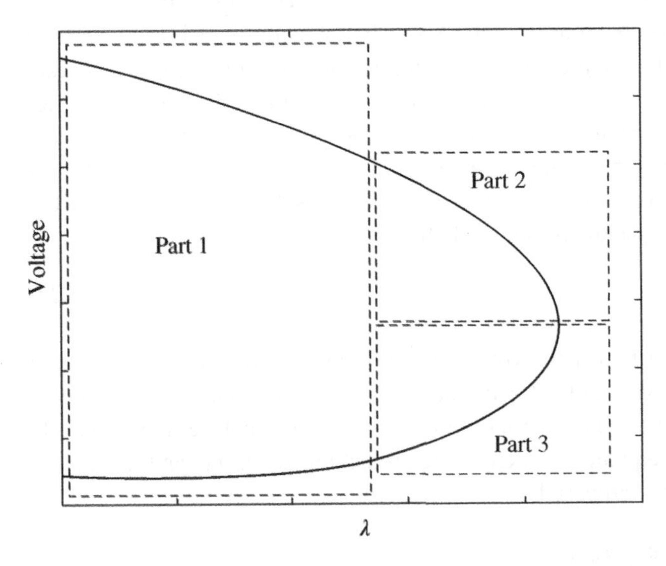

Figure 5.6 Decomposition of PV curve for prediction step size determination.

points of the first part, the tangent vector is determined by selecting a certain $\Delta\lambda$ [Eq. (5.5)]. In Ref. [8], the difference between the tangent vector length and Δs is considered as a measure of the prediction error. When this difference becomes more than a limit, Δs is reduced. For this purpose, Eq. (5.26) is used.

$$\begin{aligned} &\text{If } \left|1-\left\{\left(s_j-s_{j-1}\right)/h_0\right\}\right| > \beta_1 \\ &\text{Then } h_1 = h_0 \cdot \alpha_1 \end{aligned} \tag{5.26}$$

where h_0 is the length of the tangent vector. If the absolute value of the left side of Eq. (5.26) becomes greater than β_1, Δs equals h_1, otherwise equals h_0. α_1 is a parameter whose value is less than 1.

Also for the parts 2 and 3 of the PV curve, the rules in Eqs. (5.27) and (5.28) are used. In these parts, Δs is selected equal to h_2 and h_3, respectively. α_2 is less than 1, but for α_3 a value greater than 1 is considered.

$$\begin{aligned} &\text{If } |\Delta x/\Delta\lambda|_{\max} > \beta_2 \quad \text{and} \quad \Delta\lambda_j \geq 0 \\ &\text{Then,} h_2 = \left(s_j - s_{j-1}\right) \cdot \alpha_2 \end{aligned} \tag{5.27}$$

$$\begin{aligned} &\text{If } |\Delta x/\Delta\lambda|_{\max} > \beta_2 \quad \text{and} \quad \Delta\lambda_j < 0 \\ &\text{Then,} h_3 = \left(s_j - s_{j-1}\right) \cdot \alpha_3 \end{aligned} \tag{5.28}$$

5.7 Comparison of Predictors

In this section, the tangent, secant, second-order nonlinear, and third-order nonlinear predictors are compared. In all these methods, the correction step is performed by the arclength parameterization method. The simulations are done on the IEEE 57-bus test system. By increasing the loading factor λ, the operating point of the system is moved from the base state to the maximum load point (MLP). λ will be increased until reaching λ_{max}.

With respect to the number of required points in different methods, the second point is determined by tangent method, the third point by tangent and secant methods, the fourth point by tangent, secant and second order nonlinear methods, and the fifth point onwards by all four methods. In tangent and secant methods, σ_0 is considered equal to 0.1, and Eq. (5.23) is used to determine σ at each step.

Before approaching the area near the PV curve nose, in the tangent method, λ is selected as the known variable. Also, in nonlinear methods, Δs is the same as the difference between the previous two points. After reaching the area near the PV curve nose, which is determined by $|\Delta x/\Delta\lambda|_{max} > \beta$, the voltage magnitude of the bus 30 that is a weak bus in the system and the PV curve is drawn at it, is considered as a known variable in the tangent method. Also in this area, in nonlinear methods, Δs is selected for the upper parts of the PV curve as $(s_j - s_{j-1}) \bullet \alpha_1$ and for the lower parts of the curve as $(s_j - s_{j-1}) \bullet \alpha_2$. After calculating each point, if the generator exceeds its reactive power generation limit, the generator bus is converted from PV to PQ, and the operating point is recalculated. Respectively, the values 0.3, 0.6, and 1.01 are assigned to the parameters β, α_1, and α_2, respectively.

Figures 5.7–5.10 show the PV curves obtained by different prediction methods. At each obtained point, the number of required iterations in the correction step

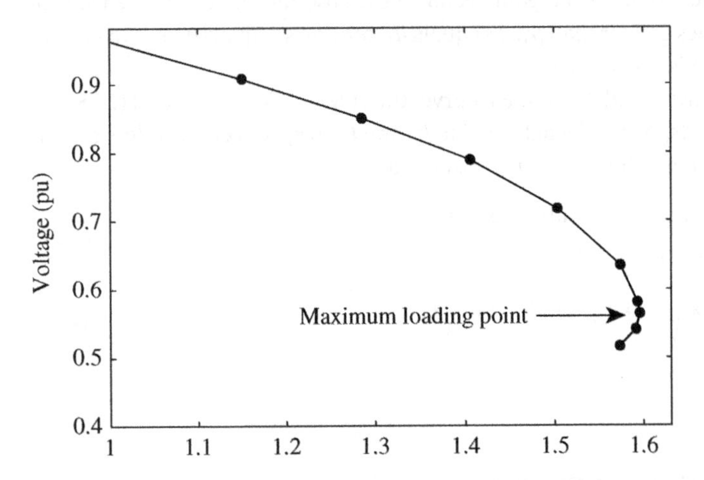

Figure 5.7 PV curve obtained using tangent predictor.

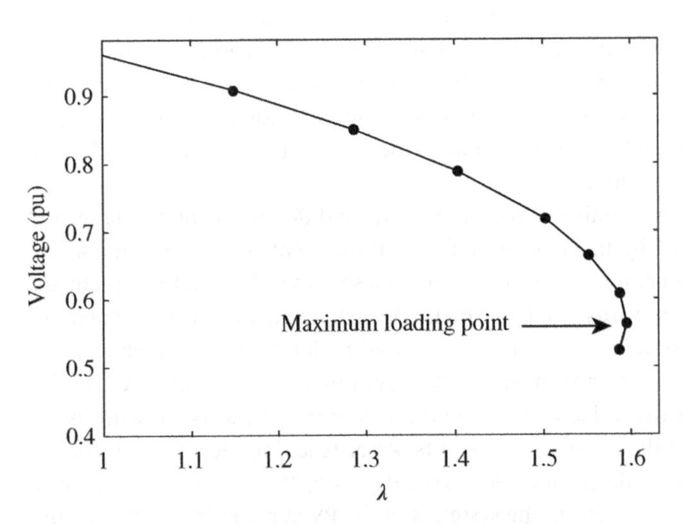

Figure 5.8 PV curve obtained using secant predictor.

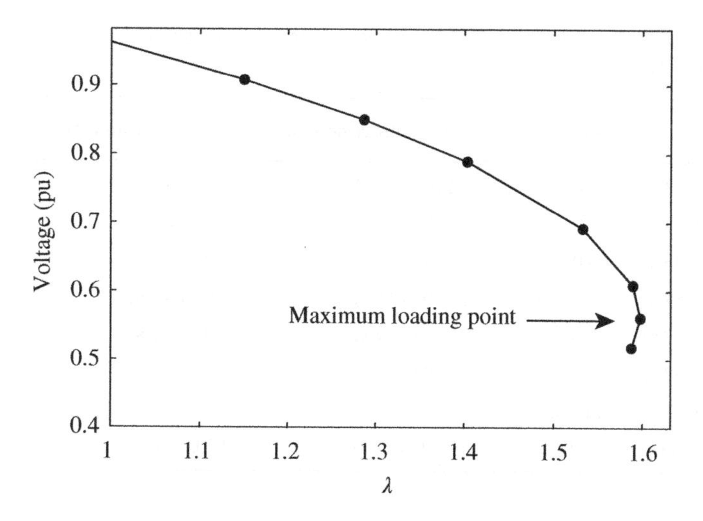

Figure 5.9 PV curve obtained using second-order nonlinear predictor.

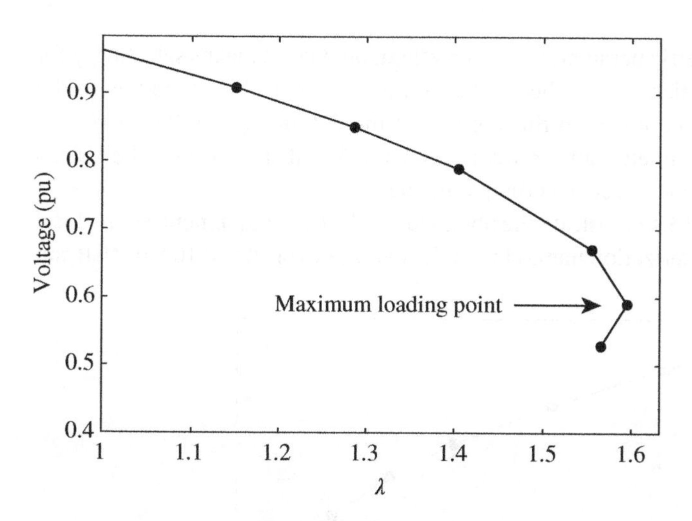

Figure 5.10 PV curve obtained using third-order nonlinear predictor.

and the solution time in seconds are given in Table 5.1. For the sake of comparison, just the points before the PV curve nose are presented in Table 5.1. λ decrement is a sign of crossing the MLP. It is observed that the nonlinear predictors need less iterations in the correction step because they provide better prediction about the actual operating point.

Table 5.1 Number of required iterations in the correction step and the solution time (s), with different predictors.

Point No.	Tangent	Secant	Second order nonlinear	Third order nonlinear
1	2 (0.0563 s)	2 (0.0564 s)	2 (0.0560 s)	2 (0.0563 s)
2	3 (0.0848 s)	3 (0.0855 s)	3 (0.0859 s)	3 (0.0857 s)
3	3 (0.0845 s)	3 (0.0847 s)	2 (0.0566 s)	2 (0.0566 s)
4	3 (0.0853 s)	3 (0.0851 s)	2 (0.0855 s)	1 (0.0286 s)
5	3 (0.0852 s)	2 (0.0563 s)	1 (0.0284 s)	1 (0.0282 s)
6	2 (0.0837 s)	2 (0.0567 s)	1 (0.0281 s)	—
7	2 (0.0531 s)	2 (0.0565 s)	—	—
Total	18 (0.5059 s)	17 (0.4812 s)	11 (0.3405 s)	9 (0.2554 s)

5.8 Simulation of Local Geometric Parameterization Method

In the local geometric parameterization method, prediction means selecting t_β for the next point, which is described in Section 5.4.3. t_β selection is done with the values of 20 and 100 for N. In this section, in the flat area of the PV curve, λ is selected as the parameter, and in the areas 2 and 3 of it (Figure 5.6), the bus 30 voltage magnitude is selected as the parameter.

Figures 5.11 and 5.12 illustrate the obtained results from implementing the local geometric parameterization method with the two values of 20 and 100 for N. It can

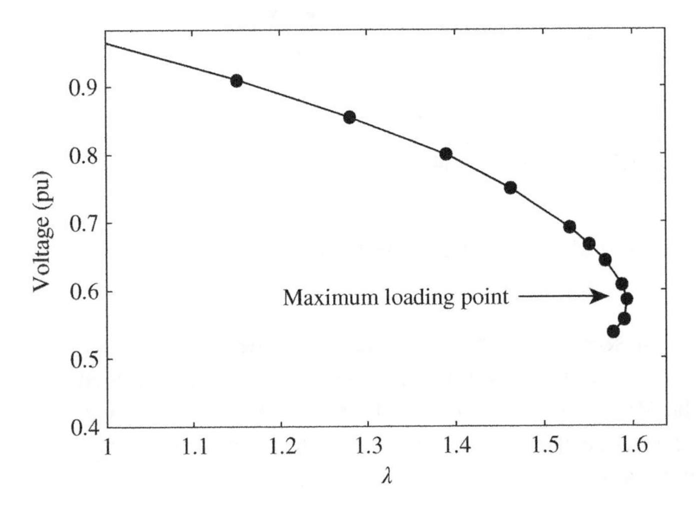

Figure 5.11 PV curve obtained using local geometric parameterization predictor with $N = 20$.

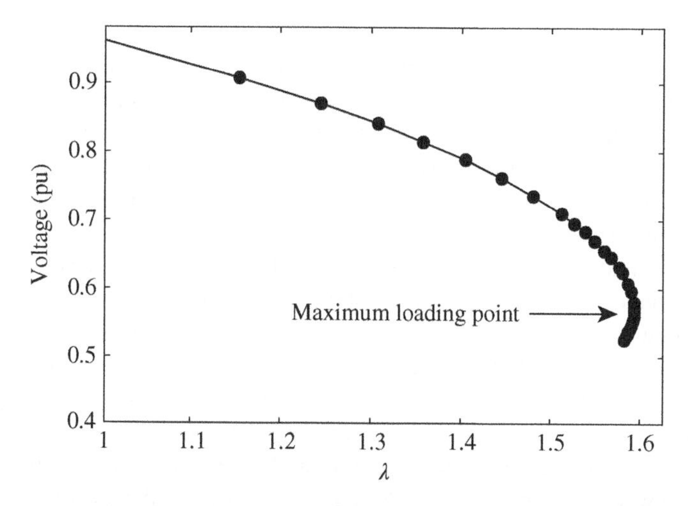

Figure 5.12 PV curve obtained using local geometric parameterization predictor with $N = 100$.

be seen that with increasing t_β, the obtained points are very close to each other at the vicinity of the nose of the PV curve, which makes it possible to accurately determine the MLP.

5.9 Some Real-world Applications of CPF

The presented literatures on CPF show, CPF has applications in real-world power systems toward various aims. In the following paragraphs, some of these applications are briefly introduced.

The paper [11] tried to assess voltage security on unbalanced multiphase distribution systems using CPF method. Wherein, a multiphase continuation power flow (MCPF) methodology with comprehensive models of most network components is used. This voltage assessment security method is a useful tool to assist utility engineers and planners in their analyses of voltage stability phenomena in real-world applications.

An application of CPF in distribution systems is introduced in [12]. Wherein, a modified version of CPF is used to determine the maximum loadability point of radial distribution systems. Using this method, the CPF remains stable even when the considered radial distribution system is close to its maximum loadability point.

In paper [13], a three-phase CPF method for voltage stability assessment of active unbalanced distribution systems is presented. The method is useful in distribution systems with high penetration of distributed energy resources (DERs). The method assesses voltage stability in both the meshed and radial distribution systems and the balanced and unbalanced three-phase distribution systems. Analyzing distribution system voltage stability with DER will allow high penetration of renewable energy necessary for the sustainability goals. The developed tool allows the voltage stability analysis to facilitate the planning, operation, control, and distribution system management.

In Ref. [14], a continuation three-phase power flow (CTPFlow) to analyze voltage stability in unbalanced three-phase power systems is presented. This application of CPF is useful when the power system is unbalanced from the viewpoints of network and load. When the network or load is unbalanced in a power system, the pattern of PV curves is interesting and different from the normal conditions. With unbalanced network or load, at least one of the PV curves at a bus is very similar to that of single-phase or balanced three-phase systems, and the tracing direction of the PV curve is clockwise while the rest of the PV curves (or curve) at the bus have the anti-clockwise tracing direction. For those PV curves with anti-clockwise tracing direction, the higher voltage portion of the PV curves is corresponding to the unstable power-flow solutions, while the lower voltage portion of the PV curves is corresponding to the stable power-flow solutions. This characteristic is unique to unbalanced three-phase power systems. In this paper, this approach is implemented on some real-world power systems with common unbalanced conditions in power systems.

In paper [15], CPF is used in a market clearing model that is modeled as a voltage stability constrained optimal power flow (OPF) problem. In the model, the distance to the closest critical power flow solution is represented by means of a loading parameter and evaluated using CPF method. This model is implemented on a simple test system as well as on a more realistic test system, successfully.

5.10 Summary

As stated, singularity of power flow Jacobian matrix at the nose of PV curves leads to divergence of iterative method to obtain the PV curves. To obtain the loadability limits, there are direct and CPF methods. Of course, the CPF methods are widely used in literature, and the major focus of this chapter is on these methods. The major four elements of CPF, i.e. parameterization, prediction, correction, and step size control are introduced in detail. Also, some real-world applications of CPF presented in literatures were introduced.

References

1 Ajjarapu, V. and Christy, C. (1992). The continuation power flow: a tool for steady state voltage stability analysis. *IEEE Trans. Power Syst.* 7 (1): 416–423. https://doi.org/10.1109/59.141737.

2 Abbasi, S. and Karbalaei, F. (2016). Quick and accurate computation of voltage stability margin. *J. Electr. Eng. Technol.* 11 (1): 1–8.

3 Pama, A. and Radman, G. (2009). A new approach for estimating voltage collapse point based on quadratic approximation of PV-curves. *Electr. Power Syst. Res.* 79 (4): 653–659. https://doi.org/10.1016/j.epsr.2008.09.018.

4 Chiang, H.D., Flueck, A.J., Shah, K.S., and Balu, N. (1995). CPFLOW: a practical tool for tracing power system steady-state stationary behavior due to load and generation variations. *IEEE Trans. Power Syst.* 10 (2): 623–634. https://doi.org/10.1109/59.387897.

5 Karbalaei, F. and Abasi, S. (2013). Quadratic approximation of PV curve path based on local phasor measurements, in presence of voltage dependent loads. *Majlesi J. Electr. Eng.* 7 (3): 8–13.

6 Ajjarapu, V. (1991). Identification of steady-state voltage stability in power systems. *International Journal of Energy Systems* 11 (1): 43–46.

7 Alves, D.A., Da Silva, L.C.P., Castro, C.A., and Da Costa, V.F. (2003). Continuation fast decoupled power flow with secant predictor. *IEEE Trans. Power Syst.* 18 (3): 1078–1085. https://doi.org/10.1109/TPWRS.2003.814892.

8 Alves, D.A., Da Silva, L.C.P., Castro, C.A., and Da Costa, V.F. (2003). Alternative parameters for the continuation power flow method. *Electr. Power Syst. Res.* 66 (2): 105–113. https://doi.org/10.1016/S0378-7796(03)00024-5.

9 Zhu, P., Taylor, G., and Irving, M. (2009). Performance analysis of a novel Q-limit guided continuation power flow method. *IET Gener. Transm. Distrib.* 3 (12): 1042–1051. https://doi.org/10.1049/iet-gtd.2008.0504.

10 Nino, E.E., Castro, C.A., Da Silva, L.C.P., and Alves, D.A. (2006). Continuation load flow using automatically determined branch megawatt losses as parameters. *IEE Proc. Gener. Transm. Distrib.* 153 (3): 300–308. https://doi.org/10.1049/ip-gtd.

11 De Araujo, L.R., Penido, D.R.R., Pereira, J.L.R., and Carneiro, S. (2015). Voltage security assessment on unbalanced multiphase distribution systems. *IEEE Trans. Power Syst.* 30 (6): 3201–3208. https://doi.org/10.1109/TPWRS.2014.2370098.

12 Dukpa, A., Venkatesh, B., and El-Hawary, M. (2009). Application of continuation power flow method in radial distribution systems. *Electr. Power Syst. Res.* 79 (11): 1503–1510. https://doi.org/10.1016/j.epsr.2009.05.003.

13 Nirbhavane, P.S., Corson, L., Rizvi, S.M.H., and Srivastava, A.K. (2021). TPCPF: three-phase continuation power flow tool for voltage stability assessment of

distribution networks with distributed energy resources. *IEEE Trans. Ind. Appl.* 57 (5): 5425–5436. https://doi.org/10.1109/TIA.2021.3088384.

14 Zhang, X.P., Ju, P., and Handschin, E. (2005). Continuation three-phase power flow: a tool for voltage stability analysis of unbalanced three-phase power systems. *IEEE Trans. Power Syst.* 20 (3): 1320–1329. https://doi.org/10.1109/TPWRS.2005.851950.

15 Milano, F., Cañizares, C.A., and Conejo, A.J. (2005). Sensitivity-based security-constrained OPF market clearing model. *IEEE Trans. Power Syst.* 20 (4): 2051–2060. https://doi.org/10.1109/TPWRS.2005.856985.

6

PV-Curve Fitting

6.1 Introduction

The process of finding a curve or a mathematical function that most closely resembles the available data about a thing or a process is called "curve fitting". Usually, a function is used to state the relationship between variables. This helps a lot to visualize data and understand the information that already exists. If the fitted curve passes through all available points (data), the fitting accuracy will have the highest value based on the available data. Often in such cases, the researcher is able to make prediction based on obtained curve to find a point as a prediction for a point that does not exist in the domain of given data set. If this point is in the domain of given data set, the prediction procedure is called "interpolation", and if this point is beyond the domain of the given set, it is called "extrapolation". Certainly, neither of these two predictions can be done without error [1].

One of the most common curve fittings is the polynomial curve fitting. It is well known that the general form of a polynomial of degree n is given as $f(x) = a_0 + a_1x + a_2x^2 + ... + + a_nx^n$. Depending on the number of available points, it is possible to use a polynomial with appropriate degree. For example, to fit a curve with n points, a polynomial of degree $n - 1$ or less than it can be used. When the degree is $n - 1$, the obtained polynomial passes through all used n points for curve fitting, i.e. the prediction error of curve fitting is zero. For polynomials with degree less than $n - 1$, depending on the how the points are distributed, this error may be not zero [1].

Now as an example, the six points (−2, 0.02), (−1, 1), (0, 2), (0.5, 2.1887), (1, 2), and (1.5, 1.3062) from the function $f(x) = 2 + 0.67x - 0.5x^2 - 0.17x^3$ are selected; the aim is finding the value of $f(x)$ at points $x = 0.5$ (as interpolation) and $x = 1.5$ (as extrapolation) based on the other four points using curve fitting. The available choices are polynomials of degrees 1, 2, and 3. Using the least square method, the obtained polynomials are in respect $f(x) = 1.6 + 0.7x$, $f(x) = 1.85 + 0.45x - 0.25x^2$, and $f(x) = 2 + 0.67x - 0.5x^2 - 0.17x^3$, depicted in Figure 6.1. The least square

Voltage Stability in Electrical Power Systems: Concepts, Assessment, and Methods for Improvement, First Edition. Farid Karbalaei, Shahriar Abbasi, and Hamid Reza Shabani.

Published 2023 by John Wiley & Sons, Inc.

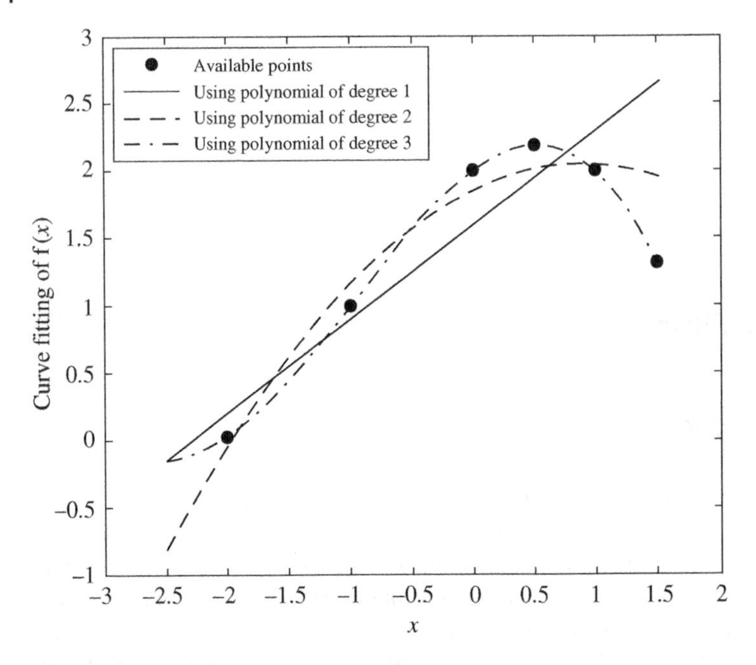

Figure 6.1 Curve fitting of $f(x)$ using polynomial.

Table 6.1 The predicted values of $f(x)$ at points $x = 0.5$ and $x = 1.5$.

At points	By polynomial of degree 1 (error %)	By polynomial of degree 2 (error %)	By polynomial of degree 3 (error %)
$x = 0.5$ (Interpolation)	1.9500 (11%)	2.0125 (8%)	2.1887 (0%)
$x = 1.5$ (Extrapolation)	2.6500 (103%)	1.9625 (50%)	1.3062 (0%)

method is a statistical procedure to find the best fit for a set of data points by minimizing the sum of the squares of the offsets. Also, the predicted values of $f(x)$ at points $x = 0.5$ and $x = 1.5$ by obtained polynomials and the prediction errors (In parenthesis) are presented in Table 6.1. As expected, in interpolation, the prediction error by polynomial of degree 3 is zero, and this error by polynomials of degrees 1 and 2 is in respect 11% and 8%. Also, the predicted point in extrapolation by polynomial of degree 3 is exact, and the prediction error by polynomials of degrees 1 and 2 is 103% and 50%, respectively. Therefore, along with increase in degree of polynomial used in curve fitting, the prediction becomes more accurate. Of course, calculation burden increases.

It is well known that PV-curves are approximately quadratic functions and become exactly quadratic in close neighborhood of the maximum loadability point [2, 3]. Giving this fact, some methods are suggested to use three, two, or one power flow solution(s) for quadratic PV-curve fitting for relatively fast, but approximate calculation of the maximum loadability point. By doing this, with minimum power flow solutions the maximum loadability point can be estimated. From the point of view of the number of power flow solutions, the proposed methods are described in the following sections.

6.2 Curve Fitting Using Three Power Flow Solutions

In Ref. [2], the PV-curve fitting is done using three power flow solutions. These solutions are found by increasing the power system load and generation using Eqs. (5.1) and (5.2) that are represented here as

$$P_i = \lambda P_{\mathrm{Gi0}} - \lambda P_{\mathrm{Li0}} = \sum_{j=1}^{n} V_i V_j y_{ij} \cos\left(\delta_i - \delta_j - \theta_{ij}\right), \quad i \in [\text{PV and PQ buses}] \tag{6.1}$$

$$Q_i = -\lambda Q_{\mathrm{Li0}} = \sum_{j=1}^{n} V_i V_j y_{ij} \sin\left(\delta_i - \delta_j - \theta_{ij}\right), \quad i \in [\text{PQ buses}] \tag{6.2}$$

For the generic bus i, the quadratic equation between λ and the bus voltage magnitude (V_i) is as follow:

$$\lambda = a_i V_i^2 + b_i V_i + c_i \tag{6.3}$$

Assuming $\lambda_1 < \lambda_2 < \lambda_3$, the available three points on the PV curve of bus i obtained by power flow solutions are (λ_1, $V_{i,1}$), (λ_2, $V_{i,2}$), and (λ_3, $V_{i,3}$). The coefficients a_i, b_i, and c_i are then computed by solving the set of linear equations:

$$\lambda_1 = a_i V_{i,1}^2 + b_i V_{i,1} + c_i \tag{6.4}$$

$$\lambda_2 = a_i V_{i,2}^2 + b_i V_{i,2} + c_i \tag{6.5}$$

$$\lambda_3 = a_i V_{i,3}^2 + b_i V_{i,3} + c_i \tag{6.6}$$

The first estimation of maximum loading factor and corresponding voltage $(\hat{\lambda}_*, \hat{V}_{i,*})$ at which $\dfrac{d\lambda}{dV} = 0$ is given by

$$\hat{V}_{i,*} = \frac{-b_i}{2a_i} \tag{6.7}$$

$$\lambda_* = a_i V_{i,*}^2 + b_i V_{i,*} + c_i = c_i - \frac{b_i^2}{4a_i} \tag{6.8}$$

To find the actual value of the voltage $V_{i,*}$, the power flow equations are solved at the estimated loading factor $\hat{\lambda}_*$. Of course, the power flow calculations may diverge at this loading level that this happens when the estimated loading factor is more than the actual one. In case of divergence of power flow calculations, the estimated $\hat{\lambda}_*$ must be modified accordingly in the following way:

$$\hat{\lambda}_*(\text{new}) = \frac{\hat{\lambda}_* + \lambda_3}{2} \tag{6.9}$$

Also, the actual value of $V_{i,*}$ is find using power flow calculations with $\hat{\lambda}_*(\text{new})$.

Now, as the second step of the PV-curve fitting, with the last obtained three points on the PV-curve i.e. $(\lambda_2, V_{i,2})$, $(\lambda_3, V_{i,3})$, and $\left(\hat{\lambda}_*, V_{i,*}\right)$, the above process is repeated to predict a new $\hat{\lambda}_*$. This procedure continues until the tolerance between the last two $\hat{\lambda}_*s$ is less than a predefined accuracy.

The algorithm of finding the maximum loadability point using three power flow solutions can be summarized as the following steps:

1) Obtain three power flow solutions at different load levels and save the points $(\lambda_1, V_{i,1})$, $(\lambda_2, V_{i,2})$, and $(\lambda_3, V_{i,3})$ for the selected bus i.
2) For the bus i, estimate the maximum loading factor and corresponding voltage $\left(\hat{\lambda}_*, \hat{V}_{i,*}\right)$ using solving Eqs. (6.4)–(6.6) and using Eqs. (6.7) and (6.8).
3) Obtain a new power flow solution at the estimated loading factor $\hat{\lambda}_*$ to find the actual value of voltage $V_{i,*}$. If in loading factor $\hat{\lambda}_*$ the power flow calculations diverge, and the $\hat{\lambda}_*$ value must be modified as Eq. (6.9).
4) Based on the last three available power solutions $(\lambda_2, V_{i,2})$, $(\lambda_3, V_{i,3})$, and $\left(\hat{\lambda}_*, V_{i,*}\right)$, repeat the steps 2–3 to obtain the next estimation of $\hat{\lambda}_*$.
5) If the difference between the last two $\hat{\lambda}_*s$ is within a given tolerance, computations terminate otherwise go to the step 2 using the last three obtained power flow solutions.

6.3 Curve Fitting Using Two Power Flow Solutions

The PV-curve fitting in Ref. [3] is based on two power flow calculations as $(\lambda_1, V_{i,1})$ and $(\lambda_2, V_{i,2})$. By using the information of the derivative of V_i with respect to λ at the second point, the coefficients a_i, b_i, and c_i are obtained, using following equations:

$$\lambda_1 = a_i V_{i,1}^2 + b_i V_{i,1} + c_i \tag{6.10}$$

$$\lambda_2 = a_i V_{i,2}^2 + b_i V_{i,2} + c_i \tag{6.11}$$

$$1 = a_i V_{i,2} \frac{dV_{i,2}}{d\lambda_2} + b_i \frac{dV_{i,2}}{d\lambda_2} \tag{6.12}$$

where $\frac{dV_{i,2}}{d\lambda_2}$ is the derivative of the bus i voltage with respect to its loading factor λ at $(\lambda_2, V_{i,2})$.

Now, similar to Eqs. (6.7) and (6.8), the first estimation of maximum loading factor and corresponding voltage are computed. Also, in the case of divergence of power flow calculations at this estimated point, the value of $\hat{\lambda}_*$ will be modified, according to Eq. (6.9), with replacing λ_3 by λ_2. Then, with the last obtained two points $(\lambda_2, V_{i,2})$ and $(\hat{\lambda}_*, V_{i,*})$, the above process is repeated to predict a new $\hat{\lambda}_*$. The procedure continues until the tolerance between the last two $\hat{\lambda}_*s$ is less an acceptable accuracy.

The important issue about the proposed method by Ref. [3] is how to find the value of $\frac{dV_{i,2}}{d\lambda_2}$. To do this, taking the first derivatives of both sides of (6.1) and (6.2) with respect to λ, the below equations are obtained:

$$\frac{d(P_i)}{d\lambda} = \frac{\partial P_i}{\partial \delta_2}\frac{d\delta_2}{d\lambda} + \cdots + \frac{\partial P_i}{\partial \delta_n}\frac{d\delta_n}{d\lambda} + \frac{\partial P_i}{\partial V_{m+1}}\frac{dV_{m+1}}{d\lambda} + \cdots + \frac{\partial P_i}{\partial V_n}\frac{dV_n}{d\lambda}, \quad i\epsilon[\text{PV and PQ buses}] \tag{6.13}$$

$$\frac{d(Q_i)}{d\lambda} = \frac{\partial Q_i}{\partial \delta_2}\frac{d\delta_2}{d\lambda} + \cdots + \frac{\partial Q_i}{\partial \delta_n}\frac{d\delta_n}{d\lambda} + \frac{\partial Q_i}{\partial V_{m+1}}\frac{dV_{m+1}}{d\lambda} + \cdots + \frac{\partial Q_i}{\partial V_n}\frac{dV_n}{d\lambda}, \quad i\epsilon[\text{PQ buses}] \tag{6.14}$$

where the indices n and m in respect denote the numbers of all buses and PV buses in the considered power system. Eqs. (6.13) and (6.14) can be calculated for other buses and finally expressed in matrix form to give (6.15), which includes the Jacobian matrix from power flow solution:

$$\begin{bmatrix} \mathbf{P_0} \\ \mathbf{Q_0} \end{bmatrix} = [\mathbf{J}] \begin{bmatrix} \frac{\mathbf{d\delta}}{\mathbf{d\lambda}} \\ \frac{\mathbf{dV}}{\mathbf{d\lambda}} \end{bmatrix} \quad \text{or} \quad \begin{bmatrix} \frac{\mathbf{d\delta}}{\mathbf{d\lambda}} \\ \frac{\mathbf{dV}}{\mathbf{d\lambda}} \end{bmatrix} = [\mathbf{J}]^{-1} \begin{bmatrix} \mathbf{P_0} \\ \mathbf{Q_0} \end{bmatrix} \tag{6.15}$$

where, $\boldsymbol{P}_0$ and $\boldsymbol{Q}_0$ are the vectors of injected active and reactive powers at base case. $\frac{d\boldsymbol{\delta}}{d\lambda}$ and $\frac{d\boldsymbol{V}}{d\lambda}$ are in respect the derivative vectors of the voltages angle and magnitude with respect to λ.

6.4 Curve Fitting Using One Power Flow Solution

The PV-curve fitting method in Ref. [4] is based on one power flow solution. The proposed method is based on this assumption that estimated PV-curve passes through the coordinate origin point. With this assumption, the coefficient c_i in Eq. (6.3) becomes zero, and the quadratic PV-curve fitting at bus i is as

$$\lambda = a_i V_i^2 + b_i V_i \tag{6.16}$$

Assuming that the known point on the PV curve of bus i is $(\lambda_1, V_{i,1})$, the coefficients a_i, b_i can be computed by solving the below two linear equations:

$$\lambda_1 = a_i V_{i,1}^2 + b_i V_{i,1} \tag{6.17}$$

$$1 = a_i V_{i,2} \frac{dV_{i,1}}{d\lambda_1} + b_i \frac{dV_{i,1}}{d\lambda_1} \tag{6.18}$$

as,

$$a_i = \frac{\left(\frac{V_{i,1}}{dV_{i,1}/d\lambda_1}\right) - \lambda_1}{V_{i,1}^2} \tag{6.19}$$

$$b_i = \frac{\lambda_1 - a_i V_{i,1}^2}{V_{i,1}} \tag{6.20}$$

In a similar manner, the first estimation of maximum loading factor and corresponding voltage are given by

$$\hat{V}_{i,*} = \frac{-b_i}{2a_i} \tag{6.21}$$

$$\hat{\lambda}_* = a_i V_{i,*}^2 + b_i V_{i,*} = -\frac{b_i^2}{4a_i} \tag{6.22}$$

When the estimated point $\hat{\lambda}_*$ is beyond the actual maximum loadability limit, the power flow calculations diverges at the loading factor $\hat{\lambda}_*$. In [4], to avoid divergence of power flow calculation at an estimated loading factor, instead of $\hat{\lambda}_*$, the $\hat{V}_{i,*}$ is inserted in the power flow equations as the known variable to find a new actual

$(\lambda_*, \hat{V}_{i,*})$. Now, having this actual point, the algorithm repeats until reaching a valid accuracy.

The proposed algorithm in this paper can be summarized as the following steps:

1) Perform one power flow and save the point $(\lambda_1, V_{i,\,1})$ for the selected bus i.
2) With this loading factor, calculate $\frac{dV_{i,1}}{d\lambda_1}$ from Eq. (6.15).
3) For the selected bus i, obtain the coefficients a_i and b_i from Eqs. (6.19) and (6.20), and estimate the maximum loadability point $(\hat{\lambda}_*, \hat{V}_{i,*})$ as Eqs. (6.21) and (6.22).
4) Insert $\hat{V}_{i,*}$ obtained from step 3 in the power flow equations as the known variable and find a new actual $(\lambda_*, \hat{V}_{i,*})$.
5) With the new point obtained in step 4, repeat the above steps until the difference between two successive $\hat{\lambda}_*$s becomes less than a given tolerance.

In [4], it is suggested that in order to reduce the number of required iterations, the initial point $(\lambda_1, V_{i,1})$ should be selected as close as possible to the maximum loadability limit. This is due to the fact that although the maximum loadability point may notably change in occurrence of various contingencies, at the voltage variations range at this point is limited. Numerous simulations that have been performed in different systems for many contingencies confirm this conclusion [4]. The voltage magnitude at the maximum loadability limit depends on the level of reactive power compensation and varies from about 0.6 pu in low compensated systems to about 0.9 pu for in highly compensated systems. Based on this, the initial point can be determined by selecting a suitable value for $V_{i,1}$ and calculating λ_1 for this voltage.

Another approach in [5] for PV-curve fitting based on one power flow solution is presented. The first- and second-order derivatives of the power flow Eqs. (6.1) and (6.2) are required in this method. Assuming that the only available point on the PV-curve at the bus i is $(\lambda_1, V_{i,1})$, the first equation to obtain the required coefficients of estimated curve is as

$$\lambda_1 = a_i V_{i,1}^2 + b_i V_{i,1} + c_i \tag{6.23}$$

Using the Eq. (6.23), the first-order derivative terms $\frac{d\lambda}{dV_i}$ and $\frac{dV_i}{d\lambda}$, as well as the second-order derivative terms $\frac{d^2V_i}{d\lambda^2}$ are found as follows:

$$\frac{d\lambda}{dV_i} = 2a_i V_i + b_i \tag{6.24}$$

$$\frac{dV_i}{d\lambda} = \frac{1}{2a_i V_i + b_i} \tag{6.25}$$

$$\frac{d^2V_i}{d\lambda^2}=\frac{d}{d\lambda}\left(\frac{dV_i}{d\lambda}\right)=\frac{\partial(dV_i/d\lambda)}{\partial V_i}\frac{dV_i}{d\lambda}+\frac{\partial(dV_i/d\lambda)}{\partial\lambda}=-\frac{2a_i}{(2a_iV_i+b_i)^2}\frac{dV_i}{d\lambda}=-2a_i\left(\frac{dV_i}{d\lambda}\right)^3 \tag{6.26}$$

Solving Eqs. (6.24)–(6.26) the coefficients a_i, b_i, and c_i can be found as

$$a_i=\frac{d^2V_i/d\lambda^2}{2(dV_i/d\lambda)^3} \tag{6.27}$$

$$b_i=\frac{1}{dV_i/d\lambda}-2a_iV_i \tag{6.28}$$

$$c_i=\lambda-a_iV_i^2-b_iV_i \tag{6.29}$$

The rest of this method is similar to what stated about the estimation and modification of loading factor in Eqs. (6.7)–(6.9).

As mentioned, the first- and second-order derivatives $\frac{dV_i}{d\lambda}$ and $\frac{d^2V_i}{d\lambda^2}$ at the available point of PV-curve are required to insert in Eqs. (6.28) and (6.27). How to calculate the first-order derivative was previously described in Eq. (6.15). To compute the second-order derivative $\frac{d^2V_i}{d\lambda^2}$, the derivatives of both sides of (6.14) are taken as

$$\frac{d}{d\lambda}\left(\frac{dP_i}{d\lambda}\right)=\frac{d}{d\lambda}(P_{Gi0}-P_{Li0})=0=\frac{d}{d\lambda}\left\{\sum_{j=2}^{n}\frac{\partial P_i}{\partial\delta_j}\frac{d\delta_j}{d\lambda}+\sum_{j=m+1}^{n}\frac{\partial P_i}{\partial V_j}\frac{dV_j}{d\lambda}\right\},$$

$$i\epsilon[\text{PV and PQ buses}] \tag{6.30}$$

$$\frac{d}{d\lambda}\left(\frac{dQ_i}{d\lambda}\right)=\frac{d}{d\lambda}(-Q_{Li0})=0=\frac{d}{d\lambda}\left\{\sum_{j=2}^{n}\frac{\partial Q_i}{\partial\delta_j}\frac{d\delta_j}{d\lambda}+\sum_{j=m+1}^{n}\frac{\partial Q_i}{\partial V_j}\frac{dV_j}{d\lambda}\right\},\quad i\epsilon[\text{PQ buses}] \tag{6.31}$$

or,

$$0=\left\{\sum_{j=2}^{n}\left(\frac{d}{d\lambda}\left(\frac{\partial P_i}{\partial\delta_j}\right)\frac{d\delta_j}{d\lambda}+\frac{\partial P_i}{\partial\delta_j}\frac{d}{d\lambda}\left(\frac{d\delta_j}{d\lambda}\right)\right)+\sum_{j=m+1}^{n}\left(\frac{d}{d\lambda}\left(\frac{\partial P_i}{\partial V_j}\right)\frac{dV_j}{d\lambda}+\frac{\partial P_i}{\partial V_j}\frac{d}{d\lambda}\left(\frac{dV_j}{d\lambda}\right)\right)\right\},$$

$$i\epsilon[\text{PV and PQ buses}] \tag{6.32}$$

$$0=\left\{\sum_{j=2}^{n}\left(\frac{d}{d\lambda}\left(\frac{\partial Q_i}{\partial\delta_j}\right)\frac{d\delta_j}{d\lambda}+\frac{\partial Q_i}{\partial\delta_j}\frac{d}{d\lambda}\left(\frac{d\delta_j}{d\lambda}\right)\right)+\sum_{j=m+1}^{n}\left(\frac{d}{d\lambda}\left(\frac{\partial Q_i}{\partial V_j}\right)\frac{dV_j}{d\lambda}+\frac{\partial Q_i}{\partial V_j}\frac{d}{d\lambda}\left(\frac{dV_j}{d\lambda}\right)\right)\right\},$$

$$i\epsilon[\text{PQ buses}] \tag{6.33}$$

The Eqs. (6.32) and (6.33) can be written in the form of the matrix equation:

$$0 = [\boldsymbol{J}]\begin{bmatrix}\dfrac{\mathbf{d}^2\boldsymbol{\delta}}{\mathbf{d}\lambda^2}\\ \dfrac{\mathbf{d}^2\boldsymbol{V}}{\mathbf{d}\lambda^2}\end{bmatrix} + [\boldsymbol{A}]\begin{bmatrix}\left(\dfrac{\mathbf{d}\boldsymbol{\delta}}{\mathbf{d}\lambda}\right)^2\\ \left(\dfrac{\mathbf{d}\boldsymbol{V}}{\mathbf{d}\lambda}\right)^2\end{bmatrix} + 2\begin{bmatrix}\text{diag}\left(\dfrac{\text{d}\delta_i}{\text{d}\lambda}\right) & 0\\ 0 & \text{diag}\left(\dfrac{\text{d}V_i}{\text{d}\lambda}\right)\end{bmatrix}[\boldsymbol{B}]\begin{bmatrix}\dfrac{\mathbf{d}\boldsymbol{\delta}}{\mathbf{d}\lambda}\\ \dfrac{\mathbf{d}\boldsymbol{V}}{\mathbf{d}\lambda}\end{bmatrix}$$
$$+2\begin{bmatrix}\text{diag}\left(\dfrac{\text{d}V_i}{\text{d}\lambda}\right) & 0\\ 0 & \text{diag}\left(\dfrac{\text{d}\delta_i}{\text{d}\lambda}\right)\end{bmatrix}[\boldsymbol{C}]\begin{bmatrix}\dfrac{\mathbf{d}\boldsymbol{\delta}}{\mathbf{d}\lambda}\\ \dfrac{\mathbf{d}\boldsymbol{V}}{\mathbf{d}\lambda}\end{bmatrix} + 2[\boldsymbol{D}]\begin{bmatrix}\dfrac{\mathbf{d}\boldsymbol{\delta}}{\mathbf{d}\lambda}\dfrac{\mathbf{d}\boldsymbol{V}}{\mathbf{d}\lambda}\\ \dfrac{\mathbf{d}\boldsymbol{V}}{\mathbf{d}\lambda}\dfrac{\mathbf{d}\boldsymbol{\delta}}{\mathbf{d}\lambda}\end{bmatrix} \tag{6.34}$$

The matrices and vectors in Eq. (6.34) are as follows:

$$[\boldsymbol{A}] = \begin{bmatrix}\boldsymbol{A}_{11} & \boldsymbol{A}_{12}\\ \boldsymbol{A}_{21} & \boldsymbol{A}_{22}\end{bmatrix};\ [\boldsymbol{A}_{11}] = \left[\frac{\partial^2 P_i}{\partial\delta_j^2}\right],\ [\boldsymbol{A}_{12}] = \left[\frac{\partial^2 P_i}{\partial V_j^2}\right],\ [\boldsymbol{A}_{21}] = \left[\frac{\partial^2 Q_i}{\partial\delta_j^2}\right],\ [\boldsymbol{A}_{22}] = \left[\frac{\partial^2 Q_i}{\partial V_j^2}\right]$$

$$[\boldsymbol{B}] = \begin{bmatrix}\boldsymbol{B}_{11} & \boldsymbol{B}_{12}\\ \boldsymbol{B}_{21} & \boldsymbol{B}_{22}\end{bmatrix};\ [\boldsymbol{B}_{11}] = \left[\frac{\partial^2 P_i}{\partial\delta_i\partial\delta_{j_{j\neq i}}}\right],\ [\boldsymbol{B}_{12}] = \left[\frac{\partial^2 P_i}{\partial\delta_i\partial V_j}\right],\ [\boldsymbol{B}_{21}] = \left[\frac{\partial^2 Q_i}{\partial V_i\partial\delta_i}\right],$$

$$[\boldsymbol{B}_{22}] = \left[\frac{\partial^2 Q_i}{\partial V_i\partial V_{j_{j\neq i}}}\right]$$

$$[\boldsymbol{C}] = \begin{bmatrix}\boldsymbol{C}_{11} & \boldsymbol{C}_{12}\\ \boldsymbol{C}_{21} & \boldsymbol{C}_{22}\end{bmatrix};\ [\boldsymbol{C}_{11}] = \left[\frac{\partial^2 P_i}{\partial V_i\partial\delta_{j_{j\neq i}}}\right],\ [\boldsymbol{C}_{12}] = \left[\frac{\partial^2 P_i}{\partial V_i\partial V_{j_{j\neq i}}}\right],$$

$$[\boldsymbol{C}_{21}] = \left[\frac{\partial^2 Q_i}{\partial\delta_i\partial\delta_{j_{j\neq i}}}\right],\ [\boldsymbol{C}_{22}] = \left[\frac{\partial^2 Q_i}{\partial\delta_i\partial V_{j_{j\neq i}}}\right]$$

$$[\boldsymbol{D}] = \begin{bmatrix}\boldsymbol{D}_{11} & \boldsymbol{D}_{12}\\ \boldsymbol{D}_{21} & \boldsymbol{D}_{22}\end{bmatrix};\ [\boldsymbol{D}_{11}] = \left[\frac{\partial^2 P_i}{\partial\delta_j\partial V_{j_{j\neq i}}}\right],\ [\boldsymbol{D}_{12}] = [0],\ [\boldsymbol{D}_{21}] = [0],\ [\boldsymbol{D}_{22}] = \left[\frac{\partial^2 P_i}{\partial V_j\partial\delta_{j_{j\neq i}}}\right]$$

The elements of the above matrices can be obtained from the power flow equations:

$$\begin{bmatrix}\dfrac{\text{d}^2\boldsymbol{\delta}}{\text{d}\lambda^2}\\ \dfrac{\text{d}^2\boldsymbol{V}}{\text{d}\lambda^2}\end{bmatrix} = \begin{bmatrix}\dfrac{\text{d}^2\delta_2}{\text{d}\lambda^2} & \cdots & \dfrac{\text{d}^2\delta_n}{\text{d}\lambda^2} & \dfrac{\text{d}^2V_{m+1}}{\text{d}\lambda^2} & \cdots & \dfrac{\text{d}^2V_n}{\text{d}\lambda^2}\end{bmatrix}^T$$

$$\begin{bmatrix} \left(\dfrac{\mathrm{d}\boldsymbol{\delta}}{\mathrm{d}\lambda}\right)^2 \\ \left(\dfrac{\mathrm{d}\boldsymbol{V}}{\mathrm{d}\lambda}\right)^2 \end{bmatrix} = \begin{bmatrix} \left(\dfrac{\mathrm{d}\delta_2}{\mathrm{d}\lambda}\right)^2 & \cdots & \left(\dfrac{\mathrm{d}\delta_n}{\mathrm{d}\lambda}\right)^2 & \left(\dfrac{\mathrm{d}V_{m+1}}{\mathrm{d}\lambda}\right)^2 & \cdots & \left(\dfrac{\mathrm{d}V_n}{\mathrm{d}\lambda}\right)^2 \end{bmatrix}^T$$

$$\begin{bmatrix} \dfrac{\mathrm{d}\boldsymbol{\delta}}{\mathrm{d}\lambda} \\ \dfrac{\mathrm{d}\boldsymbol{V}}{\mathrm{d}\lambda} \end{bmatrix} = \begin{bmatrix} \dfrac{\mathrm{d}\delta_2}{\mathrm{d}\lambda} & \cdots & \dfrac{\mathrm{d}\delta_n}{\mathrm{d}\lambda} & \dfrac{\mathrm{d}V_{m+1}}{\mathrm{d}\lambda} & \cdots & \dfrac{\mathrm{d}V_n}{\mathrm{d}\lambda} \end{bmatrix}^T$$

$$\begin{bmatrix} \dfrac{\mathrm{d}\boldsymbol{\delta}}{\mathrm{d}\lambda} \dfrac{\mathrm{d}\boldsymbol{V}}{\mathrm{d}\lambda} \end{bmatrix} = \begin{bmatrix} \dfrac{\mathrm{d}\boldsymbol{V}}{\mathrm{d}\lambda} \dfrac{\mathrm{d}\boldsymbol{\delta}}{\mathrm{d}\lambda} \end{bmatrix} = \begin{bmatrix} \dfrac{\mathrm{d}\delta_1}{\mathrm{d}\lambda} \dfrac{\mathrm{d}V_1}{\mathrm{d}\lambda} & \cdots & \dfrac{\mathrm{d}\delta_n}{\mathrm{d}\lambda} \dfrac{\mathrm{d}V_n}{\mathrm{d}\lambda} \end{bmatrix}^T$$

By replacing the elements of matrices and the first-order derivatives, the second-order derivative term $\dfrac{\mathrm{d}^2 V_i}{\mathrm{d}\lambda^2}$ is found from Eq. (6.34) to be used in Eq. (6.27).

6.5 Comparison of Different PV-Curve Fitting Methods

For the sake of comparison, the mentioned PV-curve fitting methods are simulated on the IEEE 57-bus test system. The aim of simulations is finding the maximum loadability limit at an arbitrarily selected bus that here is bus 31. This is done with and without reactive power limits of generators. The maximum loadability limits (maximum loading factor) with and without reactive power limits are in respect 1.6043 and 1.8922 in this system. The obtained results are presented in Tables 6.2 and 6.3. In each iteration, the estimated maximum loadability limit and the estimation error (In parenthesis) is presented. As can be seen, from the viewpoint of required iterations, the performance of the three point-based method is better than other's methods. The one point-based method of Ref. [5] requires the most iterations to obtain the maximum loadability limit.

6.6 Summary

This chapter is dedicated to providing methods for PV-curve fitting. The purpose of these methods is to quickly determine the maximum loadability point with the minimum number of power flow solutions. To this aim, the fact is used that PV curves are approximately quadratic functions and become exactly quadratic in close neighborhood of maximum loadability point. In this chapter, a comparison is made among the performance of methods that use one, two, and three power flow solutions.

Table 6.2 Comparison of different PV-curve fitting methods, with reactive power limits.

The no. total power flow solutions	Using three power flow solutions (error %)	Using two power flow solutions (error %)	Using one power flow solution (error %)	
			Ref. [4]	Ref. [5]
1	—	—	1.5542 (−37.9%)	1.7389 (−17.18%)
2	—	1.5324 (−11.90%)	1.5454 (−9.75 %)	1.4344 (−28.24%)
3	1.5730 (−5.1795%)	1.5326 (−11.86%)	1.5805 (−3.93%)	1.4825 (−28.09%)
4	1.5943 (−1.6548%)	1.5752 (−4.82%)	1.5953 (−1.49%)	1.5388 (−12.87%)
5	1.5999 (−0.7281%)	1.5879 (−2.71%)	1.6010 (−0.55%)	1.5642 (−8.30%)
6	1.6028 (−0.2483%)	1.5963 (−1.32%)	1.6032 (−0.18%)	1.5790 (−5.24%)
7	1.6035 (−0.13%)	1.6003 (−0.66%)	1.6039 (−0.06%)	1.5882 (−3.32%)
8	1.6039 (−0.06%)	1.6020 (−0.38%)	1.6041 (−0.03%)	1.5940 (−2.66%)
9	1.6041 (−0.03%)	1.6030 (−0.22%)	1.6043 (−0.01%)	1.5978 (−1.70%)
10	1.6043 (−0.01%)	1.6037 (−0.1%)		1.6001 (−1.07%)
11		1.6040 (−0.05%)		1.6017 (−0.69%)
12		1.6041 (−0.03%)		1.6027 (−0.43%)
13		1.6043 (−0.01%)		1.6034 (−0.26%)
14				1.6038 (−0.14%)
15				1.6041 (−0.03%)
16				1.6043 (−0.01%)

Table 6.3 Comparison of different PV-curve fitting methods, without reactive power limits.

The no. total power flow solutions	Using three power flow solutions (error %)	Using two power flow solutions (error %)	Using one power flow solution (error %)	
			Ref. [4]	Ref. [5]
1	—	—	1.5542 (−37.9%)	1.7389 (−17.18%)
2	—	1.7059 (−20.8%)	1.7892 (−11.54%)	1.8206 (−8.02%)
3	1.8578 (−3.85%)	1.8307 (−6.89%)	1.8670 (−2.8%)	1.8577 (−3.86%)
4	1.8861 (−0.68%)	1.8695 (−2.54%)	1.8864 (−0.65%)	1.8753 (−1.89%)
5	1.8919 (−0.03%)	1.8850 (−0.80%)	1.8909 (−0.14%)	1.8838 (−0.94%)
6	1.8922 ($<e-2\%$)	1.8901 (−0.23%)	1.8920 ($<e-2\%$)	1.8880 (−0.47%)
7		1.8916 (−0.06%)		1.8900 (−0.24%)
8		1.8921 ($<e-2\%$)		1.8911 (−0.12%)
9				1.8916 (−0.04%)
10				1.8920 ($<e-2\%$)

References

1 Zielesny, A. (2011). *From Curve Fitting to Machine Learning: An Illustrative Guide to Scientific Data Analysis and Computational Intelligence*. Springer.

2 Ejebe, G.C., Irisarri, G.D., Mokhtari, S. et al. (1996). Methods for contingency screening and ranking for voltage stability analysis of power systems. *IEEE Trans. Power Syst.* 11 (1): 350–356.

3 Chiang, H.D., Wang, C.S., and Flueck, A.J. (1997). Look-ahead voltage and load margin contingency selection: functions for large-scale power systems. *IEEE Power Eng. Rev.* 12 (1): 173–180.

4 Abbasi, S. and Karbalaei, F. (2016). Quick and accurate computation of voltage stability margin. *J. Electr. Eng. Technol.* 11 (1): 1–8.

5 Pama, A. and Radman, G. (2009). A new approach for estimating voltage collapse point based on quadratic approximation of PV-curves. *Electr. Power Syst. Res.* 79 (4): 653–659.

7

Measurement-Based Indices

7.1 Introduction

In the previous two chapters, some methods for accurately determining the maximum loadability limit were reviewed. In this chapter and the next one, voltage stability indices will be introduced. These indices quickly provide an estimation of the distance from the current operating point to the power system maximum loadability limit. However, due to the fact that most of these indices have nonlinear behavior, it is not possible to accurately predict maximum loadability limit with voltage stability indices. But instead these indices contain important information about weak points of the system. The weak points are buses and lines in which increasing the load at them strongly push the system toward voltage instability. Finding these points will help a lot in determining the measures to prevent voltage instability. Also, these points are usually suitable places for installing reactive power compensation equipment and distributed generation (DG) units.

Voltage stability indices can be divided into two general categories: model- and measurement based. Measurement-based indices do not require system model and are based on the fact that the measured voltage and current phasors contain sufficient information to diagnose the proximity to the system maximum loadability limit. Model-based indices are more suitable for analysis the system state in facing with various changes such as contingency occurrence and the installation of new equipment. In this chapter, measurement-based indices are presented, and the next chapter is dedicated to model-based indices.

Voltage Stability in Electrical Power Systems: Concepts, Assessment, and Methods for Improvement, First Edition. Farid Karbalaei, Shahriar Abbasi, and Hamid Reza Shabani.

Published 2023 by John Wiley & Sons, Inc.

7.2 Thevenin Equivalent-Based Index

7.2.1 Background

Thevenin equivalent (TE) circuit from the viewpoint of each bus of the power system means replacing the system connected to the bus with a voltage source and a series impedance as shown in Figure 7.1, where $\overline{E}_{\text{Th}}$ and $\overline{Z}_{\text{Th}}$ are the TE parameters. It is well known that at bus loadability limit, the magnitude of the load impedance connected to that bus and the Thevenin impedance seen from that bus become equal. Therefore, the ratio of the magnitudes of these two impedances can be considered as a measure of proximity to the loadability limit.

Kirchoff's law gives

$$\overline{V} = \overline{E}_{\text{Th}} - \overline{Z}_{\text{Th}}\overline{I} \tag{7.1}$$

The TE parameters can be calculated by consecutive measurement of $\overline{V}$ and $\overline{I}$ phasors by phasor measurement units (PMUs).

Assuming that there is no change in the system during the two measurements $(\overline{V}_1, \overline{I}_1)$ and $(\overline{V}_2, \overline{I}_2)$, the TE parameters remain constant. So,

$$\overline{V}_1 = \overline{E}_{\text{Th}} - \overline{Z}_{\text{Th}}\overline{I}_1 \tag{7.2}$$

$$\overline{V}_2 = \overline{E}_{\text{Th}} - \overline{Z}_{\text{Th}}\overline{I}_2 \tag{7.3}$$

The TE parameters can be calculated from Eqs. (7.2) and (7.3) as

$$\overline{Z}_{\text{Th}} = \frac{\overline{V}_1 - \overline{V}_2}{\overline{I}_2 - \overline{I}_1} \tag{7.4}$$

$$\overline{E}_{\text{Th}} = \frac{\overline{V}_1\overline{I}_2 - \overline{V}_2\overline{I}_1}{\overline{I}_2 - \overline{I}_1} \tag{7.5}$$

As the system conditions change, this occurs continuously, and the TE parameters may also change significantly. Therefore, calculation of the parameters must be updated consecutively in a sliding window of measured data. The goal is that the parameters of the current system be correctly determined at each

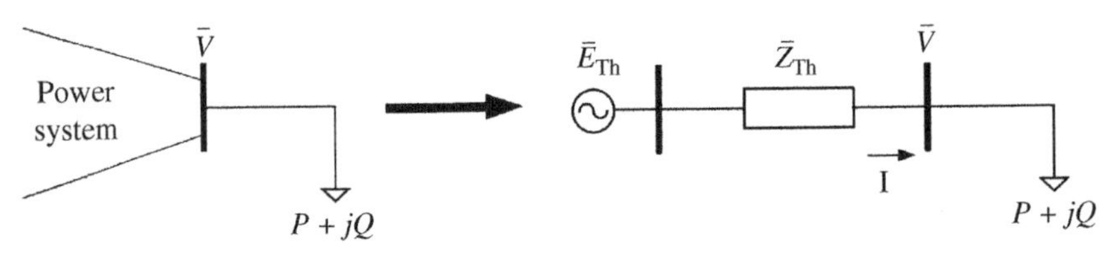

Figure 7.1 Concept of using Thevenin equivalent (TE).

calculation. For some reason, more than two measurements are usually required to determine the TE parameters at each time. One of these reasons is the change of system during two consecutive measurements. This change can be due to a switching, such as disconnecting or connecting a line or a generator, which can cause a large change in the TE parameters. The change can also be due to changes in load and generation that occur continuously. In order to reduce the probability of system change during two consecutive measurements, the time interval between the two measurements is usually reduced to one cycle [1–4]. On the other hand, although the probability of load change is higher than the probability of system change [3], but when the time interval between two consecutive measurements is reduced more, the load may remain constant during the two consecutive measurements. Under these conditions, there will be no significant difference between the measured values and the calculation of TE parameters that requires more measurements. Another reason for the need for more measurements is measurement errors. Increasing the number of measurements reduces the effect of measurement errors on the accuracy of estimating TE parameters [3, 4]. In the following, some methods for determining TE parameters and methods to reduce the impact of measurement error on parameters estimation accuracy are presented.

7.2.2 Recursive Least Square (RLS) Algorithm

By expressing the TE parameters at a time instant t_K as $\overline{E}_{\mathrm{Th}_K} = E_{\mathrm{R}_K} + jE_{I_K}$ and $\overline{Z}_{\mathrm{Th}_K} = R_{\mathrm{Th}_K} + jX_{\mathrm{Th}_K}$ and the measured phasors as $\overline{V}_K = V_{R_K} + jV_{I_K}$ and $\overline{I}_K = I_{R_K} + jI_{I_K}$, the Eq. (7.1) becomes two linear equations with four unknowns and that the matrix form of which is as follows:

$$\boldsymbol{y}_K = \boldsymbol{H}_K^T \boldsymbol{x}_K \tag{7.6}$$

where

$$\boldsymbol{y}_K = \begin{bmatrix} V_{R_K} \\ V_{I_K} \end{bmatrix}, \ \boldsymbol{x}_K = \begin{bmatrix} E_{R_K} \\ E_{I_K} \\ R_{\mathrm{Th}_K} \\ X_{\mathrm{Th}_K} \end{bmatrix}, \ \boldsymbol{H}_K^T = \begin{bmatrix} 1 & 0 & -I_{R_K} & I_{I_K} \\ 0 & 1 & -I_{I_K} & -I_{R_K} \end{bmatrix} \tag{7.7}$$

Using measurements before t_K, the dimensions of the vector $\boldsymbol{y}_K$ and matrix $\boldsymbol{H}_K^T$ increase. If more than two measurements are considered, the number of equations becomes more than the number of unknowns. For example, with m measurements,

$$\boldsymbol{y} = \boldsymbol{H}\boldsymbol{x}_K \tag{7.8}$$

where

$$\begin{aligned} \boldsymbol{y} &= \begin{bmatrix} y_K^T & y_{K-1}^T & \cdots & y_{K-m+1}^T \end{bmatrix}^T \\ \boldsymbol{H} &= \begin{bmatrix} H_K & H_{K-1} & \cdots & H_{K-m+1} \end{bmatrix}^T \end{aligned} \tag{7.9}$$

When the system changes during the measurements or the measurements have error, there will be no answer to these overdetermined equations. The purpose of the RLS algorithm is to determine the parameters $\boldsymbol{x_K}$ so that the two sides of the Eq. (7.8) become as possible as close. This purpose is achieved by having previous measurements and by using forgetting factor λ to minimize the following objective function.

$$J_1(x_K) = \sum_{i=1}^{K} \lambda^{K-i} \left\| y_i - H_i^T x_K \right\|_2 \tag{7.10}$$

The forgetting factor $\lambda < 1$ causes more weight to be assigned to the recent measurements. In the objective function of (7.20), all previous measurements are considered. If $\hat{x}_K$ and $\hat{x}_{K-1}$ are the vectors of the estimated parameters in moments t_K and t_{K-1}, respectively, the RLS algorithm is as Eqs. (7.11)–(7.14) [5]:

$$\hat{x}_K = \hat{x}_{K-1} + L_K e_K \tag{7.11}$$

$$e_K = y_K - H_K^T \hat{x}_{K-1} \tag{7.12}$$

$$L_K = P_{K-1} H_K \left[\lambda I + H_K^T \quad P_{K-1} H_K \right]^{-1} \tag{7.13}$$

$$P_K = \frac{\left[I - L_K H_K^T \right] P_{K-1}}{\lambda}, \quad \boldsymbol{P}_0 = P_0 I \tag{7.14}$$

where $P_K \in R^{n \times n}$, n is the dimension of the parameters' vector, I is an identity matrix, and P_0 is a large positive number.

In order to increase the speed of tracing the variable parameters of the power system, the estimation of the parameters should be done only based on recent measurements. To do this, the objective function (7.10) is changed as follows:

$$J_2(x_K) = \sum_{i=K-m+1}^{K} \lambda^{K-i} \left\| y_i - H_i^T x_K \right\|_2 \tag{7.15}$$

where $\boldsymbol{m}$ is the length of the measurement window, i.e. the number of considered measurements. In this case, the RLS algorithm is as Eqs. (7.16)–(7.22) [6]:

$$\hat{x}_K = \hat{V}_{K-1} + L_K e'_K \tag{7.16}$$

$$e'_K = y_K - H_K^T \hat{V}_{K-1} \tag{7.17}$$

$$L_K = P_K H_K \tag{7.18}$$

$$P_K = \frac{1}{\lambda} \left[P'_{K-1} - \left(P'_{K-1} H_K H_K^T P'_{K-1} \right) \left(\lambda I + H_K^T P'_{K-1} H_K \right)^{-1} \right] \tag{7.19}$$

$$\hat{V}_{K-1} = \hat{x}_{K-1} - \lambda^{m-1} P'_{K-1} H_{K-m} e''_K \tag{7.20}$$

$$e''_K = y_{K-m} - H^T_{K-m} \hat{x}_{K-1} \tag{7.21}$$

$$P'_{K-1} = P_{K-1} + \left[P_{K-1} H_{K-m} H^T_{K-m} P_{K-1}\right] \left[\lambda^{-m+1} I - H^T_{K-m} P_{K-1} H_{K-m}\right]^{-1}$$
$$m \geq 2, \quad \boldsymbol{P}_0 = P_0 I \tag{7.22}$$

The value of $\hat{x}_0$ is selected arbitrary. Although reducing the measurements used increases the speed of estimating the parameters, however, it increases the effect of measurement errors. In [7], an robust RLS estimator is proposed to reduce measurement errors. Also in [8] a robust classical least square method is proposed in which the parameters are determined by solving the following optimization problem.

$$\min_{x_k} \max_{\substack{\|\Delta H\|_2 \leq \rho H \\ \|\Delta Y\|_2 \leq \rho Y}} \|(\boldsymbol{H} + \Delta \boldsymbol{H}) x_k - (\boldsymbol{Y} + \Delta \boldsymbol{Y})\|_2 \tag{7.23}$$

where $\boldsymbol{\Delta Y}$ and $\boldsymbol{\Delta H}$ represent the effect of measurement errors on the vector $\boldsymbol{Y}$ and the matrix $\boldsymbol{H}$, respectively. $\boldsymbol{\rho Y}$ and $\boldsymbol{\rho H}$ are also used to specify the range of these errors.

7.2.3 Calculation of X_{Th} Assuming E_{Th} as a Free Variable

Assuming $R_{\mathrm{Th}} \ll X_{\mathrm{Th}}$ and $R_{\mathrm{Th}} \approx 0$, the relationship between the TE parameters and the measured phasors becomes as Eq. (7.24) [1].

$$\overline{E}_{\mathrm{Th}} = \overline{V} + jX_{\mathrm{Th}} \overline{I} \tag{7.24}$$

Assuming $\overline{I} = I\angle 0$, $\overline{V} = V\angle\theta$, and $\overline{E}_{\mathrm{Th}} = E_{\mathrm{Th}}\angle\beta$, the phasor diagram of the equivalent Figure 7.1 is as Figure 7.2 [1]:

Separation of Eq. (7.24) leads to two parts, real and imaginary:

$$E_{\mathrm{Th}} \cos\beta = V\cos\theta \tag{7.25}$$

$$E_{\mathrm{Th}} \sin\beta = X_{\mathrm{Th}} I + V\sin\theta \tag{7.26}$$

If $\hat{\overline{E}}_{\mathrm{Th}}$ and $j\hat{X}_{\mathrm{Th}}$ are the phasors of estimated parameters:

$$\hat{\overline{E}}_{\mathrm{Th}} = \overline{V} + j\hat{X}_{\mathrm{Th}} \overline{I} \tag{7.27}$$

By subtracting the two Eqs. (7.24) and (7.27) from each other,

$$\Delta\hat{\overline{E}}_{\mathrm{Th}} = j\Delta\hat{X}_{\mathrm{Th}} \overline{I} \tag{7.28}$$

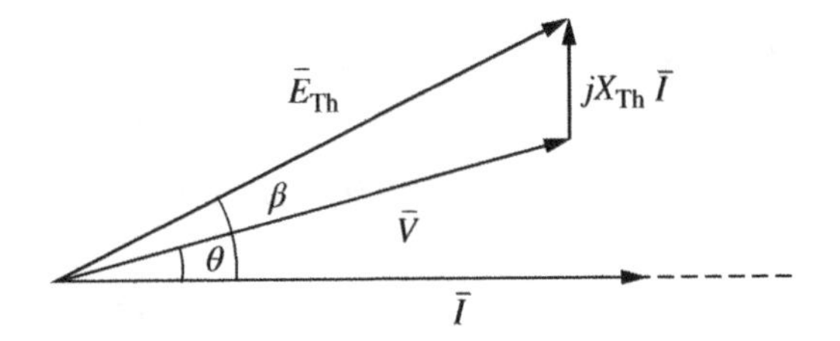

Figure 7.2 Phasor diagram of two-bus equivalent circuit [1].

Assuming that the angle $\bar{I}$ is zero, $\bar{I} \to I$

$$\Delta\bar{\hat{E}}_{\text{Th}} = j\Delta\hat{X}_{\text{Th}}I = j\Delta\hat{E}_{\text{Th}} \tag{7.29}$$

It can be seen from Eq. (7.29) that the error of the estimated parameters $\left(\Delta\hat{E}_{\text{Th}}, \Delta\hat{X}_{\text{Th}}\right)$ is same sign. That means both are larger or smaller than the actual values. The aim is to converge the estimated parameters into the actual parameters as quickly as possible during consecutive measurements. By keeping $\hat{E}_{\text{Th}}$ constant at the two sampling times t_{k-1} and t_k, values $\hat{X}_{\text{Th}_{K-1}}$ and $\hat{X}_{\text{Th}_K}$ are obtained from Eqs. (7.25) and (7.26). Assuming that in the time interval between these two samplings, the system equivalent parameters remain unchanged,

$$j\left(\Delta\hat{X}_{\text{Th}_K} - \Delta\hat{X}_{\text{Th}_{K-1}}\right) = j\Delta\hat{E}_{\text{Th}}\left(\frac{1}{I_K} - \frac{1}{I_{K-1}}\right) \tag{7.30}$$

Equation (7.30) indicates that how $\hat{E}_{\text{Th}}$ should be changed to converge $\hat{X}_{\text{Th}}$ to its actual value faster. Of course, convergence of $\hat{X}_{\text{Th}}$ to its actual value occurs when $\hat{E}_{\text{Th}}$ also converges to its actual value. For example, if the electrical current decreases during two consecutive measurements $\left[\left(\frac{1}{I_K} - \frac{1}{I_{K-1}}\right) > 0\right]$ but the estimated reactances increase $\left[\left(\Delta\hat{X}_{\text{Th}_K} - \Delta\hat{X}_{\text{Th}_{K-1}}\right) > 0\right]$, $\Delta\hat{E}_{\text{Th}}$ becomes greater than zero, indicating that $\hat{E}_{\text{Th}}$ is greater than the actual value. In this condition, to converge the parameters to the actual value, $\hat{E}_{\text{Th}}$ must be reduced. The initial value of E_{Th0} is chosen as Eq. (7.31):

$$\hat{E}_{\text{Th0}} = \left(\frac{E_{\text{Th}}^{\max} - E_{\text{Th}}^{\min}}{2}\right) \tag{7.31}$$

At each operating point, assuming the load is inductive, $E_{\text{Th}}^{\min} = V$. Also, assuming that the system operation is in the conditions of maximum loadability, the value of $E_{\text{Th}}^{\max}$ is obtained from Eqs. (7.25) and (7.26). In this condition, $Z_{\text{L}} = X_{\text{Th}}$ (Assuming $R_{\text{Th}} = 0$), and

$$E_{\text{Th}}^{\max} = \frac{V\cos\theta}{\cos\hat{\beta}} \tag{7.32}$$

$$\tan\hat{\beta} = \frac{Z_L I + V\sin\theta}{V\cos\theta} \tag{7.33}$$

According to what mentioned above, the algorithm to determine X_{Th} is as follows:

Step 1) calculate $\hat{E}_{Th_0}$ and $\hat{\beta}_0$, respectively, from Eqs. (7.31) and (7.25).
Step 2) determine $\hat{X}_{\text{Th}_0}$ from Eq. (7.26). Set $K = 1$
Step 3) Determine $\hat{E}_{\text{Th}_K}$ according to the following conditions.

If $I_K > I_{K-1}$

If $\left(\hat{X}^*_{\text{Th}_K} - \hat{X}^*_{\text{Th}_{K-1}}\right) < 0$, then $\hat{E}_{\text{Th}_K} = \hat{E}_{\text{Th}_{K-1}} - \epsilon_E$

If $\left(\hat{X}^*_{\text{Th}_K} - \hat{X}^*_{\text{Th}_{K-1}}\right) > 0$, then $\hat{E}_{\text{Th}_K} = \hat{E}_{\text{Th}_{K-1}} + \epsilon_E$

($\hat{X}^*_{\text{Th}_K}$ is the same as $\hat{X}_{\text{Th}_K}$, but in calculating it, β_{K-1} and $E_{\text{Th}_{K-1}}$ are used instead of β_K and E_{Th_K})

If $I_K < I_{K-1}$

If $\left(\hat{X}^*_{\text{Th}_K} - \hat{X}^*_{\text{Th}_{K-1}}\right) < 0$ then $\hat{E}_{\text{Th}_K} = \hat{E}_{\text{Th}_{K-1}} + \epsilon_E$

If $\left(\hat{X}^*_{\text{Th}_K} - \hat{X}^*_{\text{Th}_{K-1}}\right) > 0$ then $\hat{E}_{\text{Th}_K} = \hat{E}_{\text{Th}_{K-1}} - \epsilon_E$

If $I_K = I_{K-1}$

$\hat{E}_{\text{Th}_K} = \hat{E}_{\text{Th}_{K-1}}$

Step 4) Obtain $\hat{\beta}_K$ and $\hat{X}_{\text{Th}_K}$ from Eq. (7.25) and (7.26), respectively.
Step 5) Increase K and return to step 3.

The criterion of identification X_{Th} is that its estimated value in consecutive sampling steps varies around a value in a small range. The value ϵ_E is obtained from Eq. (7.34).

$$\epsilon_E = \min\left(\epsilon_{\inf}, \epsilon_{\sup}, \epsilon_{\lim}\right) \tag{7.34}$$

where

$$\begin{aligned} \epsilon_{\inf} &= \left|E_{\text{Th}_{K-1}} - V_K\right| \\ \epsilon_{\sup} &= \left|E_{\text{Th}_{K-1}} - E^{\max}_{\text{Th}_K}\right| \\ \epsilon_{\lim} &= \left|E_{\text{Th}_{K-1}} \times h\right| \end{aligned} \tag{7.35}$$

where h is a parameter used for limiting identification error.

7.2.4 Reduction of Parameter Estimation Errors

Errors in estimating TE parameters occur for different reasons. One of these reasons is system change during a sliding window of data measurements. In each bus, the difference between the measured values in different measurements can occur due to the changes in both sides of the load and the system. The smaller the system changes during the measurements compared to the load changes, the less errors in estimating the system parameters [2, 3]. In [3], it is shown that if during two measurements, the load remains constant and the system changes, and the estimated parameters represent the TE parameters of the load side and not the system. System changes can include small changes such as load and generation changes that occur constantly and usually do not cause much change in TE parameters during measurements made in a sliding window. But larger changes, such as disconnecting or connecting different equipment such as generators and lines, can cause large changes in parameters. In these situations, parameter estimation error occurs when some of the measurements in the sliding window are related to before the system change and some to the after that. The best way to reduce error is to eliminate the measurements taken before the system change.

Another cause of parameter estimation error is PMU errors. This error can be both in the magnitude and angle of the phasors in one measurement and also includes the error in the difference in the angle of the phasors in two consecutive measurements, which occurs due to change in the system frequency. This change, although small, causes it to slip between the system frequency and the PMU sampling frequency. This slip causes the consecutive measurements to be asynchronous and the measured phase angle to be drifted [2].

In [2], a way to synchronize measurements is provided. In this method, it is assumed that the nonsynchronization of three consecutive measurements has caused drifts ϵ_1 and ϵ_2 in the angle of the measured phasors $(\overline{V}_1, \overline{I}_1)$ and $(\overline{V}_2, \overline{I}_2)$ relative to the phasors $(\overline{V}_0, \overline{I}_0)$. Therefore, if the measurements are synchronized, the second and third measured phasors change to $(\overline{V}_1 e^{-j\epsilon_1}, \overline{I}_1 e^{-j\epsilon_1})$ and $(\overline{V}_2 e^{-j\epsilon_2}, \overline{I}_2 e^{-j\epsilon_2})$. It is shown that if the system does not change during three measurements, there is Eq. (7.36) among the phasors.

$$\det \begin{bmatrix} 1 & 1 & 1 \\ \overline{V}_0 & \overline{V}_1 e^{-j\epsilon_1} & \overline{V}_2 e^{-j\epsilon_2} \\ \overline{I}_0 & \overline{I}_1 e^{-j\epsilon_1} & \overline{I}_2 e^{-j\epsilon_2} \end{bmatrix} = 0 \tag{7.36}$$

From these equations, ϵ_1 and ϵ_2 are obtained as Eqs. (7.37) and (7.38).

$$\epsilon_1 = \pi + \delta_A - \gamma - \delta_B \tag{7.37}$$

$$\epsilon_2 = \pi + \delta_A + \alpha - \delta_C \tag{7.38}$$

where δ_A, δ_B, and δ_C are the angles of the complex numbers A, B, and C, respectively.

where,

$$A = \det\begin{bmatrix} \overline{V}_1 & \overline{V}_2 \\ \overline{I}_1 & \overline{I}_2 \end{bmatrix}, \quad B = \det\begin{bmatrix} \overline{V}_2 & \overline{V}_0 \\ \overline{I}_2 & \overline{I}_0 \end{bmatrix}, \quad C = \det\begin{bmatrix} \overline{V}_0 & \overline{V}_1 \\ \overline{I}_0 & \overline{I}_1 \end{bmatrix} \tag{7.39}$$

Also,

$$\alpha = \cos^{-1}\left(\frac{|A|^2 + |C|^2 - |B|^2}{2|A||C|}\right) \tag{7.40}$$

$$\gamma = \cos^{-1}\left(\frac{|A|^2 + |B|^2 - |C|^2}{2|A||C|}\right) \tag{7.41}$$

Equations (7.37) and (7.38) are obtained with the assumption that the error in the measured phasors is only due to the nonsynchronization of the measurements. Of course, if there are the same errors in the magnitude and angle of the phasors in the three measurements, Eqs. (7.37) and (7.38) also are established [2].

Ref. [2] also provides a method for diagnosis faulty data as well as detecting big changes in the system. This method is based on the phasor diagram of Figure 7.3, which shows the relationship between the data measured at a measuring point and the TE parameters. In this diagram, the current is considered as the reference.

Using Figure 7.3, the relationship between the measured data and the TE parameters is in accordance with Eq. (7.42).

$$(V_x + IR_{\text{Th}})^2 + (V_y + IX_{\text{Th}})^2 = E_{\text{Th}}^2 \tag{7.42}$$

Dividing both sides by I^2,

$$\left(R_{\text{Th}} + \frac{V_x}{I}\right)^2 + \left(X_{\text{Th}} + \frac{V_y}{I}\right)^2 = \left(\frac{E_{\text{Th}}}{I}\right)^2 \tag{7.43}$$

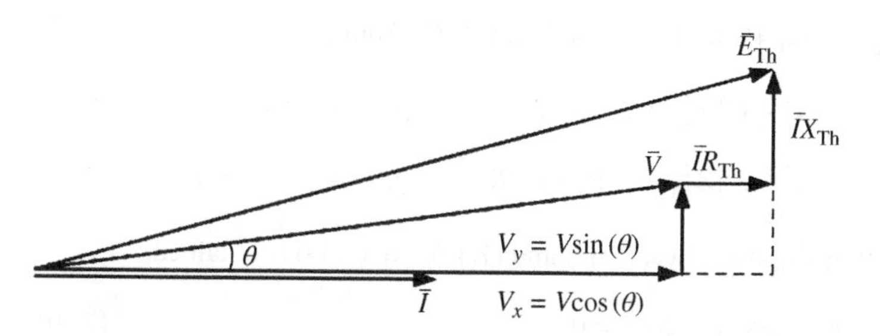

Figure 7.3 Relationship between the measured data and the TE parameters.

Equation (7.43) represents a circle on the complex plane $\overline{Z}_{Th}$ whose center is $\left(\frac{-V_x}{I}, \frac{-V_y}{I}\right)$ and radius is $\left(\frac{E_{Th}}{I}\right)$. For each measurement, a circle is obtained. If, during the measurements, the system has constant TE parameters, these circles intersect at two points, representing the R_{Th} and X_{Th} of the system. To achieve this, the centers of all circles must be along a line. This result can be used to diagnose faulty data as well as detect big changes in the system. For this purpose, a straight line must be fitted to the measured data. Data that is too far from this line is identified as faulty data or changed system data. It should be noted that the point obtained from each measurement $\left(\frac{-V_x}{I}, \frac{-V_y}{I}\right)$ depends only on the magnitude of the voltage and current as well as the angle between the measured voltages and currents. Hence, the obtained points are not affected by nonsynchronization of consecutive measurements. In [4], it is shown that assuming the power factor of load is constant, the geometric location of the voltage phasors in the complex voltage plane also becomes a circle whose radius and center are related to the TE parameters. Of course, since the points of the circle are composed of measured voltage phasors, synchronicity of the measurements is effective.

Ref. [3] presents a method in which TE parameters are obtained by measuring the magnitude of voltage and current as well as active and reactive powers. Therefore, there is no need for consecutive measurements to be synchronous. Using Eq. (7.42) and the phasor diagram in Figure 7.3,

$$E_{Th}^2 = V^2 + I^2 Z_{Th}^2 + 2PR_{Th} + 2QX_{Th} \tag{7.44}$$

With three different measurements and assuming that the system parameters are constant,

$$E_{Th}^2 = V_1^2 + I_1^2 Z_{Th}^2 + 2P_1 R_{Th} + 2Q_1 X_{Th} \tag{7.45}$$

$$E_{Th}^2 = V_2^2 + I_2^2 Z_{Th}^2 + 2P_2 R_{Th} + 2Q_2 X_{Th} \tag{7.46}$$

$$E_{Th}^2 = V_3^2 + I_3^2 Z_{Th}^2 + 2P_3 R_{Th} + 2Q_3 X_{Th} \tag{7.47}$$

Subtracting the Eq. (7.46) from (7.45) and (7.47) from (7.46),

$$V_1^2 - V_2^2 + (I_1^2 - I_2^2) Z_{Th}^2 + 2(P_1 - P_2) R_{Th} + 2(Q_1 - Q_2) X_{Th} = 0 \tag{7.48}$$

$$V_2^2 - V_3^2 + (I_2^2 - I_3^2) Z_{Th}^2 + 2(P_2 - P_3) R_{Th} + 2(Q_2 - Q_3) X_{Th} = 0 \tag{7.49}$$

By eliminating Z_{Th} from Eqs. (7.48) and (7.49), Eq. (7.50) is obtained.

$$2\Delta P R_{Th} + 2\Delta Q X_{Th} + \Delta V^2 = 0 \tag{7.50}$$

where

$$\Delta P = \det\begin{bmatrix} 1 & 1 & 1 \\ P_1 & P_2 & P_3 \\ I_1^2 & I_2^2 & I_3^2 \end{bmatrix},$$

$$\Delta Q = \det\begin{bmatrix} 1 & 1 & 1 \\ Q_1 & Q_2 & Q_3 \\ I_1^2 & I_2^2 & I_2^2 \end{bmatrix} \text{ and} \tag{7.51}$$

$$\Delta V^2 = \det\begin{bmatrix} 1 & 1 & 1 \\ V_1^2 & V_2^2 & V_3^2 \\ I_1^2 & I_2^2 & I_3^2 \end{bmatrix},$$

Equation (7.50) shows that with a set of three different measurements, a line can be created that expresses the relation between X_{Th} and R_{Th}. Using this line and one of the Eqs. (7.48) or (7.49), X_{Th} and R_{Th} can be obtained. Of course, there may be no solution when the system change is significant compared to the load change. Also, when two measurements are equal to or very close to each other, the coefficients in Eq. (7.50) become zero, in which case there is no solution. Therefore, it is better to consider more than three measurements. With more measurements, the effect of measurement error can be reduced. Of course, a large number of measurements, although creating the required redundancy, increase the likelihood that the system will change during the measurements. This causes the measurements to belong to different systems, which may lead to many errors in estimating the system parameters. Having more measurements, more lines similar to Eq. (7.50) will be created. If N is the number of measurements, the number of C_3^N lines will be created. As the number of lines increases, the number of intersection points will increase. If the system remains constant during the measurements (its parameters remain constant) and there is no measurement error, all the lines intersect each other at a single point, which is $\overline{Z}_{Th}$.

In view of the above, Ref. [3] has proposed five measurements. With this number of measurements, there will be 10 lines and a maximum of 45 intersection points, resulting in 45 $\hat{\overline{Z}}_{Th}$. The estimated impedance with the highest frequency is selected as the Thevenin impedance of the system. It is also possible to obtain the probability of R_{Th} and X_{Th} being in a certain range. Also, by checking the intersection points, high-error measurements can be identified. These measurements lead to lines that notably change the intersection points. Of course, it should be noted that each of these measurements affects a large number of intersection points. Therefore, identifying them requires identifying all unusual intersection points and their intersection lines.

7.2.5 Simulations

In order to evaluate the capability of the TE-based method in diagnosis of proximity to the maximum loadability limit, it is first assumed that there is no error in the measurement of phasors, and with two measurements the Thevenin parameters are obtained according to Eqs. (7.4) and (7.5). In the first part of the simulation, the loading factor λ is increased in all PQ and PV buses until reaching the loadability limit. In order to investigate the effect of system variation during the two measurements, two states are considered. In the first state, it is assumed that in the time interval between the two measurements, λ increases by 1%, and in the second state, it is assumed that this increase is 3%.

Figures 7.4 and 7.5 show the changes of $|\overline{Z}_{Th}|/|\overline{Z}_L|$ for a number of weak buses in the IEEE 57-bus test system. In addition to the nonlinear changes of the index, the important issue is that in none of the states, the index value reaches one at the maximum loadability limit. The reason is that with load changing during the two

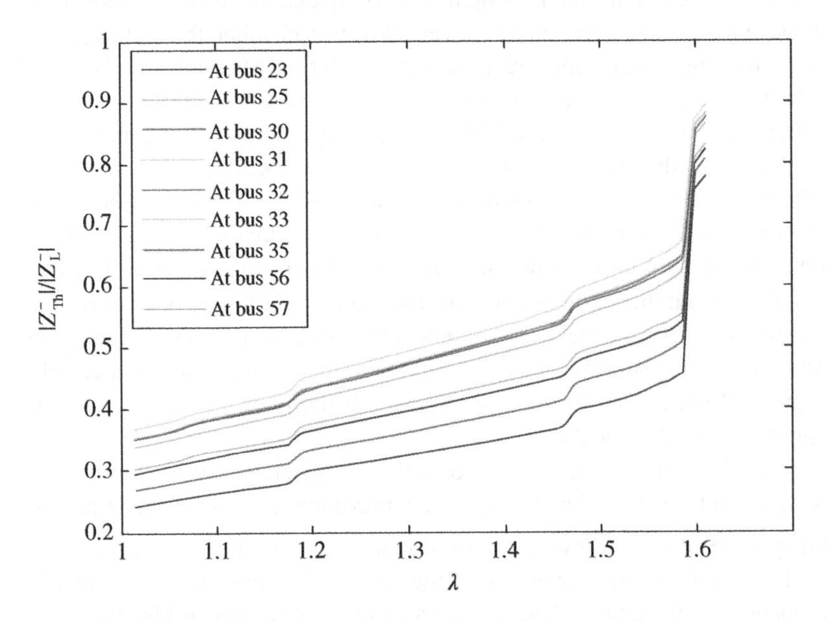

Figure 7.4 The $|\overline{Z}_{Th}|/|\overline{Z}_L|$ index variations when λ increases by 1%.

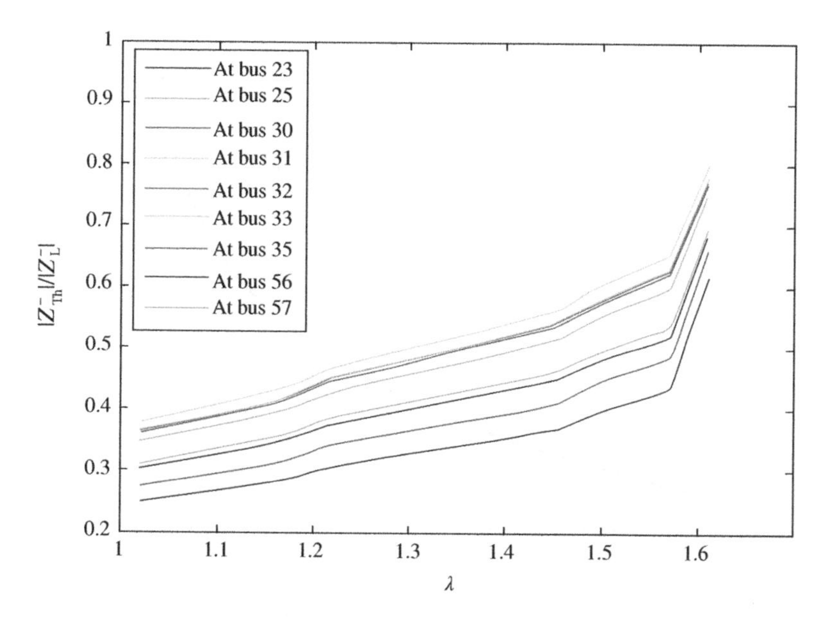

Figure 7.5 The $|\bar{Z}_{Th}|/|\bar{Z}_L|$ index variations when λ increases by 3%.

measurements, in fact each measurement is related to a different system. This is more obvious when the system load change becomes greater between the two measurements. The observed jumps in the curves are due to the generators encountering their reactive power generation limits.

In another part, it is assumed that the load increases only at the bus in which the index is calculated. In other words, the λ value does not change at other load buses as well as at PV buses. λ increases at the desired bus by 1% steps. In this state, the index variation is as Figure 7.6. It is observed that the value of the index gets closer to one at the loadability limit because the system changes less during the two measurements. In this state, the slight changes of the system are due to the voltage variations when the load increases and consequently the impedance variations due to their constant power characteristic. Similar results are obtained for other load buses.

Figures 7.7 and 7.8 show the effect of measurement error. It is assumed that there is a random error at each measurement. The maximum value of this error in Figure 7.7 is 2% and in Figure 7.8 is 5%. In both figures, the load factor is increased at all buses in steps of one percent. It is observed that measurement error increases the unpredictable changes in the index.

Now to test the performance of the RLS algorithm, the index variations are obtained using this algorithm with the assumption that the measurement error is 5%. Figures 7.9 and 7.10 show the results for two forgetting factor of 0.4 and

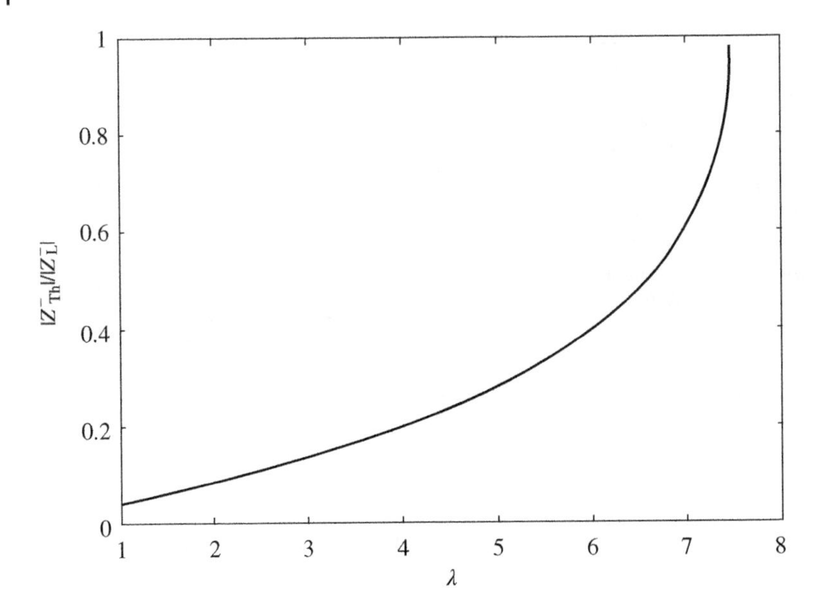

Figure 7.6 The $|\bar{Z}_{Th}|/|\bar{Z}_L|$ index variations at bus 30 with load increment at bus 30.

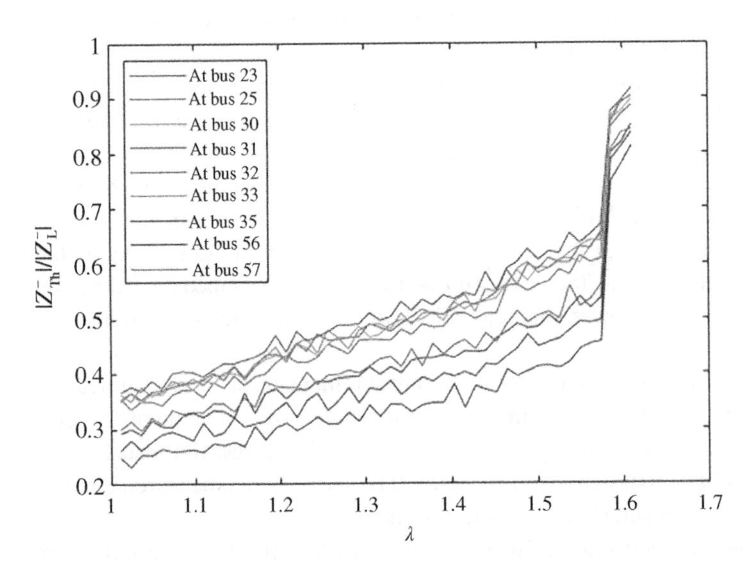

Figure 7.7 Impact of measurement error on the $|\bar{Z}_{Th}|/|\bar{Z}_L|$ index variations when λ increases by 1%, error is 2%.

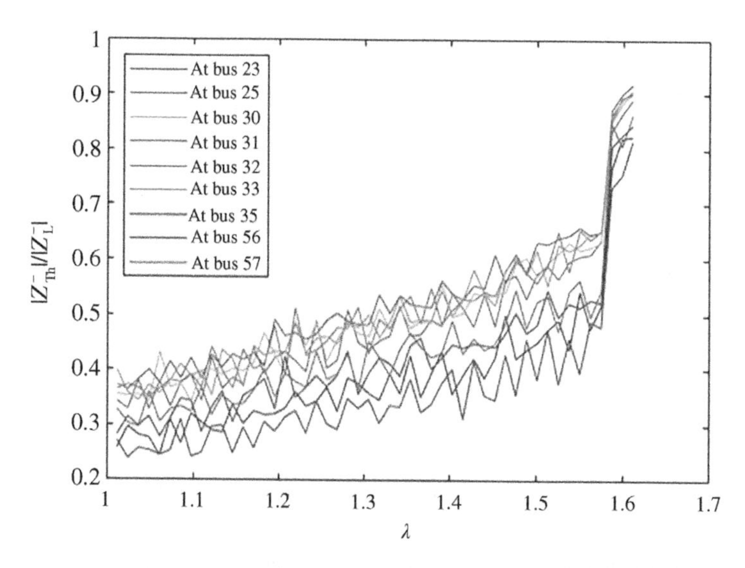

Figure 7.8 Impact of measurement error on the $|\bar{Z}_{Th}|/|\bar{Z}_L|$ index variations when λ increases by 1%, error is 5%.

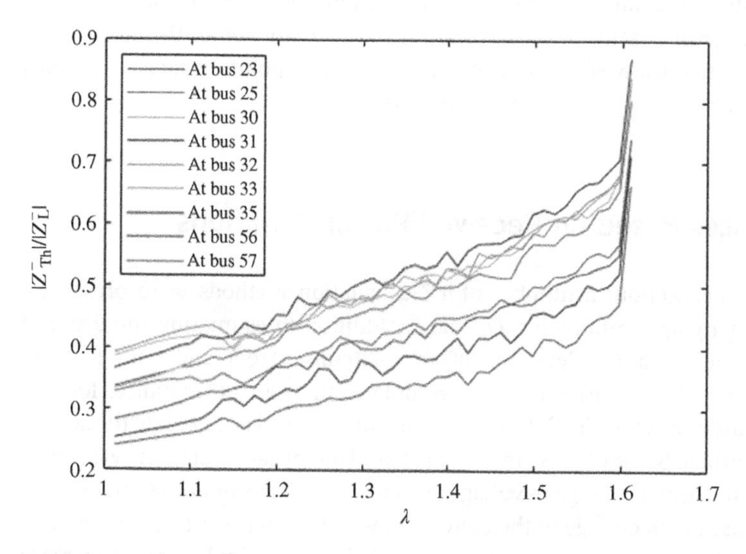

Figure 7.9 Use of RLS algorithm with forgetting factor 0.4 when λ increases by 1%, error is 5%.

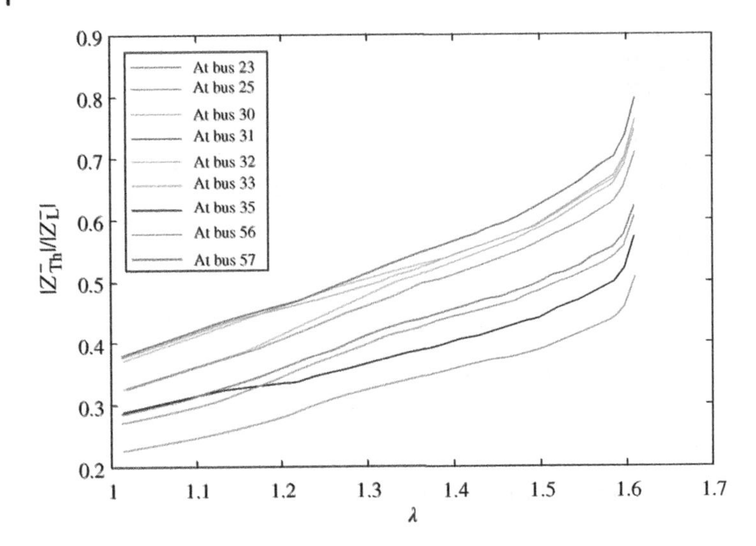

Figure 7.10 Use of RLS algorithm with forgetting factor 0.8 when λ increases by 1%, error is 5%.

0.8. It is observed that with the forgetting factor of 0.8, since more measured points are considered in estimating the Thevenin impedance, the error effect is less; but due to the consideration of previous measured points that actually belong to another system, the value of the index is far from one at the loadability limit. But using forgetting factor of 0.4 has less impact on reduction of the error effect. However, due to the low effect of the previous measured points, it causes the index value to be closer to one at the loadability limit.

7.3 Indices Based on Received Power Variations

In the previous section, a number of TE calculation methods were presented. There are many applications for TE that in addition to estimating the distance to the loadability limit, calculation of short-circuit current and fault location are from them [3]. In addition to TE-based indices, there are other indices for estimating the distance to the loadability limit that do not need to calculate the system equivalent circuit. Some of these indices are based on the fact that in the vicinity of the loadability limit, a change in voltage and current of a bus or at the end of a line does not cause much change in the delivery power. In other words, an increase in current that is expected to increase the power will be ineffective by a more decrease in voltage. This can be detected by two consecutive measurements of voltage and

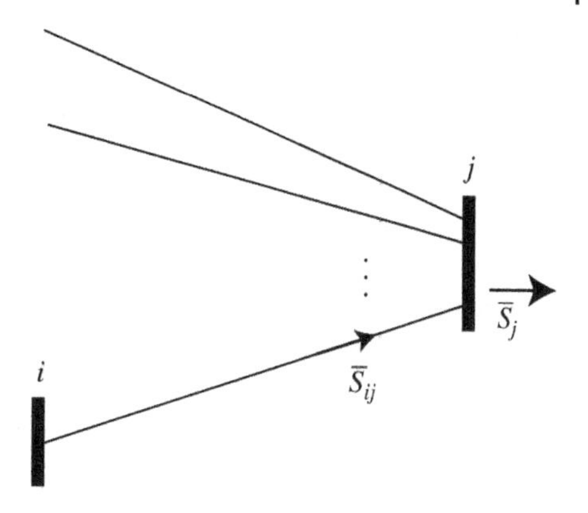

Figure 7.11 Complex power at bus j and in line $i - j$.

current phasors. This method was first used to detect the proximity to the loadability limit of a line [9, 10]. According to Figure 7.11, the flowing complex power in line $i - j$ and in the vicinity of bus j can be expressed as Eq. (7.52).

$$\overline{S}_{ij} = \overline{V}_j \overline{I}_{ij}^* \tag{7.52}$$

By increasing the power flowing in line $i - \text{j}$ during the two measurements at t_K and t_{K+1} moments, the following equation can be written:

$$\begin{aligned}\overline{S}_{ij}^{(K+1)} &= \overline{S}_{ij}^{(K)} + \Delta\overline{S}_{ij}^{(K+1)} = \left(\overline{V}_j^{(K)} + \Delta\overline{V}_j^{(K+1)}\right)\left(\overline{I}_{ij}^{(K)} + \Delta\overline{I}_{ij}^{(K+1)}\right)^* \\ &= \overline{S}_{ij}^{(K)} + \Delta\overline{V}_j^{(K+1)}\overline{I}_{ij}^{(K)*} + \overline{V}_j^{(K)}\Delta\overline{I}_{ij}^{(K+1)*} + \underbrace{\Delta\overline{V}_j^{(K+1)}\Delta\overline{I}_{ij}^{(K+1)*}}_{\approx 0}\end{aligned} \tag{7.53}$$

With high sampling frequency, the term $\Delta\overline{V}_j\Delta\overline{I}_{ij}^*$ has a very small value and can be ignored. In the vicinity of the loadability limit, increase in power in the sending end of the line does not lead to increase in the receiving end of the line. Therefore,

$$\Delta\overline{S}_{ji}^{(K+1)} = \Delta\overline{V}_j^{(K+1)}\overline{I}_{ij}^{(K)*} + \overline{V}_j^{(K)}\Delta\overline{I}_{ij}^{(K+1)*} = 0 \tag{7.54}$$

It should be noted that the diagnosis of approaching the loadability limit is only possible by studying the complex power; due to the presence of reactive power compensators, an increase in flowing active power may be associated with a decrease in flowing reactive power, so that the apparent power amount remains constant without approaching the loadability limit.

Eq. (7.54) can be rewritten as follows:

$$1 + \frac{\Delta\overline{V}_j^{(K+1)}\overline{I}_{ij}^{(K)*}}{\overline{V}_j^{(K)}\Delta\overline{I}_{ij}^{(K+1)*}} = 1 + ae^{j\varphi} = 0 \tag{7.55}$$

Eq. (7.55) is valid in the following two states:

$$\Delta \overline{V}_j^{(K+1)} = 0 \quad \text{and} \quad \Delta \overline{I}_{ij}^{(K+1)} = 0 \tag{7.56}$$

or,

$$a = 1 \quad \text{and} \quad \varphi = \pm \pi \tag{7.57}$$

State (7.56) appears when there is no significant change in current and voltage between the two measurements. Of course, this state is not a sign of proximity to voltage instability and should not be taken into account when calculating the index. Using Eq. (7.55), the S different criterion (SDC) is defined as Eq. (7.58).

$$\text{SDC} = \left|1 + ae^{j\varphi}\right| \tag{7.58}$$

In light loading of a line, the SDC value is more than zero, and at the loadability limit, its value becomes zero. The SDC calculation will be done when, first, in distance between the two measurements $|\Delta I| > \Delta I_{\min}$, and second the line consumes reactive power. When the delivery power is less than the line natural loading (surge impedance loading), the line generates reactive power. In these conditions, there is a large distance to the line loadablity limit. The value $\Delta I_{\min}$ depends on the sampling frequency. This value is chosen 0.01 pu in [10, 11] and 0.015 pu in [12].

The SDC index is used to detect proximity to loadability limits of lines. Reaching loadability limit of a line does not necessarily mean reaching loadability limit of its end bus because there may be other lines for delivering power to the bus. To detect proximity to the loadability limit of a bus, voltage, and current must be measured at that bus (instead of its connected lines). No change in received power by a bus despite change in its voltage and current is an alarm of reaching that bus loadability limit. In this situation, similar to Eq. (7.54) [12],

$$\Delta \overline{S}_j^{(K+1)} = \Delta \overline{V}_j^{(K+1)} \overline{I}_j^{(K)*} + \overline{V}_j^{(K)} \Delta \overline{I}_j^{(K+1)*} = 0 \tag{7.59}$$

Leads to

$$\left|\frac{\overline{V}_j^{(K)}}{\overline{I}_j^{(K)*}}\right| = \left|\frac{\Delta \overline{V}_j^{(K+1)}}{\Delta \overline{I}_j^{(K+1)*}}\right| \tag{7.60}$$

Now using Eq. (7.60), the bus apparent-power criterion (BSDC) index is defined as follows [12]:

$$\text{BSDC} = \frac{\left|\overline{V}_j^{(K)} / \overline{I}_j^{(K)*}\right| - \left|\Delta \overline{V}_j^{(K+1)} / \Delta \overline{I}_j^{(K+1)*}\right|}{\left|\overline{V}_j^{(K)} / \overline{I}_j^{(K)*}\right|} = 1 - \frac{\left|\Delta \overline{V}_j^{(K+1)} / \Delta \overline{I}_j^{(K+1)*}\right|}{\left|\overline{V}_j^{(K)} / \overline{I}_j^{(K)*}\right|} \tag{7.61}$$

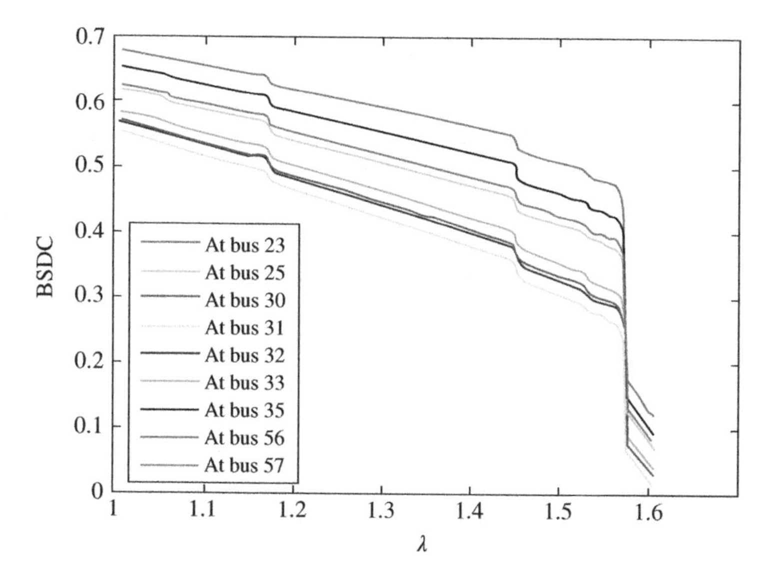

Figure 7.12 The BSDC index variations.

The performance of BSDC and SDC indices is evaluated by calculating them in IEEE 57-bus test system. For this purpose, the loading factor is increased at all buses. This increase continues by steps 1% until encountering the system loadability limit. With the assumption that the obtained phasors at each operating point are the same as the measured values, the indices are calculated based on the last two operating points. Figure 7.12 shows the BSDC index variations at some buses of the mentioned system. Also, Figure 7.13 shows the SDC index variations in a number of lines connected to these buses.

It can be seen that the value of the BSDC index at weak buses becomes very close to the expected value at the loadability limit. It can also be seen from Figure 7.13 that reaching a line to its loadability limit, which is determined when the SDC value becomes zero, may not mean that the line end bus reaches the loadability limit.

7.4 Early Detection of Voltage Instability

As mentioned, one of the applications of the voltage stability index is to estimate the distance between the stable operating point and the loadability limit. In this situation, if the system moves slowly to the loadability limit with a gradual increase in load and without occurrence of a large contingency, the indices are

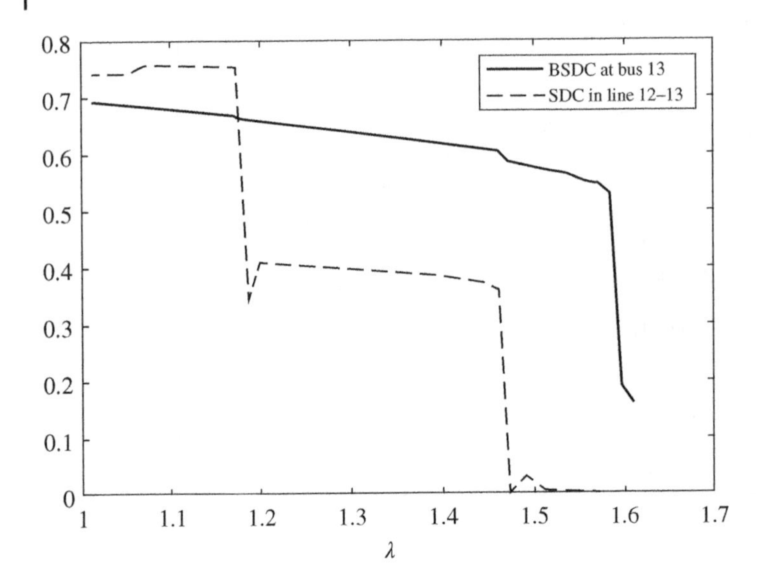

Figure 7.13 The BSDC index variations at bus 13 and the SDC index variations in line 12–13.

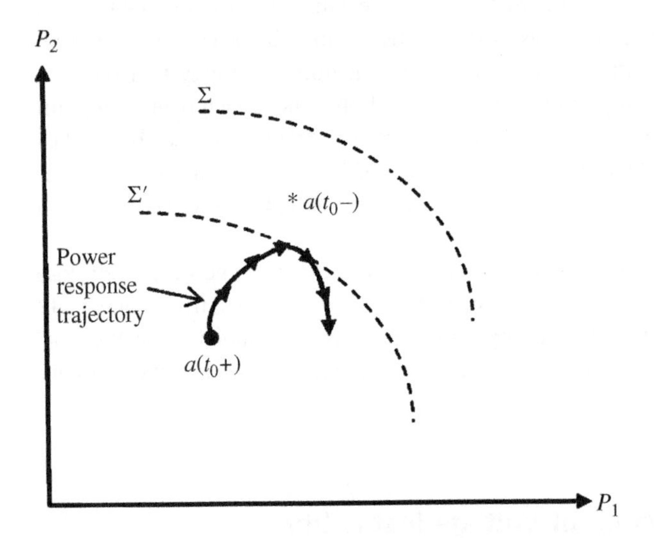

Figure 7.14 The system response to the maximum loadability limit reduction.

expected to send the necessary warning alarm to the system operator, in time. On the other hand, occurrence of a large contingency and a sudden decrease in the loadability limit can cause the demanded power to exceed this limit and the occurrence of voltage instability. Figure 7.14 shows a simple representation of this event in a two-dimensional space. In this figure, Σ and Σ' are the maximum loadability limits before and after contingency occurrence, respectively. $a(t_{0^-})$ denotes the power vector demanded by the system, which is beyond the maximum loadability limit after contingency occurrence. $a(t_{0^+})$ is the power consumption immediately after the contingency occurrence, which is due to the voltage drop caused by contingency. It is observed that the response of the system power consumption caused by the voltage change leads the operating point to move toward the operating point before the contingency. But, since it is not possible to go beyond the maximum loadability limit, it ultimately leads to reduction of the loads received power due to voltage collapse.

Early detection of above-mentioned situations is important because it reduces the emergency action needed to voltage recovery. The collision of the system trajectory with the maximum loadability limit means the definite occurrence of voltage instability. The magnitude of transmission bus voltage when the system trajectory collides with the maximum loadability limit depends on the reactive power compensation level and can be even more than 0.9 pu. Therefore, voltage magnitude cannot be a good index for early detection of voltage instability and voltage stability indices are expected to help system operators in this regard. The necessary condition for an index to be able to be used in the early detection of voltage instability is that the value and sign of the index change after colliding with the maximum loadability limit in such a way that completely indicate the collision with this limit. Figure 7.15 shows the results of this study about the BSDC

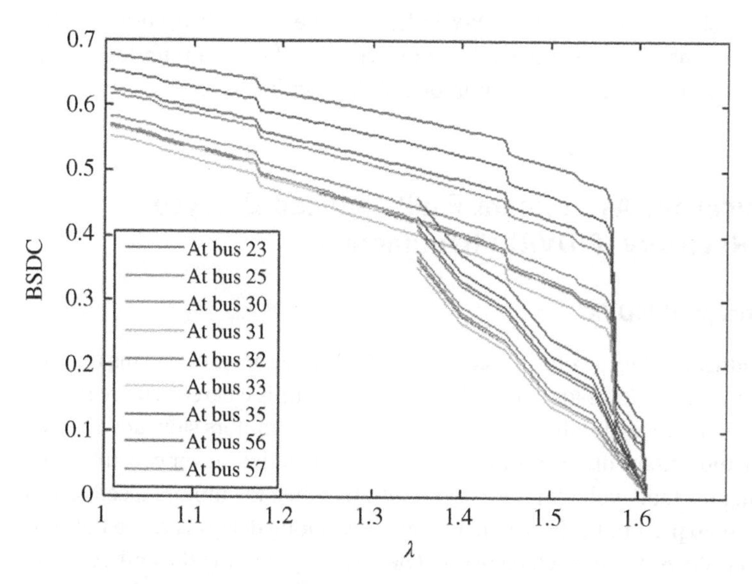

Figure 7.15 The change of BSDC index before and after colliding with the maximum loadability limit.

index. In order to investigate the points after colliding with the maximum loading limit (i.e. the lower points of the PV curves), the load model is considered as impedance according to the Eqs. (7.62) and (7.63). Because it is not possible to increase the loading factor after colliding the maximum loadability limit when the loads characteristic is constant power.

It is observed that the BSDC index after hitting the maximum loadability level is not the same as its values before hitting this level. Therefore, this index is not suitable for the mentioned aims of early detection of voltage instability.

$$P_{li} = \lambda P_{li0}\left(\frac{V_i}{V_{i0}}\right)^2 \tag{7.62}$$

$$Q_{li} = \lambda Q_{li0}\left(\frac{V_i}{V_{i0}}\right)^2 \tag{7.63}$$

In paper [13], a measurement-based index is presented which, unlike SDC and BSDC indices, does not need to measure voltage and current phasors and uses only the magnitude of variables. This index, called DSY, is based on changes in load apparent power with respect to changes in load admittance.

$$\text{DSY} = \frac{\Delta S}{V^2 \Delta y} = \frac{S_2 - S_1}{V_1^2\left(\frac{I_2}{V_2} - \frac{I_1}{V_1}\right)} \tag{7.64}$$

Figure 7.16 shows the change of DSY index at some buses. It is observed that changing the symbol of this index after colliding with the maximum loadability level allows the possibility of prediction of colliding with this level.

7.5 Indices for Assessment Fault-Induced Delayed Voltage Recovery (FIDVR) Phenomenon

7.5.1 Concept of FIDVR

The fault-induced delayed voltage recovery (FIDVR) phenomenon means that the voltage stays at a low level for a long time after the fault is cleared. The reason of FIDVR is the presence of induction motors loads. These motors slow down when fault occurs and voltage drops, resulting in increased reactive power consumption. After clearing the fault, voltage recovery occurs only when the motors speed return to values close to pre-fault values, which may be with long delays because the voltage drop itself slows down speed recovery. Therefore, by clearing the fault, the voltages do not immediately recover to pre-fault values. Deceleration and stalling of

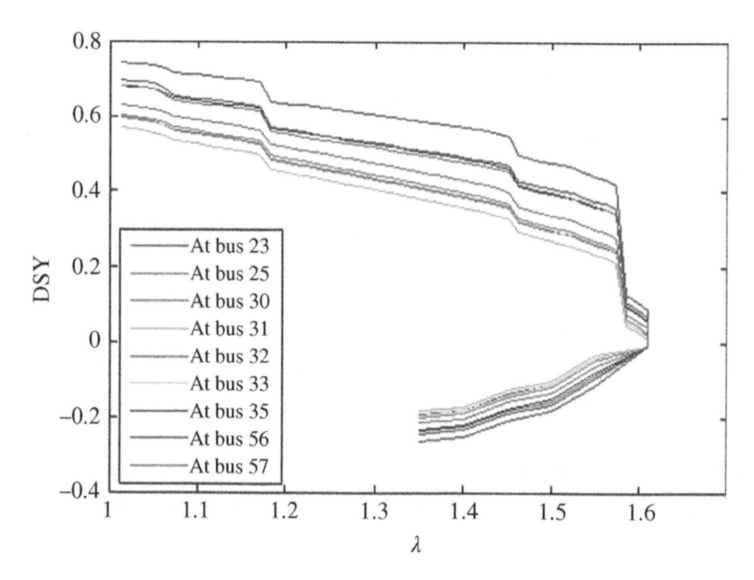

Figure 7.16 The change of DSY index at SOME before and after colliding with the maximum loadability limit.

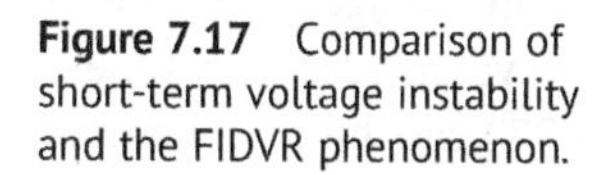

Figure 7.17 Comparison of short-term voltage instability and the FIDVR phenomenon.

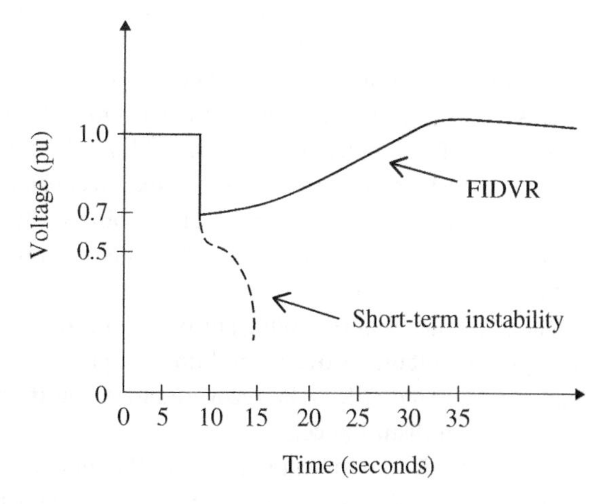

induction motors usually happen in motors with constant torque load, which are common in single-phase motors used in home air conditioning systems [14]. In Figure 7.17, a comparison between the short-term voltage instability with the FIDVR phenomenon is done. Both short-term voltage instability and the FIDVR phenomenon occur due to the presence of induction motors. In FIDVR, despite short-term voltage instability, voltage recovery is done, but with a time delay.

The amount of delay in voltage recovery depends on several factors including pre-fault operating point, fault location, fault type, fault duration, and penetration level of induction motor load [15]. Other effective factors include the use of recloser relays, which in the situation of permanent fault occurrence, and their performance will cause more delay in voltage recovery [15].

Delays in voltage recovery can lead to operation of under voltage load shedding relays and loss of load. The more important undesirable effect of delay in voltage recovery is that low voltage duration may cause outage of some DG units. Unlike load shedding, outage of generation units increases the time delay for voltage recovery and can even cause voltage collapse. Therefore, it is so important to quickly assess the severity of FIDVR so that, if necessary, there is enough time to take emergency actions, which may include load shedding. Accordingly, indices based on measuring and monitoring the trajectory of voltage variations are provided. The goal is to estimate the voltage recovery time by studying a short period of voltage variations. In the following, some of the indices are introduced.

7.5.2 FIDVR Assessment Indices

A method to estimate voltage recovery time is to use the slope of voltage variation trajectory [16]. This slope can be calculated by measuring the voltage magnitude at two consecutive sampling times. In Ref. [16], the optimum voltage recovery time is determined based on the low-voltage ride-through (LVRT) capability of generators. An example of LVRT requirements similar to that used in [16] is shown in Figure 7.18. The protection system should not disconnect generator from the grid as long as the voltage is in the dashed area. Accordingly, the voltage recovery conditions should be such that the operating points of the generators be not outside of the dashed area, because in these conditions, the protection system is allowed for outage of the generators.

Due to the fact that the voltage recovery in generators is done earlier than loads, the optimal voltage recovery conditions in [16] are considered so that voltage at load buses should at least be recovered to 0.8 pu in maximum four seconds after occurrence of disturbance.

The slope calculation begins 1.5 seconds after occurrence of disturbance according to Eq. (7.65). The reason for the 1.5 seconds delay is to ensure passing from the minimum voltage point after fault clearance and reaching the positive slope of voltage variations. In this equation, S_i is the voltage recovery slope at the bus i, and V_i^K and V_i^{K+1} are the voltage magnitudes at bus i at t_K and t_{K+1} times.

$$S_i = \frac{V_i^{K+1} - V_i^K}{t_{K+1} - t_K} \tag{7.65}$$

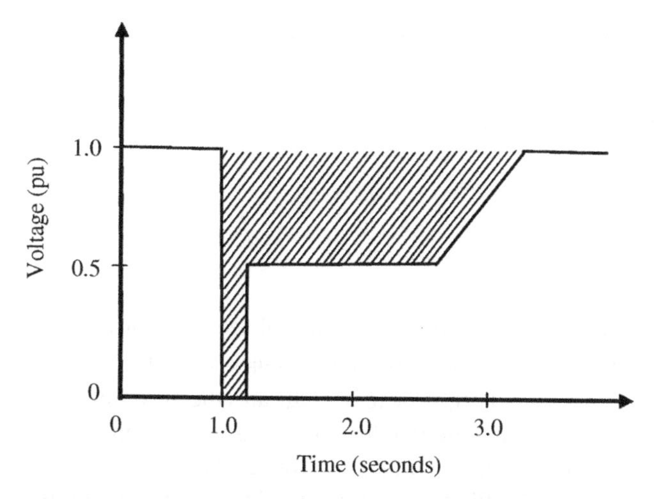

Figure 7.18 An example of FIDVR requirement [16].

The slope calculation is done continuously and compared with the minimum value of the desired slope $S_{\min}$. If the calculated slope is less than $S_{\min}$, load shedding will be performed in several steps to achieve the desired voltage recovery. The value $S_{i\ \min}$ at any time t_K is obtained from Eq. (7.66).

$$S_{i\ \min} = \frac{V_{\mathrm{R}} - V_i^K}{t_{\mathrm{R}} - t_K} \tag{7.66}$$

where V_{R} is the target voltage (here 0.8 pu) and t_{R} is the desired time to recover to the target voltage (here four seconds).

In [17], first an index called instantaneous voltage index (VI) is defined [Eq. (7.67)], which is based on the difference between the magnitude of the voltages after fault clearance with the voltage magnitude before fault at different times.

$$\mathrm{VI}_i = \frac{V_i^0 - V_i^t}{V_i^0}, \quad t \in [t_{\mathrm{cl}} \quad t_{\mathrm{a}}] \tag{7.67}$$

In this equation, V_i^0 is the voltage magnitude of bus i before occurrence of fault, t_{cl} is the fault clearing time, and t_{a} is the time that the voltage recovery assessment is considered up to it.

Then, an index called the dynamic voltage index (DVI) per window is defined, which indicates the maximum voltage deviation that remains in a sliding time window. The length of window τ, based on NERC (North American Electric Reliability Council) criteria, is equal to 20 cycles [18]. To calculate this index, first the minimum voltage deviation at each time window is obtained according to Eq. (7.68).

$$\mathrm{DVI}_i^{\Delta t_{m,n}} = \min \mathrm{VI}_i^t, \quad t \in [t_m, t_n] \\ t_m \in [t_{\mathrm{cl}}, t_{a-\tau}] \\ t_n \in [t_{\mathrm{cl}+\tau}, t_a] \tag{7.68}$$

$$\tau = \frac{20}{f_{n,m}} = \Delta t_{m,n} = t_n - t_m \tag{7.69}$$

It should be noted that $\mathrm{DVI}_i^{\Delta t_{m,n}}$ denotes the minimum voltage deviation at a time window. Hence, at all moments of a time window, the voltage deviation is equal to or more than this value. Therefore, $\mathrm{DVI}_i^{\Delta t_{m,n}}$ actually indicates the maximum voltage deviation that remains at that time window. The reason for emphasizing remaining the voltage drop is that duration of a low voltage determines the performance of protection relays. The FIDVR situation at a bus is now determined based on the maximum value $\mathrm{DVI}_i^{\Delta t_{m,n}}$.

$$\mathrm{DVI}_i = \max \mathrm{DVI}_i^{\Delta t_{m,n}} \tag{7.70}$$

In Ref. [17], the maximum acceptable value for DVI_i is chosen 0.2. Buses with DVI value more than 0.2 are known as buses with FIDVR problem. To ensure the maximum value $\mathrm{DVI}_i^{\Delta t_{m,n}}$ is reached, monitoring of voltage during 1.5 seconds is recommended. Therefore, in this reference, as in [16], determination of the FIDVR situation after a fault occurrence will be done with 1.5 seconds delay. This delay may lead to more needed emergency actions. A learning algorithm has been used to relieve this problem in [19]. For this purpose, an index called transient voltage severity index (TVSI) is used according to Eq. (7.71).

$$\mathrm{TVSI} = \frac{\sum_{i=1}^{N_{\mathrm{b}}} \sum_{t=t_{\mathrm{cl}}}^{T} \mathrm{TVDI}_{i,t}}{N_{\mathrm{b}} \times (T - t_{\mathrm{cl}})} \tag{7.71}$$

In this equation, N_{b} is the number of system buses and T is a deadline time that it is expected that the voltage recovers before it.

The transient voltage deviation index (TVDI) similar to Eq. (7.67) indicates the voltage eviation from its pre-fault value, defined as follows:

$$\mathrm{TVDI}_{i,t} = \begin{cases} \frac{|V_i^t - V_i^0|}{V_i^0} & \text{if } \frac{|V_i^t - V_i^0|}{V_i^0} \geq \mu \\ 0 & \text{otherwise} \end{cases}, \forall t \in [t_{\mathrm{cl}}, T] \tag{7.72}$$

The value μ is selected from Ref. [18] equal to 0.1. Instead of monitoring voltage magnitudes up to the time T, the TVSI value is estimated by much more limited periodic measurement of voltages using the RVFL learning algorithm and probabilistic prediction techniques. In each estimation, the upper and lower limits of the index variation and the probability of the presence of the index in this bound are found. In Ref. [19], TVSI less than 1 is known as the acceptable FIDVR, with 99% probability that the upper limit of the predicted index is less than 1 (acceptable state) or with 99% probability that the lower limit is greater than 1 (unacceptable state).

7.6 Some Real-World Applications of Measurement-based Indices

In paper [12], an algorithm to implement the BSDC index in real-world power system is presented. The algorithms is successfully tested on the Belgian–French 32-bus test system to alarm the vicinity of voltage collapse.

The authors of paper [20] introduced a measurement-based method to monitor voltage stability from a real-world transmission bus feeding a weak area of a system. This type of bus is common in approximately all real-world electrical power systems. The method generalizes both the impedance matching technique and the local identification of voltage emergency situations (LIVES) method based on LTC transformers. The required data are taken from PMU installed at the considered bus.

In paper [21], a parallel optimization method to increase voltage stability margin and minimizing reactive power losses is presented. The method uses the PMU data and can identify the weak buses from the voltage stability margin viewpoint. The used indices are in fact sensitivities of the active and reactive powers to voltage angle and magnitude at buses. The method is implemented online on real-world power systems.

7.7 Summary

In this chapter, a number of important measurements-based indices were introduced, including indices based on the equivalent Thevenin circuit, BSDC, SDC, and DSY. Also in this chapter, the FIDVR phenomenon was described, and indices were introduced to evaluate it. The BSDC index at weak buses is very close to the expected value at the loadability limit, but does not show the collision with the loadability limit. The DSY index can be used for this purpose. The methods of determining Thevenin impedance and methods for reducing the effect of measurement errors were also described in this chapter.

References

1 Corsi, S. and Taranto, G.N. (2008). A real-time voltage instability identification algorithm based on local phasor measurements. *IEEE Trans. Power Syst.* 23 (3): 1271–1279. https://doi.org/10.1109/TPWRS.2008.922586.

2 Abdelkader, S.M. and Morrow, D.J. (2012). Online tracking of Thévenin equivalent parameters using PMU measurements. *IEEE Trans. Power Syst.* 27 (2): 975–983. https://doi.org/10.1109/TPWRS.2013.2254331.

3 Abdelkader, S.M. and Morrow, D.J. (2015). Online Thévenin equivalent determination considering system side changes and measurement errors. *IEEE Trans. Power Syst.* 30 (5): 2716–2725. https://doi.org/10.1109/TPWRS.2014.2365114.

4 Abdelkader, S.M., Eladl, A.A., Saeed, M.A., and Morrow, D.J. (2018). Online Thévenin equivalent determination using graphical phasor manipulation. *Int. J. Electr. Power Energy Syst.* 97: 233–239. https://doi.org/10.1016/j.ijepes.2017.11.013.

5 Milŏsević, B. and Begović, M. (2003). Voltage-stability protection and control using a wide-area network of phasor measurements. *IEEE Trans. Power Syst.* 18 (1): 121–127. https://doi.org/10.1109/TPWRS.2002.805018.

6 Ma, J. and Ding, R. (2014). Recursive computational formulas of the least squares criterion functions for scalar system identification. *Appl. Math. Model.* 38 (1): 1–11. https://doi.org/10.1016/j.apm.2013.05.059.

7 Zhao, J., Wang, Z., Chen, C., and Zhang, G. (2017). Robust voltage instability predictor. *IEEE Trans. Power Syst.* 32 (2): 1578–1579. https://doi.org/10.1109/TPWRS.2016.2574701.

8 Su, H.Y. and Liu, T.Y. (2018). Robust Thevenin equivalent parameter estimation for voltage stability assessment. *IEEE Trans. Power Syst.* 33 (4): 4637–4639. https://doi.org/10.1109/TPWRS.2018.2821926.

9 Verbič, G. and Gubina, F. (2003). Fast voltage-collapse line-protection algorithm based on local phasors. *IEE Proc. Gener. Transm. Distrib.* 150 (4): 482–486. https://doi.org/10.1049/ip-gtd2003046.

10 Verbič, G. and Gubina, F. (2004). A new concept of protection against voltage collapse based on local phasors. *IEEE Trans. Power Deliv.* 19 (2): 576–581. https://doi.org/10.1109/ICPST.2000.897151.

11 Verbič, G. and Gubina, F. (2004). A novel scheme of local protection against voltage collapse based on the apparent-power losses. *Int. J. Electr. Power Energy Syst.* 26 (5): 341–347. https://doi.org/10.1016/j.ijepes.2003.11.001.

12 Šmon, I., Pantoš, M., and Gubina, F. (2008). An improved voltage-collapse protection algorithm based on local phasors. *Electr. Power Syst. Res.* 78 (3): 434–440. https://doi.org/10.1016/j.epsr.2007.03.012.

13 Parniani, M. and Vanouni, M. (2010). A fast local index for online estimation of closeness to loadability limit. *IEEE Trans. Power Syst.* 25 (1): 584–585. https://doi.org/10.1109/TPWRS.2009.2036460.

14 Glavic, M., Novosel, D., Heredia, E. et al. (2012). See it fast to keep calm. *IEEE Power Energy Mag.* 10: 43–55.

15 Glidewell, J.D. and Patel, M.Y. (2012). Effect of high speed reclosing on fault induced delayed voltage recovery. *IEEE Power and Energy Society General Meeting*, San Diego, CA (22–26 July 2012). pp. 1–6. IEEE. https://doi.org/10.1109/PESGM.2012.6344608.

16 Halpin, S.M., Harley, K.A., Jones, R.A., and Taylor, L.Y. (2008). Slope-permissive under-voltage load shed relay for delayed voltage recovery mitigation. *IEEE Trans. Power Syst.* 23 (3): 1211–1216. https://doi.org/10.1109/TPWRS.2008.926409.

17 Pinzón, J.D. and Gracielacolome, D. (2018). Fault-induced delayed voltage recovery assessment based on dynamic voltage indices. *IEEE PES Transmission & Distribution Conference and Exhibition-Latin America (T&D-LA)*, Lima, Peru (18–21 September 2018), pp. 1–5. https://doi.org/10.1109/TDC-LA.2018.8511712.

18 Shoup, D.J., Paserba, J.J., and Taylor, C.W. (2004). A survey of current practices for transient voltage dip/sag criteria related to power system stability. *2004 IEEE PES Power Systems Conference and Exposition*, New York, NY (10–13 October 2004), vol. 2, pp. 1140–1147. https://doi.org/10.1109/psce.2004.1397688.

19 Zhang, Y., Xu, Y., Dong, Z.Y., and Zhang, P. (2019). Real-time assessment of fault-induced delayed voltage recovery: a probabilistic self-adaptive data-driven method. *IEEE Trans. Smart Grid* 10 (3): 2485–2494. https://doi.org/10.1109/TSG.2018.2800711.

20 Vournas, C.D., Lambrou, C., and Mandoulidis, P. (2017). Voltage stability monitoring from a transmission bus PMU. *IEEE Trans. Power Syst.* 32 (4): 3266–3274. https://doi.org/10.1109/TPWRS.2016.2629495.

21 Li, H., Bose, A., and Venkatasubramanian, V.M. (2016). Wide-area voltage monitoring and optimization. *IEEE Trans. Smart Grid* 7 (2): 785–793. https://doi.org/10.1109/TSG.2015.2467215.

8

Model-Based Indices

8.1 Introduction

Since the system model is used in model-based indices, these indices can be used to assess the state of the system in the event of probable contingencies. These indices can also be used to find weak buses and lines, and generators are effective in the occurrence of voltage instability. Determining suitable locations for installing reactive power compensators is another application of model-based indices. In addition to the indices defined for the transmission network, some indices are also defined that take the distribution network structure into the account. Similar to measurement-based indices, model-based indices are defined based on system characteristics in the loadability limit. One of these characteristics is singularity of the Jacobian matrix of equilibrium equations at the loadability limit. This matrix contains useful information about the power system. Another part of the indices is based on the existence of the solution for the governing equations of the system. A number of indices are also based on the values of load buses voltage and generators reactive power.

8.2 Jacobian Matrix-Based Indices

8.2.1 Background

The eigenvalues and eigen and singular vectors of the Jacobin matrix contain important information about the state of the power system and can be used in voltage stability studies. First, an introductory description of eigenvalues of a matrix is given.

Voltage Stability in Electrical Power Systems: Concepts, Assessment, and Methods for Improvement, First Edition. Farid Karbalaei, Shahriar Abbasi, and Hamid Reza Shabani.

Published 2023 by John Wiley & Sons, Inc.

Assuming that $\boldsymbol{A}$ is an $n \times n$ matrix, λ is an eigenvalue of $\boldsymbol{A}$ if there is a nonzero vector $\boldsymbol{x}$ such that

$$\boldsymbol{A}\boldsymbol{x} = \lambda\boldsymbol{x} \tag{8.1}$$

or

$$(\boldsymbol{A} - \lambda\boldsymbol{I})\boldsymbol{x} = 0 \tag{8.2}$$

where I is the identity matrix [1]. The vector $\boldsymbol{x}$ is the right eigenvector corresponding to the eigenvalue λ. The row vector $\boldsymbol{y}$ is called the left eigenvector if,

$$\boldsymbol{y}\boldsymbol{A} = \lambda\boldsymbol{y} \tag{8.3}$$

Equation (8.2) has a nonzero solution if and only if,

$$\det(\boldsymbol{A} - \lambda\boldsymbol{I}) = 0 \tag{8.4}$$

Equation (8.4) is a polynomial with degree n in terms of λ known as the characteristic polynomial of matrix $\boldsymbol{A}$. The roots of this polynomial are the n eigenvalues of $\boldsymbol{A}$.

The left and right eigenvectors corresponding to different eigenvalues are orthogonal [1], i.e., if λ_i does not equal λ_j,

$$y_j x_i = 0 \tag{8.5}$$

For eigenvectors corresponding to an eigenvalue,

$$y_i x_i = C_i \tag{8.6}$$

where C_i is nonzero. Due to eigenvectors can be multiplied by a number or divided by a non-zero number, it is common to normalize them so that the following equation is satisfied:

$$y_i x_i = 1 \tag{8.7}$$

So,

$$\boldsymbol{Y} = \boldsymbol{X}^{-1} \tag{8.8}$$

where $\boldsymbol{X}$ is a matrix whose columns are right eigenvectors of matrix $\boldsymbol{A}$. $\boldsymbol{Y}$ is also a matrix that its rows are left eigenvectors of matrix $\boldsymbol{A}$. When all eigenvalues of $\boldsymbol{A}$ are different, there is the following equation:

$$\boldsymbol{X}^{-1}\boldsymbol{A}\boldsymbol{X} = \boldsymbol{\Omega} \tag{8.9}$$

where $\boldsymbol{\Omega}$ is a diagonal matrix whose main diagonal elements are the eigenvalues of $\boldsymbol{A}$. Also, using eigenvalues and eigenvectors, the matrix $\boldsymbol{A}$ can be decomposed as follows:

$$\boldsymbol{A} = \boldsymbol{X}\boldsymbol{\Lambda}\boldsymbol{X}^{-1} \tag{8.10}$$

In addition to eigenvalues and eigenvectors, decomposition can also be done using singular values and vectors, as given in Eq. (8.11) [2, 3].

$$A = U\Sigma V^T = \sum_{i=1}^{n} \sigma_i u_i v_i^T \quad (8.11)$$

In Eq. (8.11), U and V are $n \times n$ dimension orthogonal matrices; an orthogonal matrix is a matrix whose inverse and transposition are the same. u_i and v_i are, respectively, the columns of U and V matrices. Σ is a diagonal matrix whose elements denoted by σ_i are the singular values of matrix A. These elements are nonnegative and are usually arranged in such a way that $\sigma_1 \geq \sigma_2 \geq \cdots \geq \sigma_n \geq 0$. u_i and v_i are, respectively, the left and right singular vectors corresponding to the singular value σ_i. About the singular vectors, there are,

$$Av_i = \sigma_i u_i \quad (8.12)$$

$$A^T u_i = \sigma_i v_i \quad (8.13)$$

Assuming that the rank of A is r ($r \leq n$), its singular values are equal to root square of the r positive eigenvalues of the matrix A^TA (or AA^T). Also, the right and left singular vectors of matrix A are equal to the right eigenvectors of A^TA and AA^T matrices, respectively.

For a square matrix, the rank is the maximum number of linearly independent columns or rows in the matrix. If $r < n$, the matrix is called singular. At least one eigenvalue and singular value of a singular matrix is zero.

When a real matrix is symmetric, its eigenvalues are real and equal to the singular values. Also in this case, the singular and eigenvectors are the same.

8.2.2 Singularity of Jacobian Matrix at the Loadability Limit

In Chapter 5, the continuation power flow (CPF) methods to determine the loadability limit was presented. Another way to determine the loadability limit is to solve an optimization problem as follows [4]:

$$\text{Maximize } \lambda \quad (8.14)$$

Subject to,

$$\bar{f}(\bar{x}, \lambda) = 0 \quad (8.15)$$

where λ is the loading factor, $\bar{x}$ is the vector of power flow variables, and $\bar{f}$ denotes the power flow equations.

To use Karush–Kuhn–Tucker (KKT) conditions [5], the Lagrange function is first defined as follows:

$$L = \lambda + \overline{\boldsymbol{W}}^T \overline{\boldsymbol{f}}(\overline{\boldsymbol{x}}, \lambda) \tag{8.16}$$

where $\overline{\boldsymbol{W}}$ is the vector of Lagrange coefficients. Now, with respect to the KKT conditions,

$$\nabla_{\overline{W}} L = 0 \Rightarrow \overline{\boldsymbol{f}}(\overline{\boldsymbol{x}}, \lambda) = 0 \tag{8.17}$$

$$\nabla_{\lambda} L = 0 \Rightarrow 1 + \overline{\boldsymbol{f}}_{\lambda}^T \overline{\boldsymbol{W}} = 0 \tag{8.18}$$

$$\nabla_{\overline{x}} L = 0 \Rightarrow 1 + \overline{\boldsymbol{f}}_{\overline{x}}^T \overline{\boldsymbol{W}} = 0 \tag{8.19}$$

where $\overline{\boldsymbol{f}}_{\lambda}$ is the vector of derivative $\overline{\boldsymbol{f}}$ with respect to λ. Also, $\overline{\boldsymbol{f}}_{\overline{x}}$ is the matrix of derivative $\overline{\boldsymbol{f}}$ with respect to $\overline{\boldsymbol{x}}$ that is the power flow Jacobin matrix shown $\boldsymbol{J}$ hereafter.

According to Eq. (8.18), the vector $\overline{\boldsymbol{W}}$ is nonzero. So, the Eq. (8.19) is valid only if the columns of the matrix $\overline{\boldsymbol{f}}_{\overline{x}}^T$ (or rows of the matrix $\overline{\boldsymbol{f}}_{\overline{x}}$) be linearly dependent, which means that the Jacobin matrix is singular at the loadability limit. Therefore, the minimum singular value and the minimum magnitude of the Jacobin matrix eigenvalues become zero at the maximum loadability. Hence, these values have been used as indices to assess the distance to the loadability limit.

8.2.3 Singular Values and Vectors

Let us assume the linearized power flow equations are as follows:

$$\begin{bmatrix} \Delta \boldsymbol{P} \\ \Delta \boldsymbol{Q} \end{bmatrix} = \boldsymbol{J} \begin{bmatrix} \Delta \theta \\ \Delta \boldsymbol{V} \end{bmatrix} \tag{8.20}$$

where $\Delta \boldsymbol{P}$ and $\Delta \boldsymbol{Q}$ are the vectors of active and reactive powers changes, $\Delta\theta$ and $\Delta \boldsymbol{V}$ are the vectors of changes of voltages angle and magnitude at buses, and $\boldsymbol{J}$ is the Jacobin matrix. Due to the orthogonality of the $\boldsymbol{U}$ and $\boldsymbol{V}$ matrices, the inverse decomposition of the matrix $\boldsymbol{J}$ in terms of singular values and vectors leads to Eq. (8.21),

$$\begin{bmatrix} \Delta \theta \\ \Delta \boldsymbol{V} \end{bmatrix} = \boldsymbol{V} \boldsymbol{\Sigma}^{-1} \boldsymbol{U} \begin{bmatrix} \Delta \boldsymbol{P} \\ \Delta \boldsymbol{Q} \end{bmatrix} = \sum_{i=1}^{n} \sigma_i^{-1} \boldsymbol{v}_i \boldsymbol{u}_i^T \begin{bmatrix} \Delta \boldsymbol{P} \\ \Delta \boldsymbol{Q} \end{bmatrix} \tag{8.21}$$

In the vicinity of loadability limit, since σ_n has a value close to zero, Eq. (8.21) can be re-written as follows [2]

$$\begin{bmatrix} \Delta\theta \\ \Delta V \end{bmatrix} \simeq \sigma_n^{-1} \boldsymbol{v}_n \boldsymbol{u}_n^T \begin{bmatrix} \Delta \boldsymbol{P} \\ \Delta \boldsymbol{Q} \end{bmatrix} \tag{8.22}$$

These can be concluded from Eq. (8.22) that

- Elements with the largest values in $\boldsymbol{v}_n$ represent the buses that by changing the active and reactive powers their voltage magnitude and angle have the highest changes. Hence, these elements represent the weak buses of the system.
- Elements with the largest values in $\boldsymbol{u}_n$ represent the most sensitive directions for changes in active and reactive powers. In other words, these elements illustrate in which buses the increase in power causes the highest voltage drop in the system.

8.2.4 Simulation

Figures 8.1 and 8.2 show the changes of σ_n with respect to loading factor λ in the IEEE 57-bus test system in two states of with and without reactive power generation limits of generators. The loadability limit ($\lambda_{\max}$) in these two states is 1.61 and 1.99. As expected, this index reaches zero at the loadability limit. The sudden changes in Figure 8.1 are due to encountering the generators with their reactive

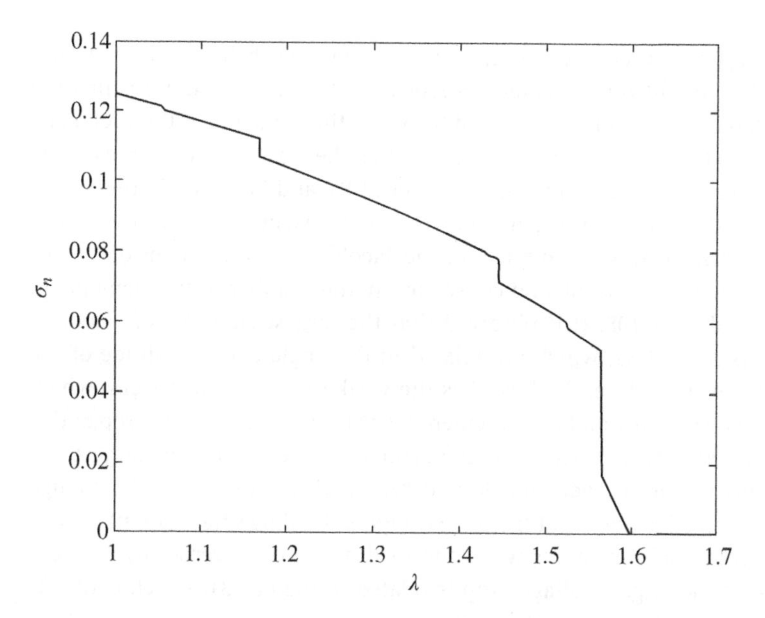

Figure 8.1 σ_n changes vs loading factor (λ) with considering reactive power generation limits of generators.

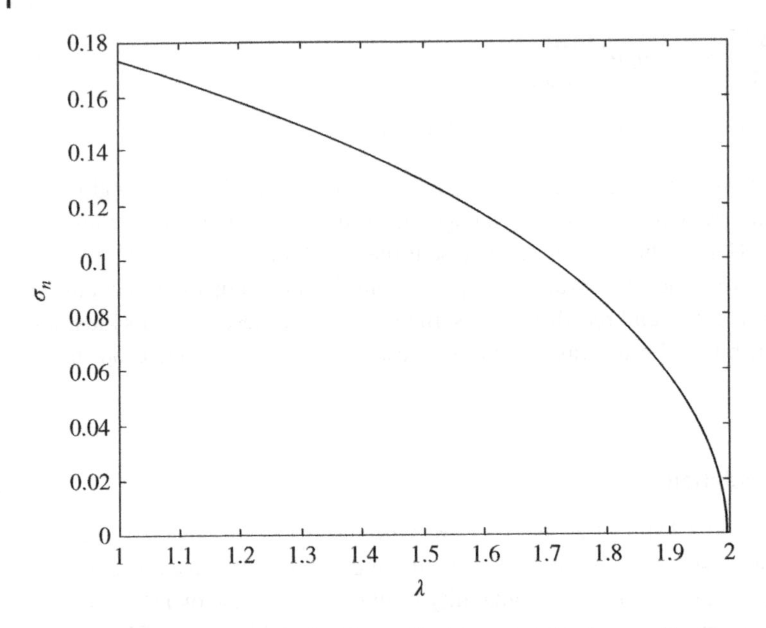

Figure 8.2 σ_n changes vs loading factor (λ) without considering reactive power generation limits of generators.

power generation limits. These jumps and the nonlinear behavior of the index, which are further illustrated in Figure 8.2, do not allow more accurate estimation of the loadability limit. Figures 8.3 and 8.4 show the elements of the vectors u_n and v_n for $\sigma_n = 0.1$, which occur at $\lambda = 1.246$ when the reactive power limits of the generators are considered. Given the number of PV and PQ buses in the 57-bus test system and the number of generators that have reached their reactive power generation limits at this loading factor, the Jacobian matrix dimension at this load level becomes 108×108 and consequently the number of the elements in vectors u_n and v_n is 108. It is observed that the largest elements of vector v_n are elements 30 and 82, which are related to the angle and magnitude of the bus 31 voltage. Therefore, the bus 31 is the weakest bus of the system, which means that power consumption increment leads to the most voltage drop at this bus. This can also be detected using the elements of vector u_n. The largest elements of this vector are elements 30 and 82, which are related to the change of active and reactive powers at the bus 31. Figure 8.5 shows the amount of voltage drop across different buses when the loading factor increases by 1%. It is observed that the largest voltage drop is related to the bus 31, which confirms the results obtained from singular vectors.

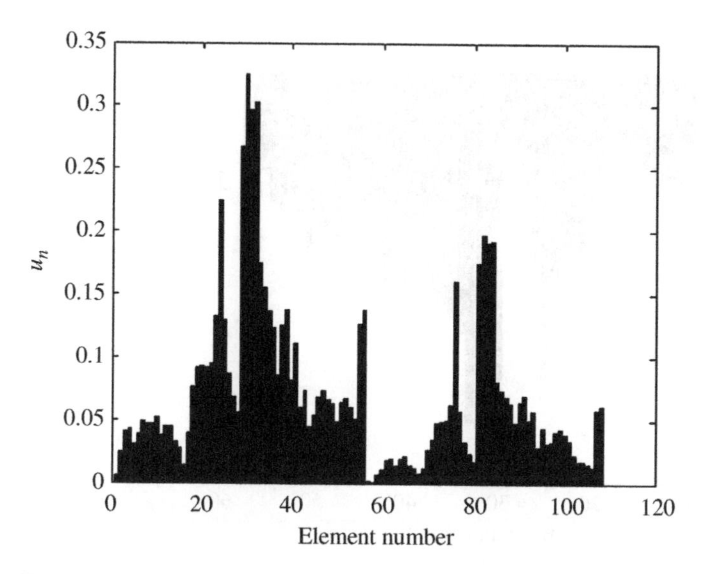

Figure 8.3 u_n elements when $\sigma_n = 0.1$ ($\lambda = 1.246$).

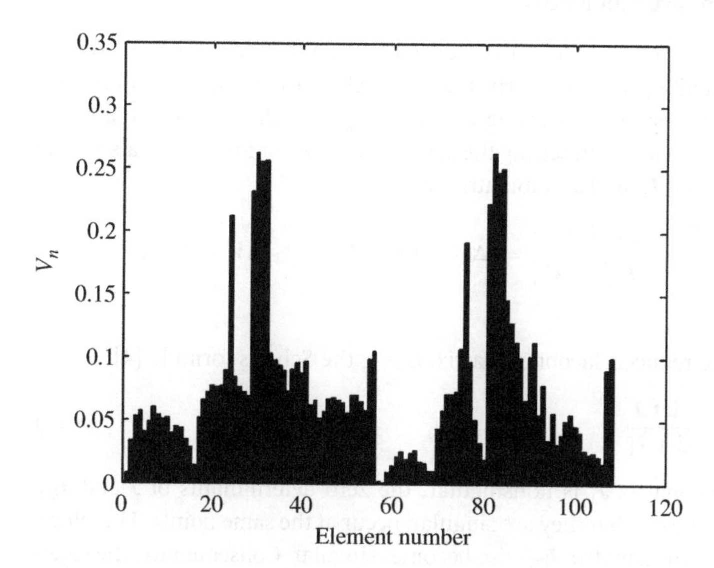

Figure 8.4 v_n elements when $\sigma_n = 0.1$ ($\lambda = 1.246$).

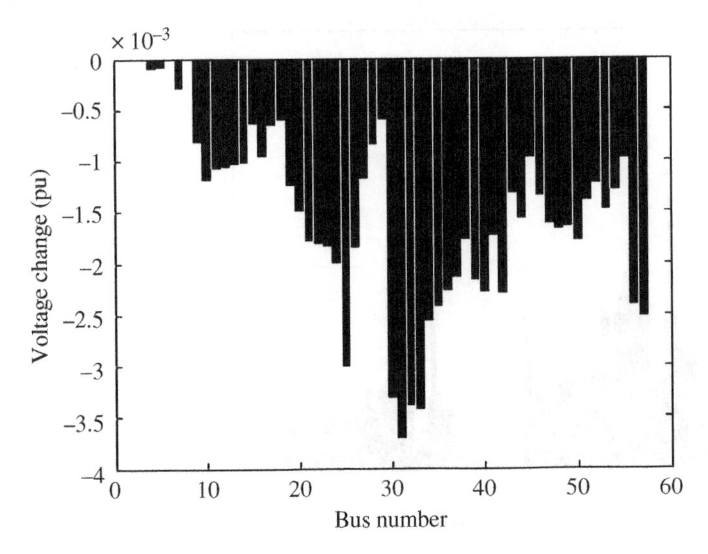

Figure 8.5 Voltage changes at $\lambda = 1.246$ after incremental step of 1% at all buses together.

8.2.5 Reduced Jacobian Matrix

In addition to the Jacobian matrix, the reduced Jacobin matrix is also used to evaluate voltage stability [2]. This matrix shows the changes in reactive power with respect to voltage changes, assuming zero active power changes. To obtain this matrix, Eq. (8.20) is used with setting the active power vector to zero and also using the following $\boldsymbol{J}_1$, $\boldsymbol{J}_2$, $\boldsymbol{J}_3$, and $\boldsymbol{J}_4$ submatrices:

$$\begin{bmatrix} 0 \\ \Delta \boldsymbol{Q} \end{bmatrix} = \begin{bmatrix} \boldsymbol{J}_1 & \boldsymbol{J}_2 \\ \boldsymbol{J}_3 & \boldsymbol{J}_4 \end{bmatrix} \begin{bmatrix} \Delta \boldsymbol{\theta} \\ \Delta \boldsymbol{V} \end{bmatrix} \Rightarrow \Delta \boldsymbol{Q} = \left(\boldsymbol{J}_4 - \boldsymbol{J}_3 \boldsymbol{J}_1^{-1} \boldsymbol{J}_2 \right) \Delta \boldsymbol{V} = \boldsymbol{J}_{QV} \Delta \boldsymbol{V} \tag{8.23}$$

where $\boldsymbol{J}_{QV}$ is the reduced Jacobian matrix. Using the Schur's formula [4],

$$\det \boldsymbol{J}_{QV} = \frac{\det \boldsymbol{J}}{\det \boldsymbol{J}_1} \tag{8.24}$$

Assuming that matrix $\boldsymbol{J}_1$ is nonsingular, the zero determinants of $\boldsymbol{J}$ and $\boldsymbol{J}_{QV}$ matrices, which means that they are singular, occur at the same points. Therefore, at the loadability limit, matrix $\boldsymbol{J}_{QV}$ also becomes singular. Consequently, the smallest singular value and the magnitude of the smallest eigenvalue of this matrix can also be used as voltage stability indices. Figure 8.6 shows the changes of the magnitude of the smallest eigenvalue of matrix $\boldsymbol{J}_{QV}$ in the IEEE 57-bus test system. It can be seen that the behavior of this index is similar to that of the smallest singular value of the Jacobian matrix.

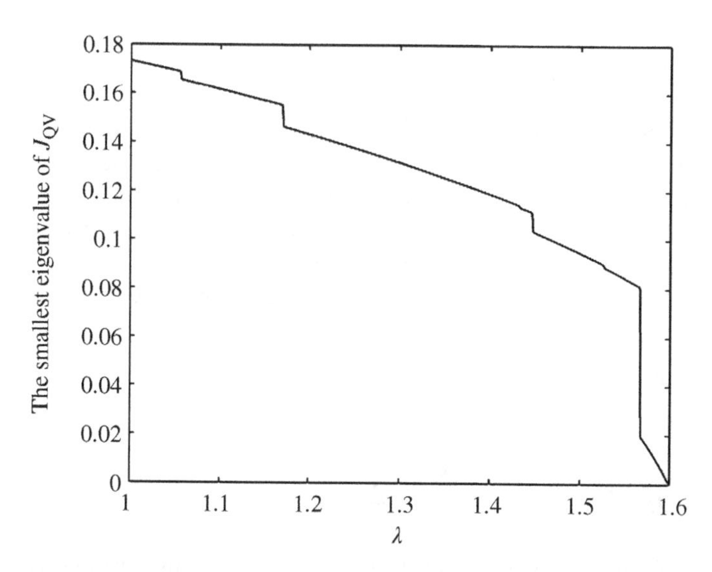

Figure 8.6 Changes of the magnitude of the smallest eigenvalue of J_{QV}.

8.2.6 Eigenvalues and Eigenvectors

The matrix $\boldsymbol{J_{QV}}$ is almost symmetric. Hence, its eigenvalues and eigenvectors are probably real. According to Eq. (8.10), decomposition of this matrix based on its eigenvalues and eigenvectors is as follows:

$$\boldsymbol{J_{QV}} = \boldsymbol{X\Omega Y} \tag{8.25}$$

$$\boldsymbol{J_{QV}^{-1}} = \boldsymbol{X\Omega^{-1}Y} \Rightarrow \Delta\boldsymbol{V} = \boldsymbol{X\Omega^{-1}Y}\Delta\boldsymbol{Q} \tag{8.26}$$

$$\Delta\boldsymbol{V} = \sum_i \frac{x_i y_i}{\lambda_i} \Delta\boldsymbol{Q} \tag{8.27}$$

Each eigenvalue λ_i, together with the corresponding eigenvectors, forms the mode i of the system. The changes of ith modal reactive power are defined as follows [6].

$$\Delta Q_{m_i} = K_i x_i \tag{8.28}$$

In general,

$$K_i^2 \sum_j x_{ji}^2 = 1 \tag{8.29}$$

Which causes the magnitude of ΔQ_{m_i} to be equal to 1. By substituting ΔQ_{m_i} in Eq. (8.27) and using the Eqs. (8.5) and (8.7), the changes of ith modal voltage will be obtained as

$$\Delta V_{m_i} = \frac{1}{\lambda_i} \Delta Q_{m_i} \tag{8.30}$$

It can be seen that when the reactive power changes are in the x_i direction, the voltage changes are also in that direction. The lower λ_i, the weaker the mode because with a slight decrease in the modal reactive power generation (increasing the modal reactive power consumption), the modal voltage decreases greatly. It should be noted that here Q is the generated reactive power.

If $\Delta Q = e_k$, that e_k is a vector whose elements are zero except kth element, which is 1:

$$\Delta \boldsymbol{V} = \sum_i \frac{y_{ik} x_i}{\lambda_i} \tag{8.31}$$

$$\Delta V_k = \sum_i \frac{y_{ik} x_{ki}}{\lambda_i} \Rightarrow \frac{\partial V_k}{\partial Q_k} = \sum_i \frac{y_{ik} x_{ki}}{\lambda_i} = \sum_i \frac{P_{ki}}{\lambda_i} \tag{8.32}$$

The system is stable only when all eigenvalues are positive. Because just in this case the voltage sensitivity to reactive power is positive at all buses, i.e. with increasing reactive power generation (or decreasing reactive power consumption) at a bus, its voltage increases. It has previously been observed that the Jacobin matrix becomes singular at the loadability limit, and therefore one of its eigenvalues becomes zero. According to Eq. (8.24), at the loadability limit, also one of the eigenvalues of the matrix $\boldsymbol{J_{QV}}$ becomes zero. In these conditions, according to Eq. (8.30), the voltage sensitivity to reactive power at all load buses tends to infinity. Figure 8.7 shows the changes of reactive power in terms of voltage with constant active power at the bus 31. This curve is called the VQ curve. Unlike the PV curve, which is plotted at a constant power factor, this curve is plotted at a constant active power (not a constant power factor). Positive and negative values of Q denote generation and consumption of reactive power consumption, respectively. The minimum point of the curve states how much the maximum reactive power consumption can be for a given active power, which represents one of the maximum loading surface points. It is obvious that at the minimum point, the voltage sensitivity to reactive power is infinite. In fact, by increasing the reactive power consumption and moving toward the maximum loadability point, this sensitivity increases and reaches an infinitely positive value when it collides with the maximum loadability surface. After passing this surface, the sensitivity will take a negative value. At first, the magnitude of this negative value is very large, but as it moves away from the maximum loadability surface, its value decreases. The *VQ* curve is plotted with placing a voltage source with zero active power generation at the bus. By changing the voltage of this source and performing power flow, its generated reactive power is obtained.

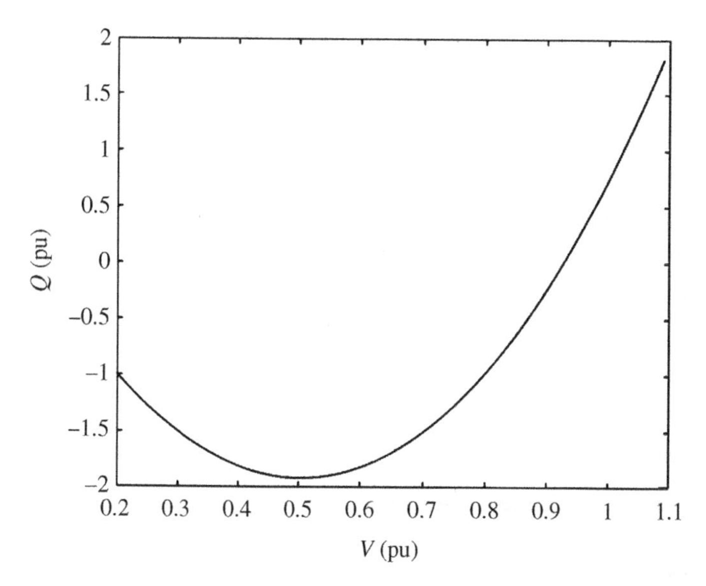

Figure 8.7 The *VQ* curve at bus 31.

In Eq. (8.32), P_{ki} is the participation factor of bus k in mode i, which in fact indicates the participation value of the ith eigenvalue in V–Q sensitivity at the bus k. The higher value of P_{ki}, the greater share of ith eigenvalue in the bus k sensitivity. For all small eigenvalues, the participation factors indicate the areas prone to voltage instability [6]. Assuming the injected reactive power changes is ΔQ_{m_i}, the value ΔV_{m_i} is obtained from Eq. (8.30) and the modal angle changes from Eq. (8.23).

$$\Delta\theta_{m_i} = -J_1^{-1} J_2 \Delta V_{m_i} \tag{8.33}$$

Having $\Delta\theta_{m_i}$ and ΔV_{m_i}, the changes of reactive power consumption of lines and the changes of the generated reactive power are obtained. If $\Delta Q_{i_{max}}$ is the maximum change of reactive power consumption of lines for mode i and ΔQ_{kji} is the change of reactive power consumption in line kj for mode i, the participation factor of line kj in mode i is [6]:

$$P_{k_{ji}} = \frac{\Delta Q_{kji}}{\Delta Q_{i_{\max}}} \tag{8.34}$$

This factor indicates that in each mode, in which line, there is the most reactive power consumption change. Branches with high participation factor are the main cause of mode i weakness. This factor tells the operator which lines need to be expanded or rerouted. They are also useful for selecting critical contingencies.

In addition to the lines participation factors, the generators participation factors are also defined. This factors are as follows [6]:

$$P_{g_{ki}} = \frac{\Delta Q_{g_{ki}}}{\Delta Q_{g_{\max\ i}}} \tag{8.35}$$

where $\Delta Q_{g_{ki}}$ is the generated reactive power change of generator k in mode i and $\Delta Q_{g_{\max\ i}}$ is the maximum change of generated reactive power in mode i. This factor indicates which generator has the most output reactive power change in each mode. This generator is important in maintaining the stability of mode i, because this generator is likely to reach its reactive power generation limit.

8.2.7 Test Function

In Ref. [7], a test function is defined as Eq. (8.36):

$$t_{Lk} \triangleq \boldsymbol{e}_L^T \boldsymbol{J}\boldsymbol{J}_{LK}^{-1} \boldsymbol{e}_L \tag{8.36}$$

In this function, $\boldsymbol{J}$ is the Jacobin matrix. $\boldsymbol{e}_L$ is a vector whose Lth element is equal to 1 and the other elements are zero. $\boldsymbol{J}_{LK}$ is also defined as follows:

$$\boldsymbol{J}_{LK} \triangleq \left(\boldsymbol{I} - \boldsymbol{e}_L \boldsymbol{e}_L^T\right)\boldsymbol{J} + \boldsymbol{e}_L \boldsymbol{e}_K^T \tag{8.37}$$

where $\boldsymbol{I}$ is the identity matrix. It can be shown that $\boldsymbol{J}_{LK}$ is obtained by substituting the Lth row of the matrix $\boldsymbol{J}$ by the row vector $\boldsymbol{e}_K^T$. If L and K are selected equal to one of the non-zero values of the eigenvectors corresponding to the zero eigenvalue, the matrix $\boldsymbol{J}_{LK}$ will become singular at the loadability limit [8]. If $L = K = C$, that C is the number of the weakest bus (critical bus) of the system, the value of the test function becomes zero at the loadability limit. In this conditions, the test function is as follows, which is called the critical test function.

$$\boldsymbol{t}_{CC} = \boldsymbol{e}_C^T \boldsymbol{J} \boldsymbol{J}_{CC}^{-1} \boldsymbol{e}_C \tag{8.38}$$

Selection of the weakest bus can be done, for example, by using the buses' participation factors for the smallest eigenvalue of the matrix $\boldsymbol{J}_{QV}$.

Changes of t_{CC} with respect to loading factor λ can be fitted with 2 or 4 degree curves. The 2 degree curve is for the case that the load increment is done only in active or reactive power of a bus. In this case, $\lambda_{\max}$ is estimated as follows [7]:

$$\hat{\lambda}_{\max} \simeq \lambda - \frac{1}{2}\frac{t_{CC}(\overline{\boldsymbol{x}}, \lambda)}{\frac{\mathrm{d}t_{CC}}{\mathrm{d}\lambda}(\overline{\boldsymbol{x}}, \lambda)} \tag{8.39}$$

where λ and $\overline{\mathbf{x}}$ are the loading factor and vector of the power flow variables at the current operating point, respectively. Also, $\hat{\lambda}_{\max}$ indicates the estimated value of the maximum loading factor. The derivative of the denominator of the fraction can be calculated by twice calculations of t_{CC} for two λ values close to each other. A method to calculate this derivative has been presented in [7]. For a more practical case where the load increases at several buses, the changes of t_{CC} can be fitted as a 4° curve. In this case,

$$\hat{\lambda}_{\max} \simeq \lambda - \frac{1}{4}\frac{t_{CC}(\overline{x}, \lambda)}{\frac{\mathrm{d}t_{CC}}{\mathrm{d}\lambda}(\overline{x}, \lambda)} \tag{8.40}$$

It is clear that the estimation error becomes less when λ is closer to $\lambda_{\max}$.

Figures 8.8 and 8.9 show the changes of t_{CC} for two values of C. In Figure 8.8, the C is 20, which is a noncritical bus number in the IEEE 57-bus test system. But in Figure 8.9, C is assumed to be 31, which is the weakest bus of the system. (This should be noted that the weakest bus may change with increase of λ, this means that the need to detect the weakest bus is the weakness of the test function.) It can be seen that when the value of C is a noncritical bus number, the test function value does not reach the expected value of zero at the loadability limit. Furthermore, the appeared sudden changes do not allow estimation of the test function with the second- and four-order functions.

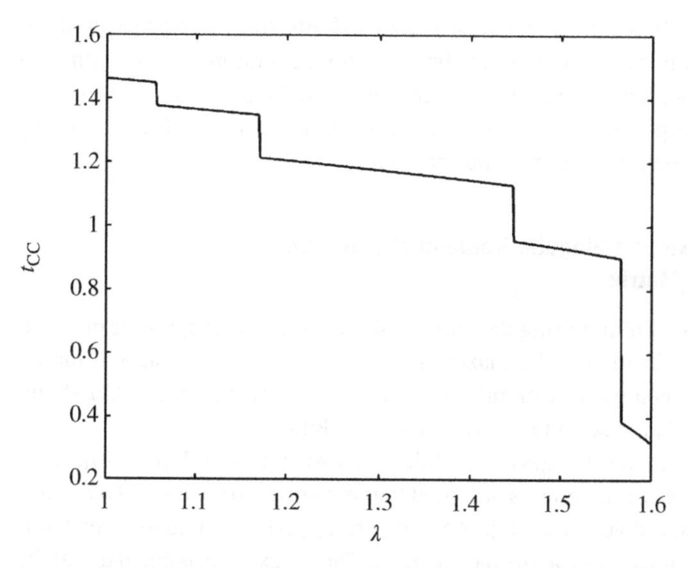

Figure 8.8 The changes of t_{CC} when $C = 20$ (bus 20 is a non-critical bus).

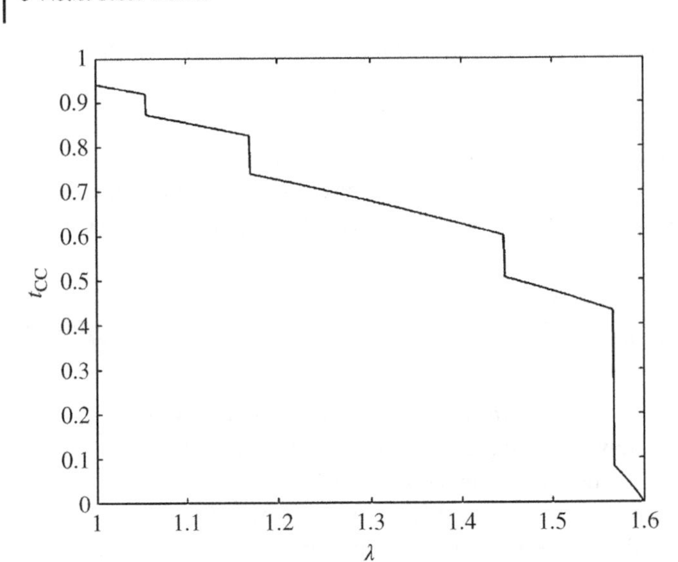

Figure 8.9 The changes of t_{CC} when C = 31 (bus 31 is the weakest bus).

In Figure 8.10, the estimated loadability margin and actual loadability margin are plotted for two states of with and without considering the reactive power limits of the generators. $\Delta\lambda$ is in fact the difference between the current values of λ with $\hat{\lambda}_{max}$ and λ^{actual}_{max}. It is observed that if the loadability limit is reached without encountering with the reactive power limits of the generators, the test function can acceptably predict the stability margin. This state is more common in weak systems in which power transmission limitation is mainly due to the lines impedances not the generators reactive power limits.

8.2.8 The Maximum Singular Value of the Inverse of the Jacobian Matrix

In addition to the minimum singular value and the minimum magnitude of eigenvalue in the Jacobin matrix, the maximum singular value of the Jacobin matrix inverse has also been used as an index for voltage stability [9]. Figure 8.11 shows the changes of this index up to reaching the loadability limit.

It can be seen that the changes of this index are also nonlinear. But in [10], using σ_{max} and its derivative, an index is proposed that is claimed to be linear. This index, called the i index, is diven in the Eq. (8.41), where C_{tot} is the system total load. i_0 is the $\sigma_{max}/(d\sigma_{max}/dC_{tot})$ value at the base loading. The index value is equal to 1 at the base loading and zero at the loadability limit. In [10], a method for calculating this

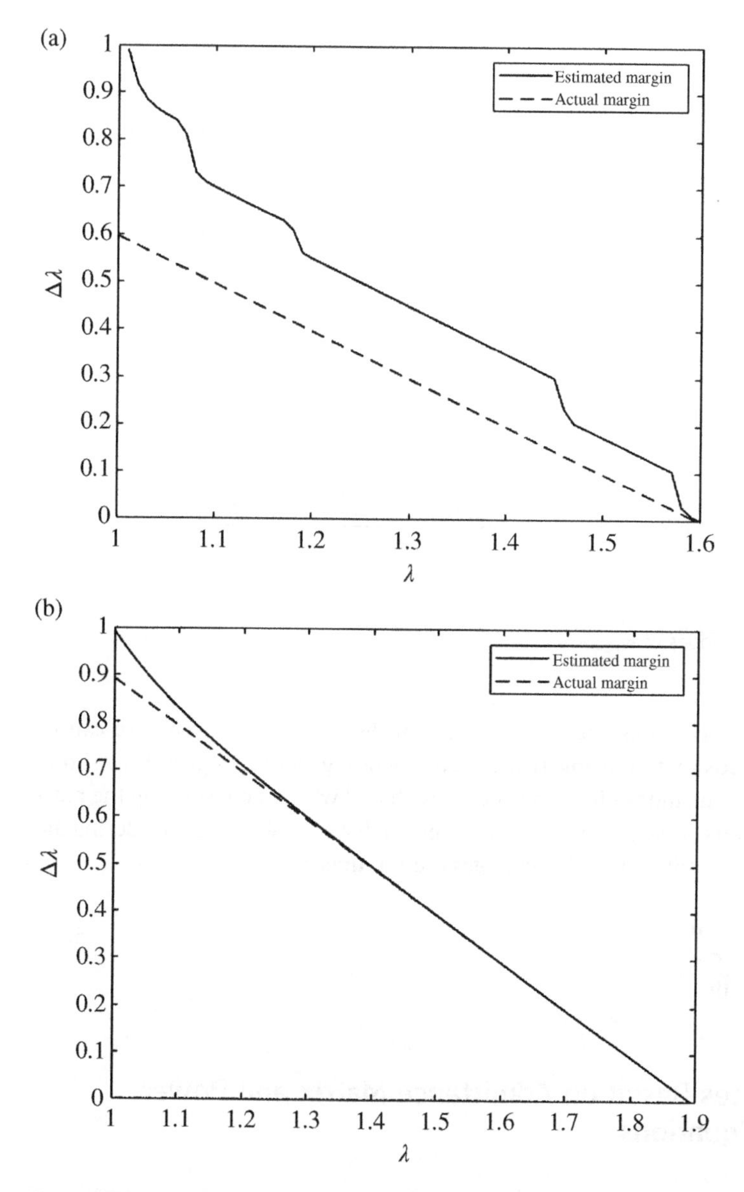

Figure 8.10 The estimation of loadability margin. (a) With considering reactive power limits of generators. (b) Without considering reactive power limits of generators.

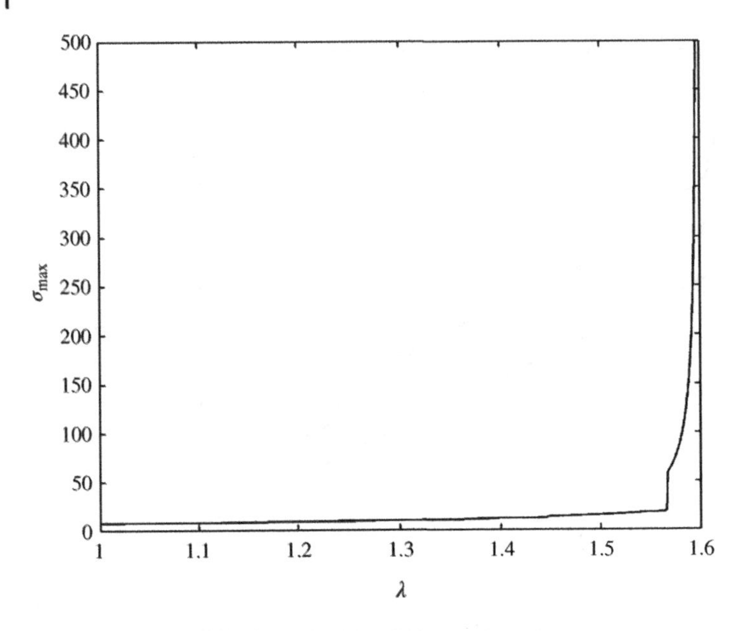

Figure 8.11 Changes of σ_{max}.

derivative has been presented. However, as in Eq. (8.39), this derivative can be obtained approximately using two closely operating points. Figure 8.12 shows the changes of the index i in both states of with and without considering the reactive power limits of the generators. It is observed that even without considering the reactive power limits, the index changes are not linear.

$$i = \frac{1}{i_0} \frac{\sigma_{max}}{\frac{d\sigma_{max}}{dC_{tot}}} \tag{8.41}$$

8.3 Indices Based on Admittance Matrix and Power Balance Equations

These indices are presented based on the necessary conditions for availability of the steady-state equilibrium point. In order to present an index, an equation is first written for one of the system variables. Then, the index is presented based on the conditions required to have a real solution for that variable. These indices can be divided into two categories: line stability indices and bus stability indices. Line indices better identify the source of voltage instability than bus indices. Because,

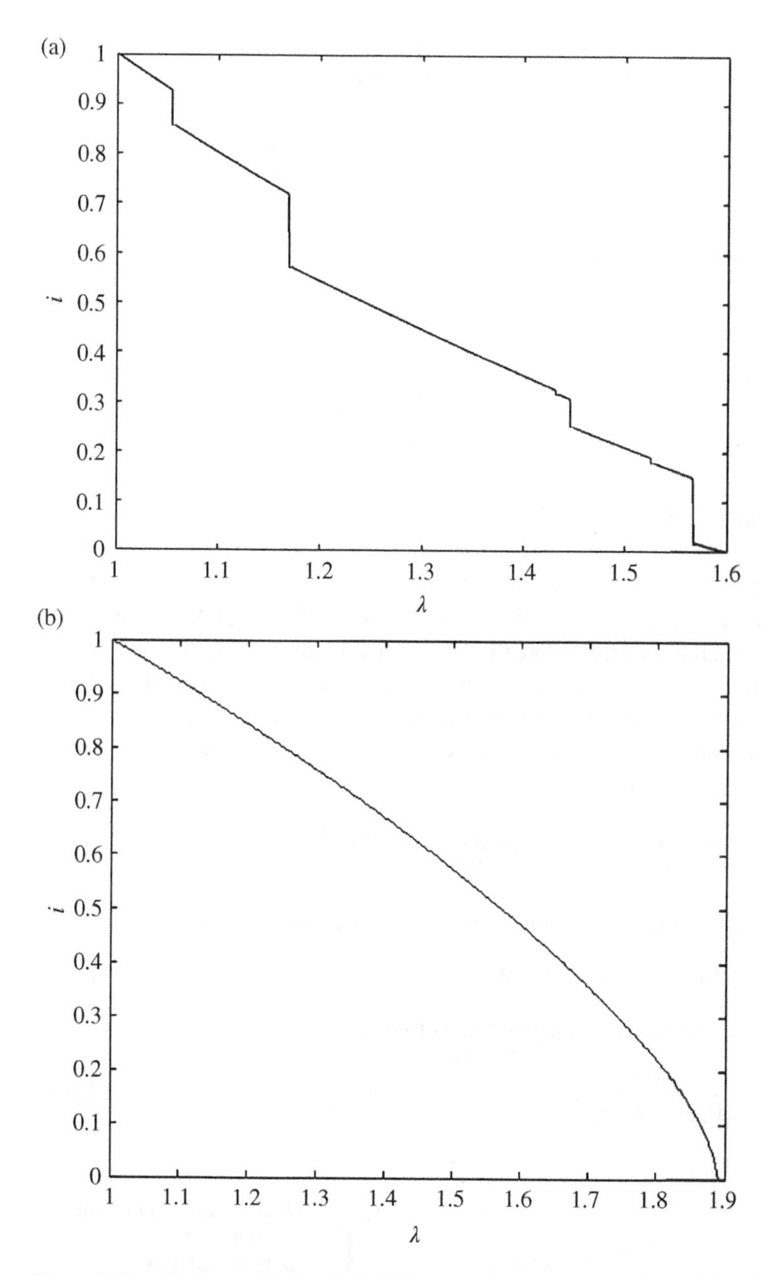

Figure 8.12 Changes of index i. (a) With reactive power limits of generators. (b) Without reactive power limits of generators.

bus indices focus on load buses; however, the source of instability may be a connecting line between two generators that is overloaded [11]. Of course, line indices can provide a pessimistic assessment about the state of system voltage stability, because a line loadability limit usually occurs before reaching the system loadability limit. In the following, line indices are introduced, first.

8.3.1 Line Stability Indices

Line indices provide an estimation of the distance from the current line flowing power to the maximum transferable power of the line. For this purpose, one of the line variables is selected, and the necessary conditions to have a real value for that variable are suggested as the line index. For powers beyond the line loadability limit, no value will be obtained for line variables.

8.3.1.1 Stability Index L_{mn}

In Ref. [11], first having the transmission line model as shown in Figure 8.13, the voltage magnitude at the receiving end of the line is obtained according to Eq. (8.42). In this equation, $\delta = \delta_1 - \delta_2$ and θ is the line impedance angle, i.e. $\theta = \tan^{-1}(X/R)$. It is worthy to note that in Figure 8.13, line capacitors are not considered. In fact, the powers S_s and S_R are the powers passing through the series impedance of the line π model. Of course, line capacitors are inserted in matrix Y_{bus} and their effect is considered in the system operating point, which is obtained by power flow calculation.

$$V_r = \frac{V_s \sin(\theta - \delta) \pm \left\{[V_s \sin(\theta - \delta)]^2 - 4XQ_r\right\}^{0.5}}{2 \sin \theta} \tag{8.42}$$

The necessary condition for the existence of a real solution for V_r is

$$\left\{[V_s \sin(\theta - \delta)]^2 - 4XQ_r\right\} \geq 0 \tag{8.43}$$

Accordingly, the index L_{mn} is defined as follows:

$$L_{mn} = \frac{4XQ_r}{[V_s \sin(\theta - \delta)]^2} \tag{8.44}$$

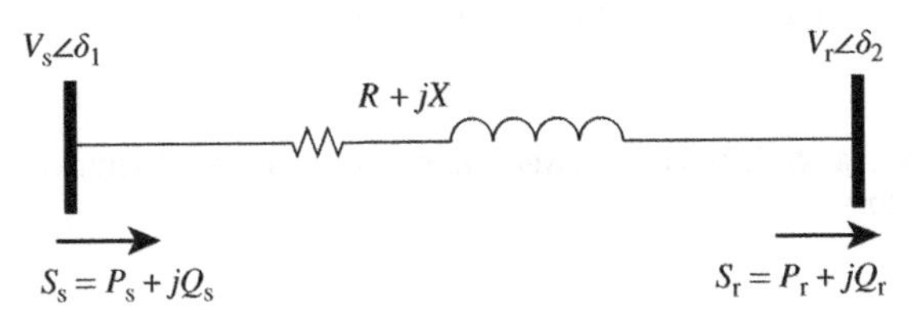

Figure 8.13 Single-line diagram of a transmission line.

Assuming that the voltage at the sending end of the line (V_s) is constant, the line loadability limit occurs at $L_{mn} = 1$.

8.3.1.2 Fast Voltage Stability Index (FVSI)

In [12], another index is proposed which is similar to L_{mn} and is based on the existence of a real solution for the voltage magnitude at the receiving end of the line. This index, called FVSI, is given as Eq. (8.45). In which, Z is the magnitude of the line impedance and the other variables are similar to Eqs. (8.42)–(8.44).

$$\mathrm{FVSI} = \frac{4Z^2 Q_r}{V_s^2 X} \tag{8.45}$$

8.3.1.3 Stability Index LQP

The index presented in [13] is based on the existence of a real value for the reactive power at the sending end of the line. This index is in the form of Eq. (8.46). All index's variables are introduced in Figure 8.13.

$$L_{QP} = 4\left[\frac{X}{V_s{}^2}\right]\left[\frac{XP_s^2}{V_s{}^2} + Q_r\right] \tag{8.46}$$

8.3.1.4 Line Collapse Proximity Index (LCPI)

In Ref. [14], an index called LCPI is introduced that uses the π model of a line as shown in Figure 8.14:

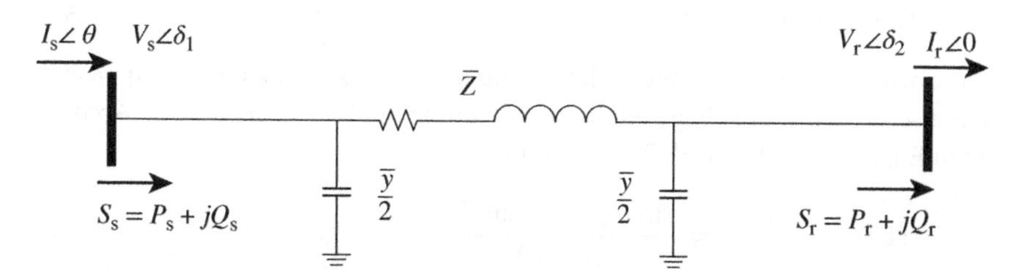

Figure 8.14 The π model of a line.

The relationship between voltages and currents at the line sending and receiving ends in terms of the ABCD constants is as follows [15].

$$\begin{bmatrix} V_s\angle\delta_1 \\ I_s\angle\theta \end{bmatrix} = \begin{bmatrix} \overline{A} & \overline{B} \\ \overline{C} & \overline{D} \end{bmatrix}\begin{bmatrix} V_r\angle\delta_2 \\ I_r\angle 0 \end{bmatrix} \tag{8.47}$$

$$\overline{A} = \overline{D} = \frac{\overline{z}\overline{y}}{2} + 1, \quad \overline{B} = \overline{z}, \quad \overline{C} = \overline{y}\left(1 + \frac{\overline{z}\overline{y}}{4}\right) \tag{8.48}$$

With the notations $\overline{A} = A\angle\alpha$, $\overline{B} = B\angle\beta$, and $\delta = \delta_1 - \delta_2$, the voltage magnitude at the receiving end of the line is [14]:

$$V_r = \frac{-V_s\cos\delta \pm \sqrt{(V_s\cos\delta)^2 - 4A\cos\delta[P_rB\cos\beta + Q_rB\sin\beta]}}{2A\cos\alpha} \tag{8.49}$$

The necessary conditions for existence of a real value for V_r lead to the following index:

$$\text{LCPI} = \frac{4A\cos\alpha(P_rB\cos\beta + Q_rB\sin\beta)}{(V_s\cos\delta)^2} \tag{8.50}$$

8.3.1.5 Integral Transmission Line Transfer Index (ITLTI)

In Ref. [16], the ratio of the current flowing power to the maximum transferable power of the line has been used as line stability index. To this aim, the relationship between the maximum transferable apparent power and the current flowing apparent power is obtained as follows:

$$S_R = S_{R(\max)}\frac{\sin(\theta'_R + \delta')\sin\delta'}{\cos(\theta'_R/2)} \tag{8.51}$$

where S_R is the apparent power at the receiving end of the line, and $S_{R(\max)}$ is the maximum amount of this power. Furthermore,

$$\delta' = \delta - \alpha \tag{8.52}$$

$$\theta'_R = 180 - (\beta - \alpha) + \theta_R \tag{8.53}$$

where δ, α, and β are the same variables and parameters defined in the Eq. (8.49). θ_R is also the angle of voltage relative to current at the line receiving end. According to Eq. (8.51), ITLTI is defined as follows:

$$\text{ITLTI} = \frac{S_R}{S_{R(\max)}} = \frac{\sin(\theta'_R + \delta')\sin\delta'}{\cos(\theta'_R/2)} \tag{8.54}$$

8.3.2 Bus Indices

8.3.2.1 Stability Index *L*

This index is one of the first indices presented to evaluate the voltage stability. The L index was first proposed for a two-bus system based on the existence of a real value for the voltage magnitude at the receiving bus, and then generalized to the n buses system. The value of this index at each load bus in an n-bus system is obtained with Eq. (8.55) [17].

$$L_j = \left| 1 - \frac{\sum_{i \epsilon \alpha_G} \overline{F}_{ji} \overline{V}_i}{\overline{V}_j} \right| \tag{8.55}$$

In this equation, L_j and $\overline{V}_j$ are the index value and the voltage phasor at the bus j. α_G denotes the set of generator buses. Matrix $\boldsymbol{F}$ is a submatrix calculated from the matrix connecting voltages and currents, as follows:

$$\begin{bmatrix} \overline{\boldsymbol{V}}_L \\ \overline{\boldsymbol{I}}_G \end{bmatrix} = \begin{bmatrix} \overline{\boldsymbol{Z}}_{LL} & \overline{\boldsymbol{F}}_{LG} \\ \overline{\boldsymbol{K}}_{GL} & \overline{\boldsymbol{Y}}_{GG} \end{bmatrix} \begin{bmatrix} \overline{\boldsymbol{I}}_L \\ \overline{\boldsymbol{V}}_G \end{bmatrix} \tag{8.56}$$

where $\overline{\boldsymbol{V}}_L$ and $\overline{\boldsymbol{I}}_L$ are the phasor vectors of voltages and currents at load buses, and $\overline{\boldsymbol{V}}_G$ and $\overline{\boldsymbol{I}}_G$ are the phasor vectors of voltages and currents at generator buses, respectively. In fact, the matrix $\boldsymbol{F}$ expresses the relationship between voltages at the load and generator buses. The L index for the whole power system is also defined as Eq. (8.57):

$$L = \max_{j \in \alpha_L} \{L_j\} \tag{8.57}$$

where α_L denotes the set of load buses. For a simple two-bus system in which impedance and voltage at the sending bus remains constant, the L index value reaches 1 at the loadability limit. Figure 8.15 shows the L index changes with

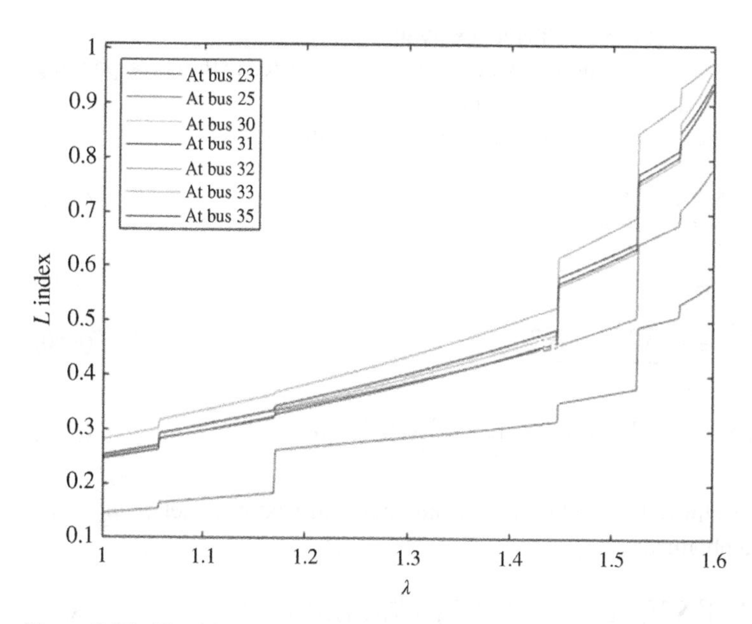

Figure 8.15 The L index changes.

respect to loading factor at a number of weak buses of the IEEE 57-bus test system. It is observed that at none of the buses, the value of L index does not reach the expected value at the loadability limit. Because, the Thevenin equivalent circuit seen from the buses is not completely similar to a two-bus system with constant impedance and voltage at sending end. This, together with the nonlinear behavior of the index, does not allow estimating the loadability margin using the L index. This index can be used to find weak buses of system; weak buses are closer to their loadability limit and have larger L index value.

Another application of the L index is contingency ranking [18]. A contingency is usually a line or generator outage. Severe contingency are the contingencies that push the system closer to the loadability limit. Due to the large number of probable contingencies, on-line system stability assessment for all contingencies is not possible. Therefore, separation of severe contingencies is necessary for more detailed studies, which is called contingency ranking. Contingencies that cause the L index to increase further are selected as severe ones. The L index can also be used for locating reactive power compensator equipment [19]. For this purpose, first the candidate buses are chosen using the L index, which is certainly the buses with higher values of this index. The compensators are then installed at the candidate buses one by one, and the system L index [Eq. (8.57)] is evaluated. The location with more decrease of this index is selected as the compensator installation place.

8.3.2.2 Improved Voltage Stability Index (IVSI)

The index proposed in [20] uses only buses directly connected to the corresponding bus. This is shown in Figure 8.16 for bus i.

Based on Figure 8.16, the Eqs. (8.58)–(8.61) are valid:

$$\overline{I}_i = \overline{V}_i \sum_{j=0}^{n} \overline{y}_{ij} - \sum_{j=1}^{n} \overline{y}_{ij} \overline{V}_j \tag{8.58}$$

$$P_i + jQ_i = \overline{V}_i \overline{I}_i^* \tag{8.59}$$

$$\frac{P_i - jQ_i}{\overline{V}_i^*} = \overline{V}_i \sum_{j=0}^{n} \overline{y}_{ij} - \sum_{j=1}^{n} \overline{y}_{ij} \overline{V}_j \tag{8.60}$$

$$P_i - jQ_i = |\overline{V}_i|^2 \sum_{j=0}^{n} |\overline{y}_{ij}| \angle \theta_{ij} - |\overline{V}_i| \sum_{j=1}^{n} |\overline{y}_{ij}| |\overline{V}_j| \angle(\theta_{ij} - \delta_{ij}) \tag{8.61}$$

By decomposing of Eq. (8.61) to its real and imaginary parts, the following equations will be obtained:

$$P_i = |\overline{V}_i|^2 \sum_{j=0}^{n} |\overline{y}_{ij}| \cos\theta_{ij} - |\overline{V}_i| \sum_{j=1}^{n} |\overline{y}_{ij}| |\overline{V}_j| \cos(\theta_{ij} - \delta_{ij}) \tag{8.62}$$

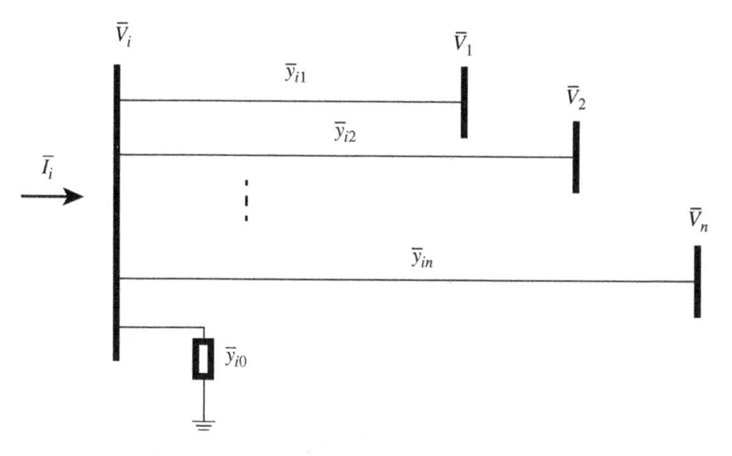

Figure 8.16 Bus *i* and its connected buses.

$$-Q_i = |\overline{V}_i|^2 \sum_{j=0}^{n} |\overline{y}_{ij}| \sin\theta_{ij} - |\overline{V}_i| \sum_{j=1}^{n} |\overline{y}_{ij}||\overline{V}_j| \sin(\theta_{ij} - \delta_{ij}) \tag{8.63}$$

From (8.62) and (8.63),

$$\sum_{j=0}^{n} (G_{ij} - B_{ij}) |\overline{V}_i|^2 - \sum_{j=1}^{n} |\overline{V}_j| \\ [G_{ij}(\cos\delta_{ij} + \sin\delta_{ij}) - B_{ij}(\cos\delta_{ij} - \sin\delta_{ij})] |\overline{V}_i| - (P_i + Q_i) = 0 \tag{8.64}$$

where G_{ij} and B_{ij} are the real and imaginary parts of $\overline{y}_{ij}$.

Given that $|\overline{V}_i|$ is real, Eq. (8.65) must be satisfied.

$$\left[\sum_{j=1}^{n} |\overline{V}_j| [G_{ij}(\cos\delta_{ij} + \sin\delta_{ij}) - B_{ij}(\cos\delta_{ij} - \sin\delta_{ij})]\right]^2 \\ + 4\sum_{j=0}^{n} (G_{ij} - B_{ij})(P_i + Q_i) \geq 0 \tag{8.65}$$

Accordingly, the improved voltage stability index (IVSI) is defined as follow,

$$\mathrm{ISVI}_i = \frac{-4\sum_{j=0}^{n} (G_{ij} - B_{ij})(P_i + Q_i)}{\left[\sum_{j=1}^{n} |\overline{V}_j| [G_{ij}(\cos\delta_{ij} + \sin\delta_{ij}) - B_{ij}(\cos\delta_{ij} - \sin\delta_{ij})]\right]^2} \tag{8.66}$$

According to the Eq. (8.65), this index is expected to reach 1 at the loadability limit for the weakest bus of the power system. Also to evaluate the voltage stability of the whole system, the following index is presented, in which N is the number of whole system buses. In [20], $ISVI_T$ is used to optimal adjustment of reactive power compensators.

$$IVSI_T = \sum_{i=1}^{N} ISVI_i \tag{8.67}$$

Figure 8.17 shows the IVSI changes at a number of weak buses in the IEEE 57-bus test system in both states of with and without generator's reactive power generation limits. It is observed that in both cases, the index value reaches the expected value at the loadability limit. Of course, like other indices, there are nonlinear changes and sudden jumps when faced with a loadability limit.

8.4 Indices Based on Load Buses Voltage and Generators Reactive Power

The purpose of these indices is not to estimate the loadability limit, other purposes such as contingency ranking, determination of weak buses, and so on are persuaded.

8.4.1 Reactive Power Performance Index (PIV)

This index is presented in order to rank the contingencies and is based on the amount of voltage change at load buses after a contingency occurrence. The criterion of ranking the contingencies is their effect on reducing the power system loadability limit. The index is established on the assumption that more severe contingencies lead to more voltage drop at load buses. The index is in the form of Eq. (8.68), in which V_i is the post-contingency voltage magnitude of the bus i. $V_{i\,min}$ and $V_{i\,max}$ are the minimum and maximum allowable voltages at the bus i, respectively. $V_{i\,nom}$ is the average value of $V_{i\,min}$ and $V_{i\,max}$, and N_{pq} is the number of load buses.

$$PIV = \sum_{i=1}^{N_{pq}} \left[\frac{2(V_i - V_{inom})}{V_{imax} - V_{imin}} \right]^2 \tag{8.68}$$

Table 8.1 lists the PIV and L values for 10 of the most severe line-outage contingencies in the IEEE 57-bus test system. The calculation of indices is done at the base load. In this table, in order to evaluate the performance of indices, the maximum loading factor (loadability limit) for each contingency is presented too. The numbers in parentheses in the last three columns are related to the rank of related contingency based on PIV, λ_{max}, and L index. It can be seen that L index has better performance in contingency ranking.

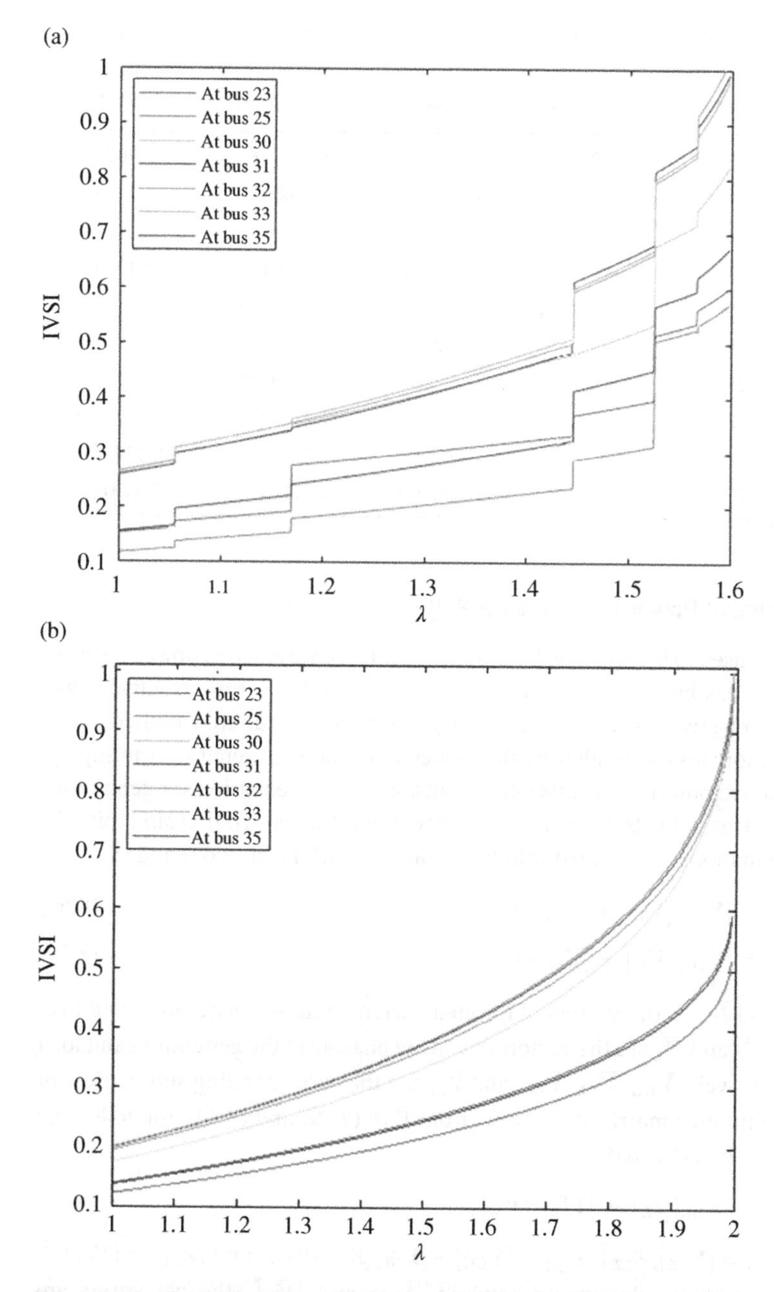

Figure 8.17 The IVSI changes. (a) With considering reactive power generation limits of generators. (b) Without considering reactive power generation limits of generators.

Table 8.1 Contingency ranking using reactive power performance index (PIV) and *L* index.

Line no.	From bus	To bus	PIV (rank)	λ_{max} (rank)	*L* index (rank)
1	37	38	21.3846 (1)	1.2870 (6)	0.6707 (5)
2	7	29	18.2189 (2)	1.2100 (4)	0.7552 (4)
3	34	35	13.7763 (3)	1.1700 (3)	0.8211 (2)
4	25	30	12.7867 (4)	1.1310 (1)	96.0106 (1)
5	36	37	10.3256 (5)	1.3760 (7)	0.5588 (8)
6	34	32	9.9628 (6)	1.1700 (2)	0.8206 (3)
7	28	29	5.4630 (7)	1.4610 (10)	0.4225 (11)
8	8	9	5.2316 (8)	1.2200 (5)	0.4330 (10)
9	9	55	4.3712 (9)	1.5150 (18)	0.3091 (17)
10	27	28	3.7462 (10)	1.5110 (15)	0.3800 (12)

8.4.2 Reactive Power Loss Index (RPLI)

This index is used to identify weak buses in order to locate reactive power compensators [21]. Weak buses here are buses at which to supply a certain reactive power value, more reactive power generation by generators is required. In other words, they cause more losses. To identify the participation value of generators in supplying the reactive power consumption of loads, in [21], an equation for generators' current in terms of loads' current and voltage was extracted. To this aim, initially, the node equations are divided into two matrix equations of (8.69) and (8.70).

$$[\overline{I}_G] = [Y_{GG}][\overline{V}_G] + [Y_{GL}][\overline{V}_G] \tag{8.69}$$

$$[\overline{I}_L] = [Y_{LG}][\overline{V}_G] + [Y_{LL}][\overline{V}_L] \tag{8.70}$$

where $\overline{I}_G$ and $\overline{I}_L$ are the vectors of injected currents into the generation and load buses, and $\overline{V}_G$ and $\overline{V}_L$ are the vectors of voltage phasors at the generation and load buses, respectively. $\overline{Y}_{GG}$, $\overline{Y}_{GL}$, $\overline{Y}_{LG}$, and $\overline{Y}_{LL}$ are the corresponding submatrices of the bus admittance matrix $\overline{Y}_{bus}$. Now, from Eqs. (8.69) and (8.70), the following equation can be extracted:

$$[\overline{I}_G] = [K_{GL}][\overline{I}_L] + [Y^e_{GG}][\overline{V}_G] \tag{8.71}$$

where $[K_{GL}] = [Y_{GL}][Z_{LL}]$, $Y^e_{GG} = [Y_{GG}] - [Y_{GL}][Z_{LL}][Y_{LG}]$, and $[Z_{LL}] = [Y_{LL}]^{-1}$.

Now, to create a relationship between $[\overline{V}_G]$ and $[\overline{V}_L]$, the generators are replaced with their equivalent admittance. For instance, this admittance for the generator connected to bus j can be calculated as

$$\overline{Y}_{Gj} = \frac{1}{\overline{V}_{Gj}}\left(\frac{-\overline{S}_{Gj}}{\overline{V}_{Gj}}\right)^{*} \tag{8.72}$$

where $\overline{S}_{Gj}$ is the generated complex power of the generator connected to bus j and $\overline{V}_{Gj}$ is the voltage of this generator. Having this admittance,

$$[\overline{\boldsymbol{I}}_G] = [\boldsymbol{Y}_G][\overline{\boldsymbol{V}}_G] \tag{8.73}$$

where $\boldsymbol{Y}_G$ is a diagonal matrix in which, except its diagonal elements which consist of the equivalent admittances of the generators, other elements are zero. By substituting $[\overline{\boldsymbol{I}}_G]$ from Eq. (8.73) to Eq. (8.69),

$$[\overline{\boldsymbol{V}}_G] = -[\boldsymbol{Y}'_{GG}]^{-1}[\boldsymbol{Y}_{GL}][\overline{\boldsymbol{V}}_L] \tag{8.74}$$

where $[\boldsymbol{Y}'_{GG}]$ is a modified submatrix of $[\boldsymbol{Y}_G]$. Having the following definition,

$$[\boldsymbol{Y}^{B}_{GL}] \triangleq -[\boldsymbol{Y}'_{GG}]^{-1}[\boldsymbol{Y}_{GL}] \tag{8.75}$$

Eq. (8.74) can be written as

$$[\overline{\boldsymbol{V}}_G] = [\boldsymbol{Y}^{B}_{GL}][\overline{\boldsymbol{V}}_L] \tag{8.76}$$

By substituting $\overline{\boldsymbol{V}}_G$ from the above equation in Eq. (8.71),

$$[\overline{\boldsymbol{I}}_G] = [\boldsymbol{K}_{GL}][\overline{\boldsymbol{I}}_L] + [\boldsymbol{Y}^{C}_{GL}][\overline{\boldsymbol{V}}_L] \tag{8.77}$$

where $[\boldsymbol{Y}^{C}_{GL}] = [\boldsymbol{Y}^{e}_{GL}][\boldsymbol{Y}^{B}_{GL}]$. To find the participation value of generators in supplying reactive power of loads, the vectors $[\overline{\boldsymbol{I}}_L]$ and $[\overline{\boldsymbol{V}}_L]$ are first replaced by the diagonal matrices. By doing so, $[\overline{\boldsymbol{I}}_G]$ converts to a matrix whose number of rows and columns, respectively, is equal to the number of generators and loads, that is denoted by $[\boldsymbol{I}_G]_{G\times L}$. Now by converting $[\overline{\boldsymbol{V}}_G]$ to a diagonal matrix and multiplying it by conjugate of $[\overline{\boldsymbol{I}}_G]$,

$$\begin{aligned}[\boldsymbol{V}_G]_{G\times G}[\boldsymbol{I}^{*}_{G}]_{G\times L} &= [\boldsymbol{S}_{gen-contrib}] \\ &= [\boldsymbol{V}_G]_{G\times G}[\boldsymbol{K}^{*}_{GL}]_{G\times L}[\boldsymbol{I}^{*}_{L}]_{L\times L} + [\boldsymbol{V}_G]_{G\times G}\left[\boldsymbol{Y}^{C\,*}_{GL}\right]_{G\times L}[\boldsymbol{V}^{*}_{L}]_{L\times L}\end{aligned} \tag{8.78}$$

Now, the participation value of the reactive power of generators in load supplying is as follows:

$$[\boldsymbol{Q}_{gen-contrib}]_{G\times L} = \mathrm{Im}\left([\boldsymbol{S}_{gen-contrib}]_{G\times L}\right) \tag{8.79}$$

Having the participation value of the generators, the reactive power loss for load bus i will be obtained from Eq. (8.80).

$$Q_{\text{loss}(i)} = \sum_{j=1}^{\text{NG}} Q_{\text{gen-contrib}(ji)} - Q_{Li} \tag{8.80}$$

where Q_{Li} is the consumed reactive power at bus i and NG is the number of generators. Obtaining reactive power loss of bus i, the reactive power loss index (RPLI) is defined as

$$\text{RPLI}_i = \text{Qloss}_{i,0}^n + \sum_{k=1}^{\text{Nc}} (\text{Qloss}_{i,k}^n \times \text{NCon}_k) \tag{8.81}$$

where $\text{Qloss}_{i,0}^n = (\text{Qloss}_{i,0} / \max(\text{Qloss}_0))$ that $\text{Qloss}_{i,\,0}$ is the reactive power loss of bus i in no-contingency (normal) state. $\max(\text{Qloss}_0)$ is the maximum reactive bus loss of buses in no-contingency state. Nc is the number of selected contingencies. $\text{Qloss}_{i,k}^n = (\text{Qloss}_{i,k} / \max(\text{Qloss}_k))$ that $\text{Qloss}_{i,\,k}$ is the reactive power loss of bus i when the kth contingency occurs. NCon_k is a coefficient that indicates the relative severity of contingency that can be determined using any of the contingency ranking methods. Buses with more RPLI value are known as weak buses, because reactive power consumption at these buses causes more reactive power loss. These buses are mainly buses that are further away from reactive power generation sources and are proper points to install reactive power compensating equipment. Figures 8.18 and 8.19 show the performance of Qloss and the L index in determining weak buses in the system. It can be seen that both indices introduce bus 31 as the weakest bus.

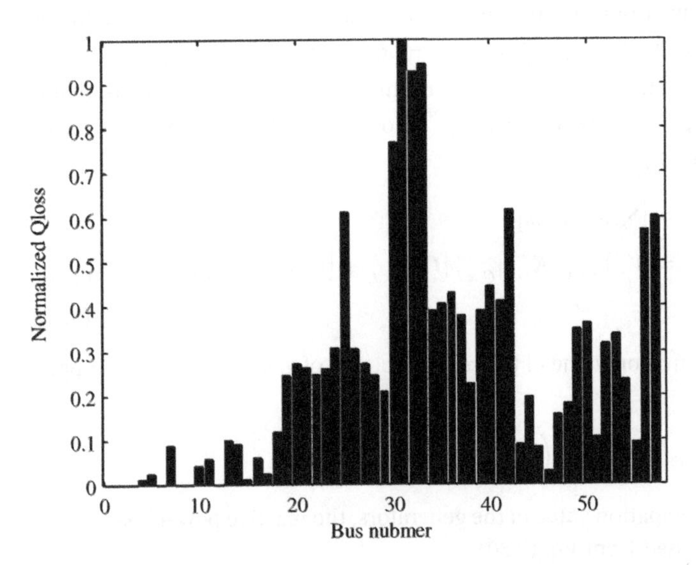

Figure 8.18 Normalized reactive power loss (Qloss) for load buses.

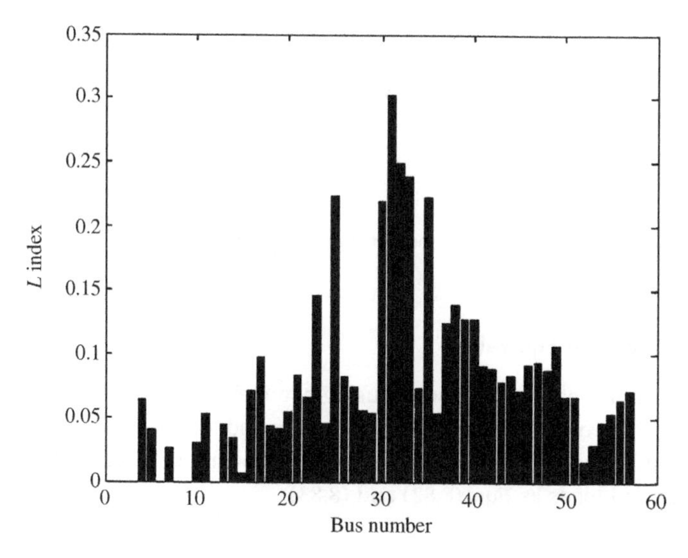

Figure 8.19 *L* index at load buses.

8.5 Indices Defined in the Distribution System

In radial distribution systems, since each consumer is fed from only one path, if each of branches reaches the loadability limit, it is no longer possible to transfer power to the downstream buses. Therefore, unlike the transmission system, the line loadability limit also indicates the bus loadability limit. Hence, voltage stability indices in distribution systems are presented based on diagnosis of proximity to the lines loadability limits. On the other hand, it was observed in the previous sections that usually at the loadability limit the line index value differs from the expected value, i.e. the index does not approach a certain value at the line loadability limit. Owing to this problem, it is not possible to predict proximity to the line loadability limit by index. The reason of that the indices does not match the expected value is that the line index is obtained assuming the voltage at the sending end of the line is constant. This assumption is only true when there is voltage control equipment at the sending end of the line, such as generators or transformers with under-load tap changer (ULTC). It is clear that in most cases, such conditions do not exist. However, in radial distribution systems, it is expected that the distribution system can be replaced with an equivalent two-bus system as shown in Figure 8.20 at which, the sending bus (bus 1) is the bus connected to the main transformer feeding the distribution network. As long as the output voltage of the main transformer is controlled by ULTC, the sending bus voltage can be considered constant. By doing this, the index value is expected to be closer to the predicted value at the loadability limit.

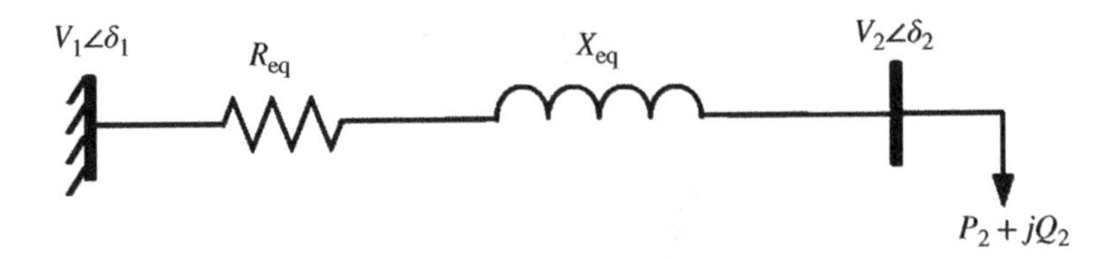

Figure 8.20 The equivalent two-bus system for distribution system.

8.5.1 Distribution System Equivalent

Distribution system equivalent is mainly done using two methods. In both methods, the values of the network parameters ($R_{eq} + jX_{eq}$) are updated at each operating point. In the first method [22–25], the parameters are obtained based on the whole system losses and loads as Eqs. (8.82) and (8.83).

$$R_{eq} = \frac{P_{loss}}{P_s^2 + Q_s^2} \tag{8.82}$$

$$X_{eq} = \frac{Q_{loss}}{P_s^2 + Q_s^2} \tag{8.83}$$

In these equations, P_{loss} and Q_{loss} are, respectively, the total active and reactive losses of the distribution system. P_s and Q_s are also active and reactive input powers of the system through the main bus. Also in Figure 8.20, the active and reactive loads connected to the ending bus (P_2,Q_2) are equal to the total active and reactive loads consumed by the distribution system. In this method, the system equivalent is not done from the viewpoint of different buses. Therefore, this equivalent method does not seem to be a suitable method for detecting weak buses. In Refs. [24, 25], in order to evaluate the voltage stability at each bus, only the load of that bus is increased and the load of the other bus is kept constant. In fact, by doing this, the effect of separate load increment at each bus on losses and consequently on the equivalent network parameters is investigated. Due to the fact that increasing the load on weak buses increases the losses more, it is expected that this method be able to identify weak buses. It is clear that with the conventional assumption of simultaneous increase of load at different buses, it is not possible to detect weak buses using the above mentioned equivalent method. In another method of equivalent, the equivalent system parameters are calculated from the viewpoint of different buses [26]. Figure 8.21 is used to explain this method. Figure 8.21a shows an n-bus distribution system. Figure 8.21b also shows the equivalent two-bus system seen from bus k. The resistance and reactance of the equivalent system are computed as Eq. (8.84).

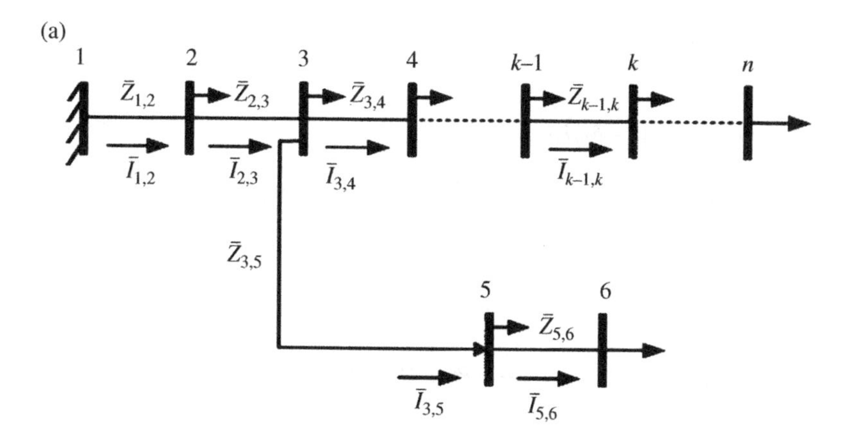

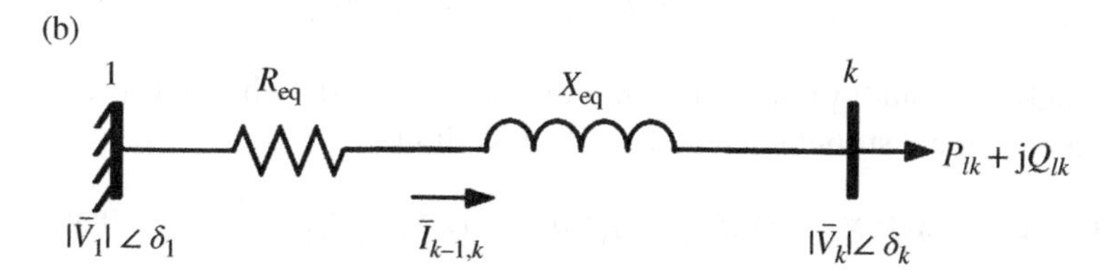

Figure 8.21 Equivalent of distribution system. (a) *n*-Bus system, (b) equivalent system [26].

$$R_{eq} + jX_{eq} = \sum_{i=2}^{k} \frac{(R_{i-1,i} + jX_{i-1,i})\bar{I}_{i-1,i}}{\bar{I}_{k-1,k}} \tag{8.84}$$

In this equation, the impedances and currents of the path branches between bus 1 (main bus) and bus k are considered. For example, for bus 6, the currents $I_{1,\,2}$, $I_{2,\,3}$, $I_{3,\,5}$, and $I_{5,\,6}$ are considered. The consuming load connected to the bus k in the equivalent system is also equal to the power consumption of the bus k plus the transmitted power from this bus to the downstream network. This power is calculated as Eq. (8.85), wherein φ_k is the angle between $\bar{V}_k$ and $\bar{I}_{k-1,k}$.

$$P_{L_k} + jQ_{L_k} = |\bar{V}_k||\bar{I}_{k-1,k}|(\cos(\varphi_k) + j\sin(\varphi_k)) \tag{8.85}$$

It is observed that in this equivalent method, from the viewpoint of each bus, a separate equivalent system will be obtained.

8.5.2 Indices

The indices presented in the distribution system are based on the conditions of existence of a real value for one of the two-bus equivalent system variables. For example, in [27], the Eq. (8.88) is first obtained from Eqs. (8.86) and (8.87).

$$\bar{I}_{12} = \frac{|\bar{V}_1|\angle\delta_1 - |\bar{V}_2|\angle\delta_2}{R_{eq} + jX_{eq}} \tag{8.86}$$

$$P_2 - jQ_2 = (|\bar{V}_2|\angle-\delta_2)\bar{I}_{12} \tag{8.87}$$

$$|\bar{V}_2|^4 - \left\{|\bar{V}_1|^2 - 2P_2R_{eq} - 2Q_2X_{eq}\right\}|\bar{V}_2|^2 + (P_2^2 + Q_2^2)\left(R_{eq}^2 + X_{eq}^2\right) = 0 \tag{8.88}$$

Now, in order to have a real value for $|\bar{V}_2|$ must,

$$\left(|\bar{V}_1|^2 - 2P_2R_{eq} - 2Q_2X_{eq}\right)^2 - 4(P_2^2 + Q_2^2)\left(R_{eq}^2 + X_{eq}^2\right) \geq 0 \tag{8.89}$$

After simplification,

$$|\bar{V}_1|^4 - 4(P_2X_{eq} - Q_2R_{eq})^2 - 4(P_2R_{eq} + Q_2X_{eq})|\bar{V}_1|^2 \geq 0 \tag{8.90}$$

Accordingly, the stability index (SI) is introduced as Eq. (8.91), which is expected to reach zero at the distribution system loadability limit.

$$\text{SI} = |\bar{V}_1|^4 - 4(P_2X_{eq} - Q_2R_{eq})^2 - 4(P_2R_{eq} + Q_2X_{eq})|\bar{V}_1|^2 \tag{8.91}$$

In a similar way, in Refs. [24, 25] based on having a real value for $|\bar{V}_2|$, indices L_P and L_V in the forms of Eqs. (8.92) and (8.93) are proposed.

$$L_P = \frac{4R_{eq}P_2}{\left[|\bar{V}_1|\cos(\theta - \delta_2 + \delta_1)\right]^2} \tag{8.92}$$

$$L_V = \frac{\left[4\sqrt{P_2^2 + Q_2^2}\cos(\varphi - \theta)\right]\left(\sqrt{R_{eq}^2 + X_{eq}^2}\right)}{\left[|\bar{V}_1|\cos(\delta_1 - \delta_2)\right]^2} \tag{8.93}$$

In these equations, θ is the impedance angle of the line of the equivalent system and φ is the power angle of bus 2 (ending bus) in the equivalent system.

$$\theta = tg^{-1}\frac{X_{eq}}{R_{eq}} \tag{8.94}$$

$$\varphi = tg^{-1}\frac{Q_2}{P_2} \tag{8.95}$$

In the Refs. [22, 23], based on the existence of a real value for the input active power of bus 1 (P_s), an index called L index is proposed; which here to avoid

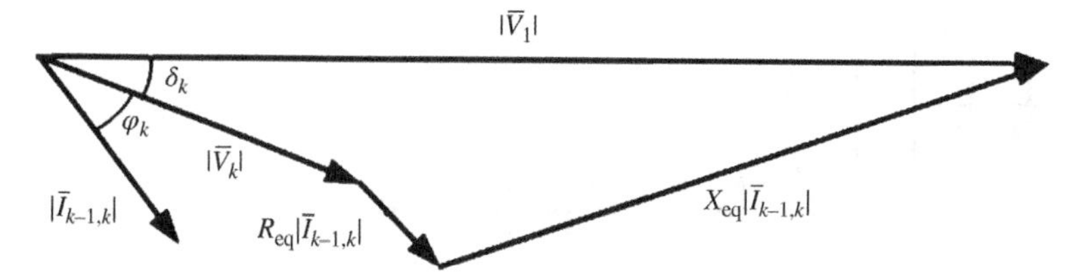

Figure 8.22 The equivalent system phasor diagram.

nominal similarity with the L index proposed in the transmission system [Eq. (8.54)], it is named L_{D}. This index that is obtained with the assumption of value 1 pu for $|\overline{V}_1|$ is as Eq. (8.96).

$$L_{\mathrm{D}} = 4\left[\left(X_{\mathrm{eq}}P_2 - R_{\mathrm{eq}}Q_2\right)^2 + X_{\mathrm{eq}}Q_2 + R_{\mathrm{eq}}P_2\right] \tag{8.96}$$

The L_{D}, L_{P}, and L_{V} indices are expected to reach 1 at the loadability limit. In the Ref. [26] using the equivalent system phasor diagram, the terms in SI [Eq. (8.21)] are obtained in terms of the voltages magnitudes and angles. The equivalent system phasor diagram is as Figure 8.22. In this figure, it is assumed that the angle δ_1 is zero.

Using the phasor diagram,

$$|\overline{V}_1|\cos\delta_k - |\overline{V}_k| = R_{\mathrm{eq}}|\overline{I}_{k-1,k}|\cos\varphi_k + X_{\mathrm{eq}}|\overline{I}_{k-1,k}|\sin\varphi_k \tag{8.97}$$

$$|\overline{V}_1|\sin\delta_k = X_{\mathrm{eq}}|\overline{I}_{k-1,k}|\cos\varphi_k - R_{\mathrm{eq}}|\overline{I}_{k-1,k}|\sin\varphi_k \tag{8.98}$$

Multiplying both sides of the equation by $|\overline{V}_k|$,

$$R_{\mathrm{eq}}P_2 + X_{\mathrm{eq}}Q_2 = |\overline{V}_k|\Delta V \tag{8.99}$$

$$X_{\mathrm{eq}}P_2 - R_{\mathrm{eq}}Q_2 = |\overline{V}_1||\overline{V}_k|\sin\delta_k \tag{8.100}$$

where

$$\Delta V = |\overline{V}_1|\cos\delta_k - |\overline{V}_k| \tag{8.101}$$

By substituting Eqs. (8.99) and (8.100) in SI, Eq. (8.102) is obtained.

$$\mathrm{E\_SI}_k = |\overline{V}_1|^4 - 4|\overline{V}_1|^2\left(|\overline{V}_k|\Delta V\right) - 4\left(|\overline{V}_1||\overline{V}_k|\sin\delta_k\right)^2 \tag{8.102}$$

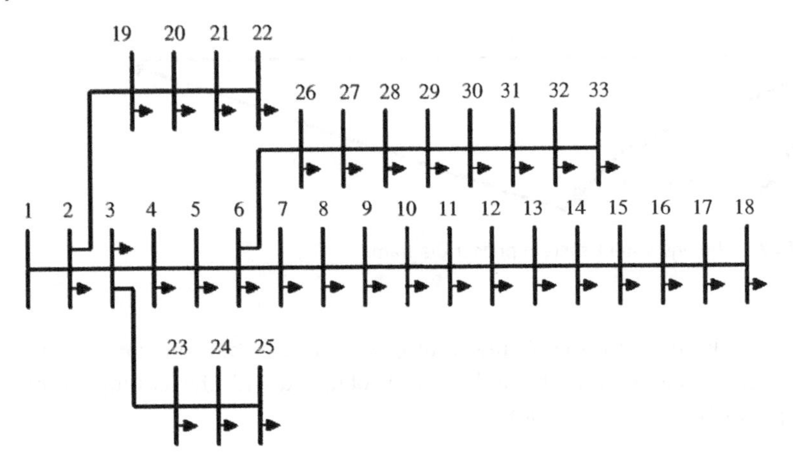

Figure 8.23 IEEE 33-bus test system.

It is observed that Eq. (8.102) is independent of the equivalent system parameters and is obtained only by measuring the voltages magnitudes and angles. Equations (8.99) and (8.100) are also applicable in the index L_D. Accordingly, the index E_L$_D$ is obtained as Eq. (8.103).

$$\mathrm{E\_L_D} = 4\left[\left(\left|\overline{V}_1\right|\left|\overline{V}_k\right|\sin\delta_k\right)^2 + \left|\overline{V}_k\right|\Delta V\right] \tag{8.103}$$

8.5.3 Simulations

To compare the distribution network indices, the IEEE 33-bus test system of Figure 8.23 is used. Equation (8.84) is used to calculate equivalent parameters. As mentioned, in this method, from each bus viewpoint, a separate equivalent network is obtained. Figures 8.24–8.27 show the changes of the indices. The indices are calculated for four end buses of the branches, which are the weakest buses of each branch. The loading factor is increased until diverging the power flow calculations. It can be seen that the changes of the indices are mostly nonlinear. At the loadability limits, the L_D and L_V indices reached the expected value, but the SI and L_P index are slightly different from the expected value. All indices have identified bus 18 as the weakest bus of the system. Table 8.2 confirms this, too. In this table, the loading factor is increased one-by-one at different buses until divergence of power flow calculations. It is observed that the lowest value of λ_{max} is related to bus 18 and then to bus 33, which is completely consistent with the results obtained from the indices.

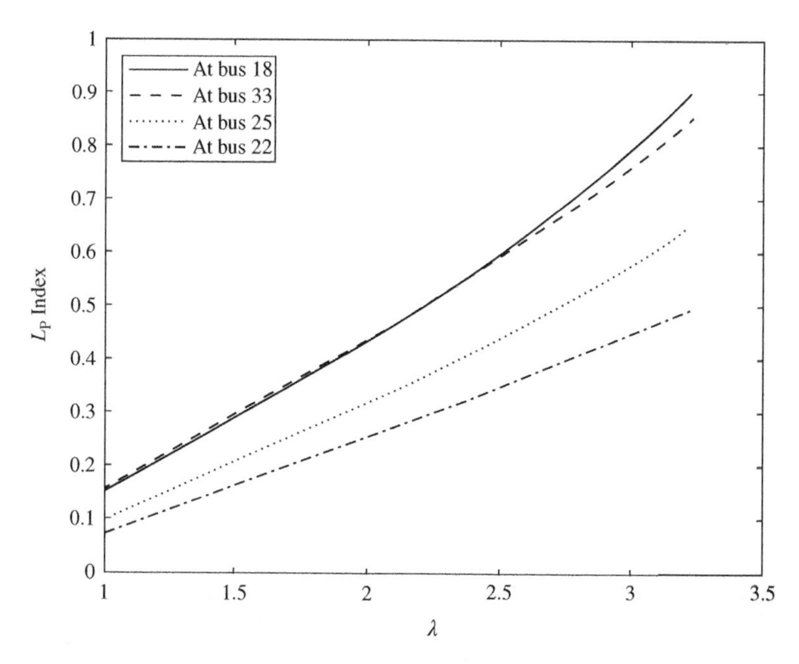

Figure 8.24 The L_P index changes.

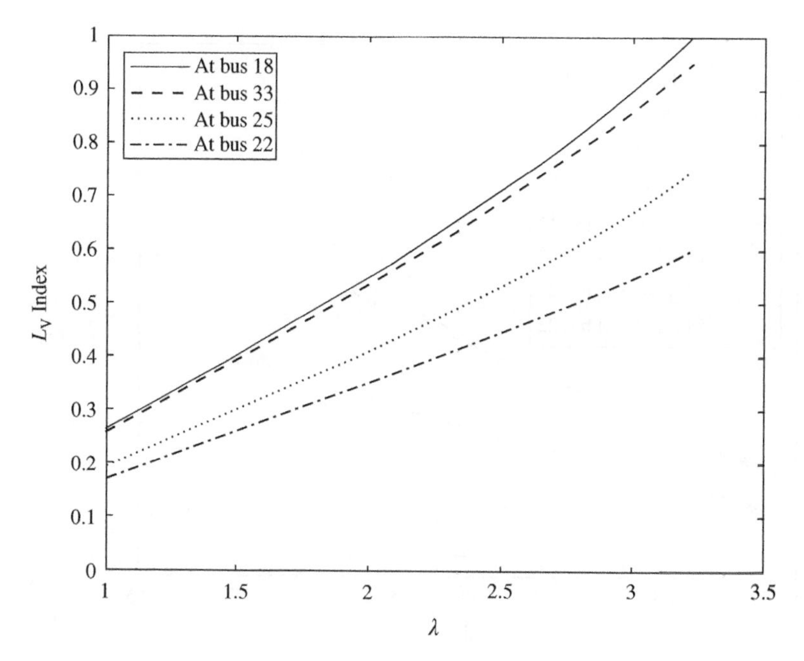

Figure 8.25 The L_V index changes.

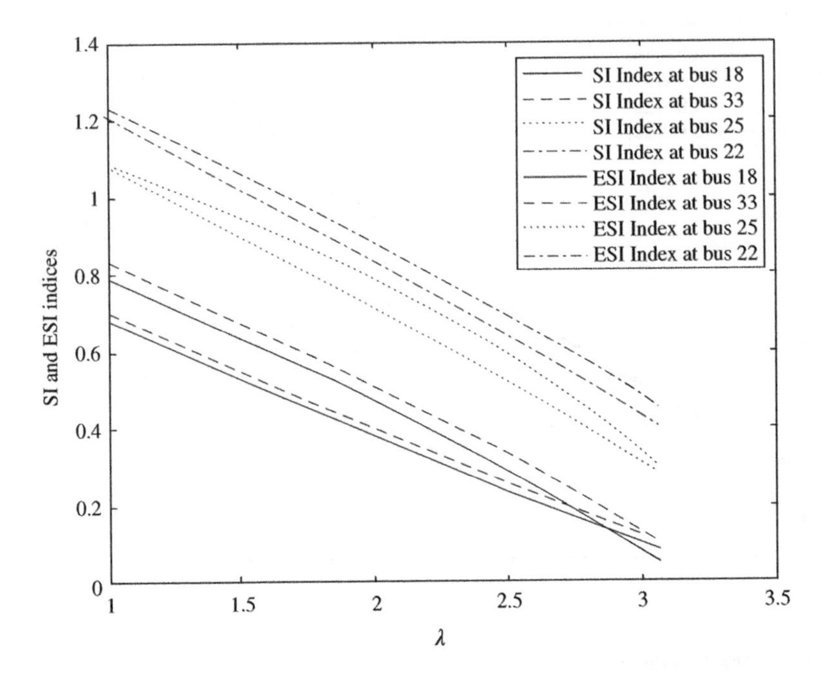

Figure 8.26 The SI and E_SI indices changes.

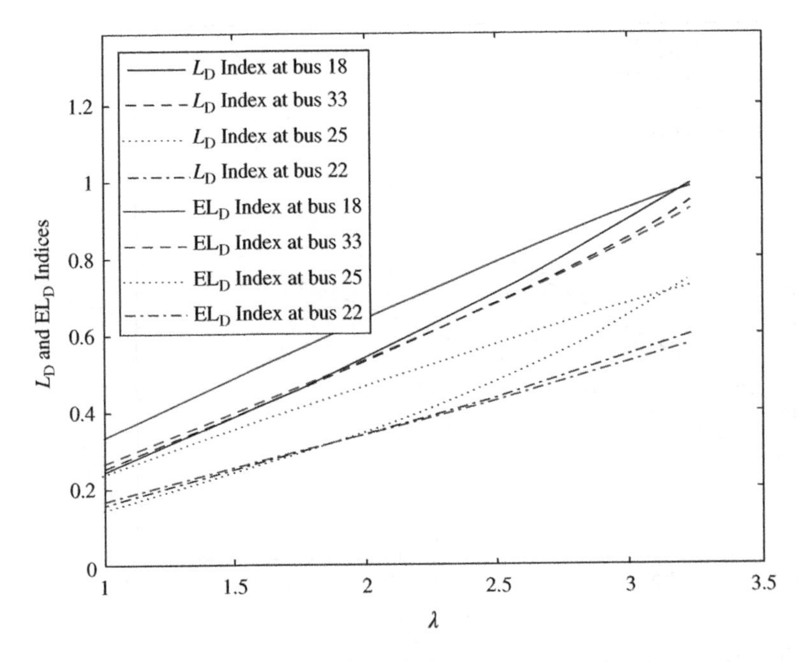

Figure 8.27 The L_D and E_L_D indices changes.

Table 8.2 Loadability limit at different buses.

	Bus 18	Bus 22	Bus 25	Bus 33
λ_{max}	23.4	100.6	50.9	23.9

8.6 Summary

In this chapter, model-based indices were presented. Some of these indices, such as indices based on the Jacobin matrix, reach the expected value at the loadability limit, but like other indices, due to nonlinear behavior and jumps in encountering the reactive power limit of the generators, they are not able to accurately estimate the loadability limit. Contingency ranking and detection of weak buses and lines are model-based index applications. In this chapter, suitable indices for evaluating voltage stability in the distribution network were also introduced.

References

1 Kundur, P., Neal, J.B., and Mark, G.L. (1994). *Power System Stability and Control*, vol. 7. New York: McGraw-Hill.

2 Löf, P.A., Andersson, G., and Hill, D.J. (1993). Voltage stability indices for stressed power systems. *IEEE Trans. Power Syst.* 8 (1): 326–335.

3 Löf, P.A., Smed, T., Andersson, G., and Hill, D.J. (1992). Fast calculation of a voltage stability index. *IEEE Trans. Power Syst.* 7 (1): 54–64.

4 Van Cutsem, T. and Vournas, C. (1988). *Voltage Stability of Electric Power Systems*. Norwell, MA: Kluwer.

5 Conejo, A.J., Castillo, E., Minguez, R., and Garcia-Bertrand, R. (2006). *Decomposition Techniques in Mathematical Programming: Engineering and Science Applications*. Springer Science & Business Media.

6 Gao, B., Morison, G.K., and Kundur, P. (1992). Voltage stability evaluation using modal analysis. *IEEE Trans. Power Syst.* 7 (4): 1529–1542.

7 Chiang, H.D. and Jean-Jumeau, R. (1995). Toward a practical performance index for predicting voltage collapse in electric power systems. *IEEE Trans. Power Syst.* 10 (2): 584–592.

8 Canizares, C.A., De Souza, A.C.Z., and Quintana, V.H. (1996). Comparison of performance indices for detection of proximity to voltage collapse. *IEEE Trans. Power Syst.* 11 (3): 1441–1450.

9 Berizzi, A. (1996). System-area operating margin assessment and security enhancement against voltage collapse. *IEEE Trans. Power Syst.* 11 (3): 1451–1462.

10 Berizzi, A. and Finazzi, P. (1998). First and second order methods for voltage collapse assessment and security enhancement. *IEEE Trans. Power Syst.* 13 (2): 543–551.

11 Moghavvemi, M. and Omar, F.M. (1998). Technique for contingency monitoring and voltage collapse prediction. *IEE Proc. Gener. Transm. Distrib.* 145 (6): 634–640.

12 Musiri, I., Rahman, A., and Khawa, T. (2002). On-line voltage stability based contingency ranking using fast voltage stability index (FVSI). *Proc. IEEE Power Eng. Soc. Transm. Distrib. Conf.* 2: 1118–1123.

13 Mohamed, A., Jasmon, G.B., and Yusoff, S. (1989). A static voltage collapse indicator using line stability factors. *J. Ind. Technol.* 7 (1): 73–85.

14 Tiwari, R., Niazi, K.R., and Gupta, V. (2012). Line collapse proximity index for prediction of voltage collapse in power systems. *Int. J. Electr. Power Energy Syst.* 41 (1): 105–111.

15 Stevenson, W.D. and Grainger, J.J. (1994). *Power System Analysis*. McGraw-Hill Education.

16 Chuang, S.J., Hong, C.M., and Chen, C.H. (2016). Improvement of integrated transmission line transfer index for power system voltage stability. *Int. J. Electr. Power Energy Syst.* 78: 830–836.

17 Kessel, P. and Glavitsch, H. (1986). Estimating the voltage stability of a power system. *IEEE Trans. Power Deliv.* 1 (3): 346–354.

18 Visakha, K., Thukaram, D., and Jenkins, L. (2004). Application of UPFC for system security improvement under normal and network contingencies. *Electr. Power Syst. Res.* 70 (1): 46–55.

19 Thukaram, D. and Lomi, A. (2000). Selection of static VAR compensator location and size for system voltage stability improvement. *Electr. Power Syst. Res.* 54 (2): 139–150.

20 Yang, C.F., Lai, G.G., Lee, C.H. et al. (2012). Optimal setting of reactive compensation devices with an improved voltage stability index for voltage stability enhancement. *Int. J. Electr. Power Energy Syst.* 37 (1): 50–57.

21 Moger, T. and Dhadbanjan, T. (2015). A novel index for identification of weak nodes for reactive compensation to improve voltage stability. *IET Gener. Transm. Distrib.* 9 (14): 1826–1834.

22 Jasmon, G.B. and Lee, L.H.C.C. (1991). Stability of loadflow techniques for distribution system voltage stability analysis. *IEE Proc. C Gener. Transm. Distrib.* 138 (6): 479–484.

23 Jasmon, G.B. and Lee, L.H.C.C. (1993). New contingency ranking technique incorporating a voltage stability criterion. *IEE Proc. C Gener. Transm. Distrib.* 140 (2): 87–90.

24 Moghavvemi, M. and Faruque, M.O. (2001). Technique for assessment of voltage stability in ill-conditioned radial distribution network. *IEEE Power Eng. Rev.* 21 (1): 58–60.

25 Hamada, M.M., Wahab, M.A.A., and Hemdan, N.G.A. (2010). Simple and efficient method for steady-state voltage stability assessment of radial distribution systems. *Electr. Power Syst. Res.* 80 (2): 152–160.

26 Beigvand, S.D., Abdi, H., and Singh, S.N. (2017). Voltage stability analysis in radial smart distribution grids. *IET Gener. Transm. Distrib.* 11 (15): 3722–3730.

27 Chakravorty, M. and Das, D. (2001). Voltage stability analysis of radial distribution networks. *Electr. Power Components Syst.* 23 (2): 129–135.

9

Machine Learning-Based Assessment Methods

9.1 Introduction

One of the important tasks of power system operators is to assess the system stability status and predict the occurrence of instability, followed by the implementation of appropriate control actions to prevent the occurrence of instability. Due to the complexities of the power system, it is not possible to achieve this goal without using appropriate tools that can assess the power system stability in a short time and using the available information [1]. Accordingly, after predicting the occurrence of contingency, which is the first task of the operator to prevent the occurrence of instability (because the instability of the power system always occurs due to occurrence of a contingency), the detection of voltage instability is the task of other power system operators. In this chapter, the main focus of the discussion is on this, and it tries to deal with various diagnostic methods using intelligent systems.

9.2 Voltage Stability Detection Based on Pattern Recognition Methods and Intelligent Systems

In recent years, the use of pattern recognition methods in detecting voltage instability is expanding. This is due to these methods have the following advantages:

- These methods can take into account the effect of all network parameters on the occurrence of instability.
- These methods have a good speed and therefore are suitable for timely evaluation of network stability.
- These methods can determine the state of voltage stability in a wide range of operating points and against various contingencies.

Voltage Stability in Electrical Power Systems: Concepts, Assessment, and Methods for Improvement, First Edition. Farid Karbalaei, Shahriar Abbasi, and Hamid Reza Shabani.

Published 2023 by John Wiley & Sons, Inc.

Stability recognition methods based on pattern extraction include two steps: the first step is classifier training and the second step is to use the classifier. The basis of pattern extraction-based methods is that several properties are predetermined, the values of these properties are applied as input to the classifier, and the output of the classifier determines the stability of the power system. It is obvious that those properties can be useful if they have enough information about the status of the power system. In fact, these characteristics determine the status of the operating point. The task of the classifier is to determine the thresholds with which the values of the properties are compared, and based on the result of this comparison, the stability status of the power system is determined.

To determine these thresholds, the so-called classifier must be trained. To do this, it is necessary to prepare a sufficient number of training samples. In fact, each training sample is a case study in which the values of the features and the stability status of the system are specified. These training samples are applied to the classifier, and the classifier determines the threshold values based on the values of the characteristics and the stability status of each sample in such a way that the classifier has the least error in determining the stability status of the training samples.

After training, the classifier is ready to use. To do this, the values of the power system properties are applied to the classifier at any given time, and the classifier determines the stability of the power system at that moment by comparing the properties values with the threshold values obtained in the training phase. In using pattern recognition methods, proper selection of these properties is very important. The more appropriate and accurate these properties are selected; the more likely it is that the stability of the system will be assessed in a wide range of workplaces [2–4].

9.2.1 The Intelligent Systems Training Approaches

These approaches are traditionally divided into three general categories, depending on the nature of the "signal" or "feedback" available to the learning system:

- Supervised learning: The computer is presented with example inputs and their desired outputs, given by a "teacher," and the goal is to learn a general rule that maps inputs to outputs.
- Unsupervised learning: No labels are given to the learning algorithm, leaving it on its own to find structure in its input. Unsupervised learning can be a goal in itself (discovering hidden patterns in data) or a means toward an end (feature learning).
- Reinforcement learning: A computer program interacts with a dynamic environment in which it must perform a certain goal (such as driving a vehicle or playing a game against an opponent). As it navigates its problem space, the program is provided feedback that is analogous to rewards, which it tries to maximize [5].

9.2.2 The Intelligent Systems Types

As mentioned before, performing intelligent systems (ISs) include creating a model, which is trained on some training data and then can process additional data to make predictions. Various models are used and researched for ISs. Some of these models are artificial neural networks (ANNs), decision trees (DTs), and support vector machines (SVMs) [4]. These models are applied widely in the literature for the case study of on-line power system voltage stability assessment [6]. As mentioned in the previous chapters, there are numerous tools that are developed to conduct a comprehensive analysis of the voltage stability assessment, such as P–V and Q–V curves, continuation power flow (CPF), and voltage stability indices. However, the developed software tools have the scarcity to be used in a real time or on-line operation as they are computationally time consuming due to its reliant on a complex mathematical modeling of a power system. The aforementioned predicament of enormous computational requirements could be resolved by utilizing the machine learning (ML) techniques such as ANNs, DTs, and SVMs. In the next sections, after describing these IS techniques, brief reviews of published research papers discussed the application of these techniques in on-line voltage stability assessment are presented [7].

9.2.2.1 Artificial Neural Networks (ANNs)

ANNs, or connectionist systems, are computing systems vaguely inspired by the biological neural networks that constitute animal brains. Such systems "learn" to perform tasks by considering examples, generally without being programmed with any task-specific rules.

In general, the ANN techniques have the same structure: input layer, hidden layer, and output layer. In fact, an ANN technique is a model based on a collection of connected units or nodes called "artificial neurons," which loosely model the neurons in a biological brain. Each connection, like the synapses in a biological brain, can transmit information, a "signal," from one artificial neuron to another. An artificial neuron that receives a signal can process it and then signal additional artificial neurons connected to it. In common ANN implementations, the signal at a connection between artificial neurons is a real number, and the output of each artificial neuron is computed by some nonlinear function of the sum of its inputs. The connections between artificial neurons are called "edges." Artificial neurons and edges typically have a weight that adjusts as learning proceeds. The weight increases or decreases the strength of the signal at a connection. Artificial neurons may have a threshold such that the signal is only sent if the aggregate signal crosses that threshold. Typically, artificial neurons are aggregated into layers. Different layers may perform different kinds of transformations on their inputs. Signals travel from the first layer (the input layer) to the last layer (the output layer), possibly after traversing the layers multiple times [7].

The original goal of the ANN approach was to solve problems in the same way that a human brain would. However, over time, attention moved to performing specific tasks, leading to deviations from biology. ANNs have been used on a variety of tasks, including computer vision, speech recognition, machine translation, social network filtering, playing board and video games, and medical diagnosis [7].

As mentioned, neural networks are actually a kind of simplistic modeling of real human nervous systems that have many applications in solving various problems. In other words, neural networks use real-time input and output data sets to use the training algorithms to find hidden connections between input and output data through weighting coefficients, biases, and functions applied to the output of each layer. One of the simplest and at the same time the most efficient layouts for use in modeling real nerve is the multilayer perceptron model. This network consists of one input layer and one output layer (containing input and output neurons) and one or more hidden layers (containing hidden neurons). The number of neurons in the input and output layers depends on the type of problem, while the number of neurons in the hidden layers is arbitrary and is calculated by trial and error. Figure 9.1 shows an multilayer perceptron (MLP) neural network with two hidden layers. In this structure, all single-layer neurons are connected to all next-layer neurons. This arrangement represents a network so-called with complete connections. It should also be noted that one of the neural network training algorithms is the post-error propagation algorithm. In this algorithm, in each step, the newly calculated output value is compared with the real value, and according to the obtained error, the weights and network biases are corrected. So that at the end of each iteration, the magnitude of error is less than obtained value in the previous iteration [8].

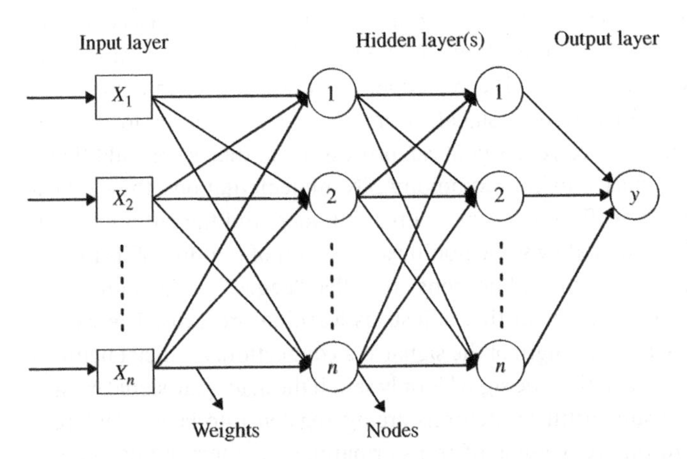

Figure 9.1 The global architecture of ANNs.

The following are some of the key motivations for using neural networks:

- Possibility of accepting unlimited inputs and outputs; this unique advantage makes neural networks more important and popular than other methods of artificial intelligence and makes them suitable for analyzing small or large data.
- Learning skills and modeling nonlinear and complex relationships artificial neural networks can manage different real applications in different contexts that are complex and nonlinear. This is a very important advantage.
- Unlike other deep learning techniques, ANNs do not require any execution constraints on input variables, such as how data is distributed.
- Deep learning skills: Neural networks can learn events and manage wise decisions by commenting to achieve better similar events.
- Multiple processing capability: ANNs can ensure numerical performance with the ability of performing multiple tasks, simultaneously.
- Error tolerance, whereby ANNs can produce output results even if some cells are damaged, and this advantage allows ANNs to tolerate errors.
- Ability to generalize: Once neural networks learn primary input relationships, they can guess unknown relationships in anonymous data, so they generalize the model and allow it to predict unknown data.

On the other hand, the following points outline the main challenges of using neural networks:

- The mysterious behavior of the network, after the ANN produced an analytical result, does not explain why or how selecting these outputs and rejecting others may make it unreliable on the network.
- Designing network architecture proper for ANNs do not have a precise rule for determining the best structure design, or in other words, the proper network structure must be obtained through trial and error.
- Hardware dependency, ANNs require powerful dual processors. These drawbacks, in fact, mean that the whole approach depends on the equipment [9].

Thus, it should be noted that the properties of neural networks make their use attractive in many engineering issues. Accordingly, the use of neural network properties in many issues related to power systems, especially the issue of dynamic security of power systems, which evaluation of them using conventional methods is complex and time consuming can be very instructive. In this regard, neural networks are widely used today to evaluate and detect voltage stability in power systems.

Accordingly, Refs. [10, 11] are examples of the proposed methods for diagnosing voltage stability, in which a neural network is used to determine the distance to the instability boundary. While in [10], only the loading amount and increment direction of load are applied as inputs to the neural network, and in [11], the angle and magnitude of the bus voltage are also used as input. In addition to the above, in

[12], the distance to the instability limit is determined using the neural network. In this paper, the active and reactive power of loads, bus voltages, and reactive power differences are generated, and the maximum reactive power that can be produced by generators is used as the input of the neural network. The disadvantage of these methods is that to determine the state of long-term voltage stability (distance to the limit of long-term voltage instability), they have used the characteristics of the system steady state and cannot be used in the transition period after contingency occurrence to determine dynamic stability of voltage. In other words, although pattern-based methods are fast, they cannot be used immediately after contingency due to the delay in extracting the properties values (because steady-state information is used as neural network input), and determination of stability status will be with delay.

A real-time continuous monitoring system (CMS) for long-term voltage stability assessment with sliding three-dimensional convolutional neural network (3D-CNN) is proposed in Ref. [13]. Given a power system, its dynamic responses and topological information are constituted the learning input of the sliding 3D-CNN. In this paper, numerical tests are carried out on the New England 10-generator-39-bus system to demonstrate the effectiveness. Simulation results also indicate the potential of the CMS in robustness to PMU measurement errors and information losses. Ref. [14] proposes a novel graph neural network (GNN) based model for the short-term voltage stability (STVS) assessment. Based on mainly the steady-state information as the inputs, both the fast voltage collapses (FVC) and fault-induced delayed voltage recovery (FIDVR) events can be identified. Also, the work of [15] presents a novel methodology for long-term voltage stability monitoring that exploits the feasibility of phasor-type information in order to estimate the long-term voltage stability status. The information regarding the current system condition is acquired through synchronized phasor measurements, and the power system is divided in subareas for improving its supervision; then, an artificial intelligence approach based on kernel extreme learning machine is used for long-term voltage stability assessment. The validation of the proposed method is performed on the 39-bus test system, obtaining feasible results. The tests confirmed that the proposed method works properly under different scenarios and system conditions, always ensuring proper voltage stability status results independently of its cause. In the same direction, in the work done in Ref. [16], a machine learning approach is presented to predict the long-term voltage stability margin. In addition, it is presented a methodology to generate training data under different operational conditions and $N-1$ contingencies to train the machine learning models. The studies were conducted on the IEEE 14-bus system and IEEE 118-bus system and led to the selection of Random Forest Regression machine learning algorithm and confirmed the accuracy and robustness of the proposed method. In work [17], a time-series deep learning framework is proposed using

1D-convolutional neural networks (1D-CNN) for real-time short-term voltage stability assessment (STVSA), which relies on a limited number of phasor measurement units (PMU) voltage samples. This work also considers DB-SCAN clustering-based fault detection and physics-based fault localization for effective short-term voltage stability assessment and remedial actions by identifying the most critical PMUs. The proposed framework is validated using the IEEE 39- and 118-bus test systems and compared against other machine learning models to demonstrate the superiority of 1D-CNN-based time-series deep learning for short-term voltage stability assessment. Finally, in Ref. [18], a deep learning technique is utilized, and a multilevel deep neural network (ML-DNN) is proposed that achieves feature fusion of the electrical parameter measurements, topology, and contingency information. Actually, to monitor the voltage stability state of complex power grid, a four-category stability classification problem that incorporates a set of serious contingencies is posed. Experiments are implemented on IEEE 39-bus test system, which demonstrates its advantage for online voltage stability monitoring.

9.2.2.2 Decision Trees (DTs)

DT learning uses a DT as a predictive model to go from observations about an item (represented in the branches) to conclusions about the item's target value (represented in the leaves). It is one of the predictive modeling approaches used in statistics, data mining, and machine learning. The DT method is shown in Figure 9.2. Tree models where the target variable can take a discrete set of values are called classification trees; in these tree structures, leaves represent class labels, and branches represent conjunctions of features that lead to those class labels [19].

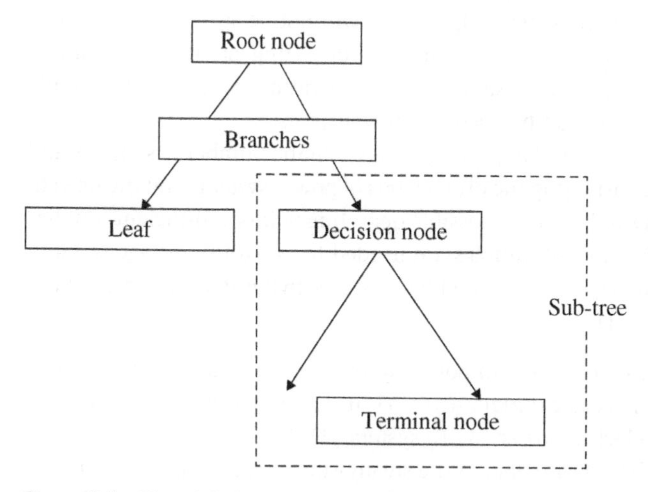

Figure 9.2 The global architecture of DTs.

In the classification process using the DT, a sample is applied to the first decision node, commonly called the root node. In each of the decision nodes, the value of one of the characteristics (properties) of the sample is compared with a threshold value (which is determined in the DT training step and is different for different decision nodes), and based on the result of this comparison, the sample moves down through interface vectors. This downward movement of the sample among the decision nodes continues until the sample reaches one of the end nodes. Each of the end nodes identifies a class that represents the class of the sample that reaches that nodes [2, 3].

In the DT training step, the structure of the DT with the characteristics of the comparison at each decision node is determined. Although there are various algorithms such as J48 and Random Forest for training DTs, in general, it can be said that at each decision node, the sample property with the highest resolution among samples will be evaluated and compared (with the threshold value).

In general, the DT has advantages over other existing classifiers. Due to this, most of the studies use the DT to determine the power systems stability that is based on statistical pattern recognition. These advantages can be described as follows:

- The DT is a suitable tool for classifying samples in situations where there are more than two classes. Hence, this classifier is more suitable than tools such as SVMs, which are normally used to separate samples of two classes from each other. Given that in system stability analysis, it is necessary to classify the stability status of the operating point based on the distance from the instability limit, and the DT can be a suitable tool for this purpose.
- For problems in which each sample belongs to only one class, classifiers such as the DT are required to be able to place each sample in a class. Obviously, in such problems, classifiers such as fuzzy algorithms will not be suitable tools. Given that in the study of the stability of different operating points, each sample belongs to only one class (for example, each operating point is either stable or unstable), and the DT can be used for this purpose.
- DT is a suitable tool to be used in problems in which the number of samples and properties is high. Considering the dimensions of power systems and the need to evaluate the system stability in all possible operating points, a sufficient number of properties and the studied samples are needed to assess the voltage stability status. Therefore, the DT can be a good tool for classifying stability status of the operating point [20, 21].

Accordingly, DT is frequently used in power system studies and of course in voltage stability assessment. As the initial uses of DT, it was used to assess voltage security of the American electric power (AEP) system [22, 23].

In Ref. [22], using the DT and data of the steady-state voltage angle and magnitude of buses applied to the DT, the voltage stability status is determined. In the DT training step, this reference determines the system stability status based on the minimum voltage of buses and the distance to the instability limit. Due to the properties used to train the DT, this method cannot be used in the transient period (to determine dynamic voltage stability status) after a contingency occurrence.

After that, DT was used for on-line voltage stability assessment based on voltage and current phasor measurements as a wide-area measurement system [24–26]. In some works, DT is combined with other algorithms such as Fuzzy logic for on-line voltage stability monitoring [27, 28]. In Ref. [27], initially, the dimension of the measured data by PMUs is reduced by principle component analysis (PCA). Then, the data are used in DT algorithm for online voltage security assessment. In Ref. [28], a combined version of DT-based PCA is used for voltage stability assessment. In which, DT-based PCA is combined with optimization algorithms namely biogeography-based optimization and invasive weed optimization.

In Ref. [29], a case study to evaluate voltage stability is presented in which PMU and DT measurement units are used. The results of simulations performed in this study show that the differences in angle between the buses voltage phasors and the transferred reactive power are good options for determining the voltage stability status. This method, similar to the above-mentioned methods, cannot be used to determine the dynamic voltage stability status.

Ref. [30] constructs DT using C4.5 algorithm for online voltage stability assessment. C4.5 algorithm is finally applied to construct DT. The case study on a practical power system demonstrates that DT can extract operating guidelines from offline voltage stability analysis results and helps system operators assess voltage stability status in real time.

Ref. [31] presents a method that performs classification of thousands of operating conditions w.r.t. power system voltage stability by using DTs. The proposed method uses a new and flexible classification criterion that allows to identify operating conditions that are near or within the region for which the system is voltage unstable, and more importantly, that can consider operational requirements. Case studies were performed using the IEEE 9-bus system for several operating conditions and different network configurations.

Also, in work [32], a novel fast machine learning-based scheme is presented for predicting the short-term voltage stability status and determining the driving force of instability, without any need to measure post-fault-clearance data. In order to evaluate the performance of the proposed scheme, the method has been tested on IEEE 39-bus test system and IEEE Nordic test system. The simulation results show the proposed scheme is highly accurate and timely.

9.2.2.3 Support-Vector Machines (SVMs)

SVMs, also known as support vector networks, are a set of related supervised learning methods used for classification and regression, as shown in Figure 9.3. Given a

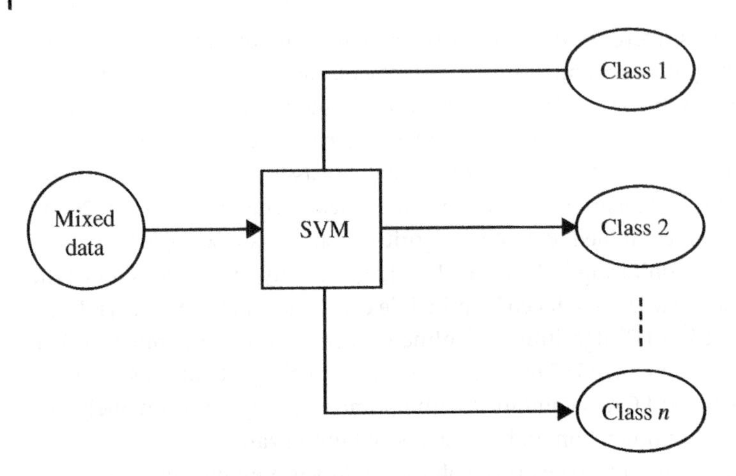

Figure 9.3 The global architecture of SVMs.

set of training examples, each marked as belonging to one of two categories, an SVM training algorithm builds a model that predicts whether a new example falls into one category or the other. An SVM training algorithm is a nonprobabilistic, binary, linear classifier, although methods such as Platt scaling exist to use SVM in a probabilistic classification setting. In addition to performing linear classification, SVMs can efficiently perform a nonlinear classification using what is called the kernel trick, implicitly mapping their inputs into high-dimensional feature spaces [33].

SVMs were expanded by Cortes and Vapnik in 1995 [34] for binary classification. The approach of this classifier can be approximately described as follows:

The classification problem can be limited to a two-class problem without losing its comprehensiveness and totality. In this case, the goal is to separate the two classes by a function that is taken from existing examples. In other words, the goal is to provide a classifier that works well on unseen samples, i.e. generalize well (Figure 9.4). There are many linear classifiers here that can separate data, but there is only one classifier that maximizes the margin (maximizes the distance between it and the nearest data point of each class). This linear classifier is called the optimal separator hyperplane. In other words, this approach basically seeks the optimal hyperplane between two classes by maximizing the margin between the nearest points of the classes. The points on the borders are the support vectors, and plane in the middle of the margin is also called the optimal separator hyperplane [35, 36].

In general, the following can be mentioned as advantages of SVR:

- SVM works relatively well when there is an obvious margin of separation between classes.
- SVM is more effective in high-dimensional spaces.

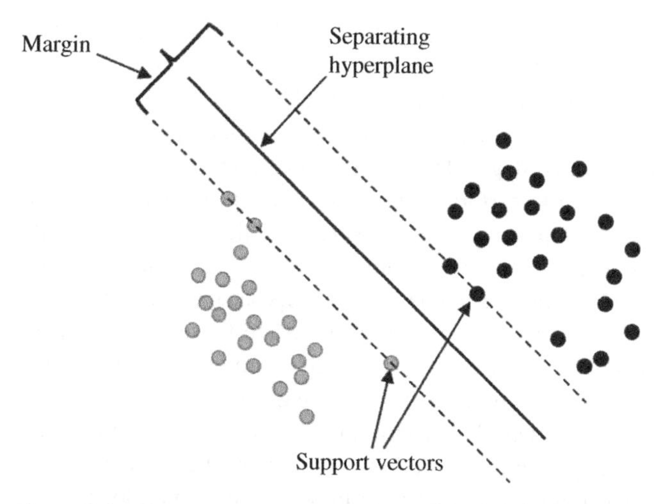

Figure 9.4 Classification (linear separable case).

- SVM is effective in cases where the number of dimensions is more than the number of samples.
- SVM is relatively memory efficient.

Of course, the following can be considered as disadvantages of this classifier, which are

- The SVM algorithm is not suitable for large data sets.
- When the set of data has more noisy, i.e. the target classes overlap the SVM does not work very well.
- In cases where the number of properties of each data point exceeds the number of instructional data samples, SVM will perform poorly.
- Since the support vector classifier works by placing data points, there is no possible explanation for classification at the top and bottom of the classifier hyperplane [37].

Given the above, SVM is a new and powerful technique for machine learning and has many applications in various engineering sciences. In voltage stability analysis and the use of this powerful classification tool, the work done in Ref. [38] can be considered as one of the first studies in this field. In which, the genetic algorithm approach based on SVM (GA-SVM) for online monitoring of long-term voltage instability is proposed. The magnitude and angle of the voltage phasor are considered as GA-SVM input properties that are assumed to be derived from the PMU. Based on this, the voltage stability margin index is estimated in the steady-state analysis. Thus, the effectiveness of the proposed approach has been investigated using the New England 39-bus test system and the real North India 246 bus network (NRPG) system. In Ref. [39], a new approach based on the SVC called least square SVM (LSSVM) is used. In fact, in this approach, a PMU-based method is proposed to predict short-term voltage instability for multibus power systems. The method is divided into two stages. The first step uses simultaneous measurements in the operation center to perform online probability analysis to produce a search table. In the second step, using local measurement, the short-term voltage instability after a contingency is identified in advance based on time-series predictions and search tables. Here, the effectiveness of the proposed method on the New England 39-bus system is confirmed, too. As one of the latest researches done on this field, the study conducted in Ref. [40] in which a comprehensive plan for online monitoring of voltage stability using the weighted least squares SVM (WLS-SVM) is proposed, in which, the effect of overcurrent protection for transmission lines is included. Also, in the work done in this study, the principal component analysis (PCA) technique has been used to reduce the dimensions of the input vector applied to WLS-SVM to make it suitable for real-time applications. The proposed approach is simulated on IEEE 39- and 118-bus systems, and the results show the efficiency, high accuracy, and better performance of the proposed method compared to other approaches.

Furthermore, as the most common application form of SVM, support vector regression (SVR) is used to evaluate voltage stability in power systems [38, 41, 42]. SVR uses the same principle as SVM, but for regression problems. The problem of regression is to find a function that approximates mapping from an input domain to real numbers on the basis of a training sample. In Ref. [41], the SVR model has been used to assess the voltage stability of power system incorporating flexible alternating current transmission systems (FACTS) devices. In Ref. [42], the ν-SVR and ε-SVR models with radial basis function (RBF) and polynomial kernel functions are used in on-line prediction of voltage stability margin. The author of [38] proposed a hybrid model combining genetic algorithm (GA) with SVR for voltage stability monitoring. It was reported that the proposed GA-SVR model has better performance.

Based on the above, the methods presented in the literature can be shown in the form of Table 9.1, in which the researches reviewed in this chapter are classified. The criteria used in this classification are the ability of these methods to identify voltage stability as well as the type of the used method along with the inputs and the test system.

The below table shows that the different machine learning techniques are applicable in analysis voltage stability of power systems. Different input(s) and outputs(s) can be considered in these techniques. These techniques are successfully implemented on various real and test power systems.

Table 9.1 Review of techniques for voltage stability monitoring.

Ref. no.	Technique used	Input(s)	Output(s)	Case study
[10]	ANN	Loading rates and loading directions	Loadability margins of power systems	IEEE two-area benchmark system and IEEE 50-machine test system
[11]	ANN	Node voltage magnitudes and the phase angles	Long-term voltage stability margin	New England 39-bus test system and a practical power system (consists of 1844 buses, 746 load buses, and 302 generator buses)
[12]	ANN	Real and reactive power loads and the voltage magnitude at the vulnerable load bus	Online voltage stability assessment	IEEE 30-bus test system
[13]	Three-dimensional convolutional neural network (3D-CNN)	Dynamic responses and topological information	Long-term voltage stability assessment	New England 10-generator-39-bus system
[14]	Graph neural network (GNN)	Steady-state information	Short-term voltage stability (STVS) assessment	New England 10-generator-39-bus system
[15]	ANN based on kernel extreme learning machine	Feasibility of phasor-type information (information regarding the current system condition)	Long-term voltage stability monitoring	39-bus test system
[16]	ANN based on machine learning	Different voltage stability indices (VSIs)	Long-term voltage stability margin (loadability margin)	IEEE 14-bus system and IEEE 118-bus system

(*Continued*)

Table 9.1 (Continued)

Ref. no.	Technique used	Input(s)	Output(s)	Case study
[17]	1D-convolutional neural networks (1D-CNN)	A limited number of phasor measurement units (PMU) voltage samples	Real-time short-term voltage stability assessment	IEEE 39- and 118-bus test systems
[18]	Multilevel deep neural network (ML-DNN)	Active/reactive power of generators and loads	Voltage stability state monitoring	IEEE-39 system
[22]	DT	PMU-related system parameters.	State of the system (secure or insecure)	American electric power system
[23]	DT	Reactive power of generators and angular difference attributes	State of the system (secure or insecure)	American electric power system
[24]	DT	Voltage and current phasor measurements	State of the system (stable, alert, unstable)	9-bus, New England 39-bus
[25]	DT	Active power flows and PMU-based voltage angle differences	State of the system (secure or insecure)	Zhejiang power grid of China
[26]	DT based on bagging and adaptive boosting methods	Active load and active generation variation	State of the system (secure or insecure)	IEEE 118-bus
[27]	Principal component analysis and DTs	Load flow convergence, loading index (LI), and profile index (PI)	State of the system (more secure, less secure, etc.)	New England 39-bus and a part of Iran power grid
[28]	Principal component analysis and DTs	PMU data	State of the system (Secure or insecure)	66-bus and Iranian power grid
[29]	DT	Voltage angle differences and transmission line reactive flows	Security from voltage collapse	8211-bus system rep- resenting summer 2003 peak load conditions

Table 9.1 (Continued)

Ref. no.	Technique used	Input(s)	Output(s)	Case study
[30]	DT	*P–V* curves analysis	Online voltage stability assessment	Practical power system that is a large load center in Southwestern China
[31]	DT	Load active powers and voltages	Classification of the degree of voltage stability	IEEE 9-bus system
[32]	DT	Using the measurements of the installed phasor measurement unit (PMU) at the generator and the induction motor (IM) buses	Short-term voltage stability	IEEE 39-bus test system and IEEE Nordic test system
[38]	Genetic algorithm based support vector machine (GA-SVM)	Voltage magnitude and phase angle obtained from phasor measurement units (PMUs)	Online monitoring of long-term voltage instability	New England 39-bus test system and the Indian Northern Region Power Grid (NRPG) 246-bus real system
[39]	Least square support vector machine (LSSVM)	PMU-based method	Short-term voltage instability	New England 39-bus system
[40]	Weighted least square support vector machine (WLS-SVM)	Collected data from Phasor measurement units (PMUs)	Online monitoring of voltage stability	IEEE 39-bus and IEEE 118-bus systems
[41]	Artificial imine system-SVM	Real and reactive power load	Loadability margin index	IEEE 30-bus, Indian 181-bus
[42]	Support vector regression (SVR)	Real and reactive power load	Loadability margin index	IEEE 30-bus, IEEE 118-bus
[34]	GA optimized support vector regression (SVR)	PMU measurements	Loadability margin index	IEEE 30-bus, IEEE 118-bus

9.3 Summary

In this chapter, the machine learning-based methods to assess and monitor voltage stability status of power systems are introduced. General topologies of these methods and their capabilities were introduced. As mentioned, these methods can be categorized into five methods. For each method, a table including input(s), output(s), used technique, and case study are presented.

References

1 Kundur, P., Paserba, J., Ajjarapu, V. et al. (2004). Classification of power system stability using support vector machines. *IEEE Trans. Power Syst.* 19 (3): 1387–1401. https://doi.org/10.1109/pes.2005.1489266.

2 Khoshkhoo, H. and Shahrtash, S.M. (2014). Fast online dynamic voltage instability prediction and voltage stability classification. *IET Gener. Transm. Distrib.* 8 (5): 957–965. https://doi.org/10.1049/iet-gtd.2013.0296.

3 Khoshkhoo, H. and Shahrtash, S.M. (2012). On-line dynamic voltage instability prediction based on decision tree supported by a wide-area measurement system. *IET Gener. Transm. Distrib.* 6 (11): 1143–1152. https://doi.org/10.1049/iet-gtd.2011.0771.

4 Mitchell, T. (1997). *Machine Learning*. New York: McGraw-Hill.

5 Bishop, C.M. (2006). *Pattern Recognition and Machine Learning*. Springer.

6 Amroune, M. (2021). Machine learning techniques applied to on-line voltage stability assessment: a review. *Arch. Comput. Methods Eng.* 28 (2): 273–287. https://doi.org/10.1007/s11831-019-09368-2.

7 Schmidhuber, J. (2015). Deep learning in neural networks: an overview. *Neural Netw.* 61: 85–117. https://doi.org/10.1016/j.neunet.2014.09.003.

8 Karami, A. (2011). Power system transient stability margin estimation using neural networks. *Int. J. Electr. Power Energy Syst.* 33 (4): 983–991. https://doi.org/10.1016/j.ijepes.2011.01.012.

9 Abdolrasol, M.G., Hussain, S.M., Ustun, T.S. et al. (2021). Artificial neural networks based optimization techniques: a review. *Electronics* 10 (21): 2689.

10 Gu, X. and Cañizares, C.A. (2007). Fast prediction of loadability margins using neural networks to approximate security boundaries of power systems. *IET Gener. Transm. Distrib.* 1 (3): 466–475.

11 Zhou, D.Q., Annakkage, U.D., and Rajapakse, A.D. (2010). Online monitoring of voltage stability margin using an artificial neural network. *IEEE Trans. Power Syst.* 25 (3): 1566–1574. https://doi.org/10.1109/TPWRS.2009.2038059.

12 Suthar, B. and Balasubramanian, R. (2007). A novel ANN based method for online voltage stability assessment. *2007 International Conference on Intelligent Systems Applications to Power Systems*, Niigata Japan Toki Messe (5–8 November 2007), pp. 1–6.

13 Cai, H. and Hill, D.J. (2022). A real-time continuous monitoring system for long-term voltage stability with sliding 3D convolutional neural network. *Int. J. Electr. Power Energy Syst.* 134: 107378.

14 Zhong, Z., Guan, L., Su, Y. et al. (2022). A method of multivariate short-term voltage stability assessment based on heterogeneous graph attention deep network. *Int. J. Electr. Power Energy Syst.* 136: 107648.

15 Villa-Acevedo, W.M., López-Lezama, J.M., Colomé, D.G., and Cepeda, J. (2022). Long-term voltage stability monitoring of power system areas using a kernel extreme learning machine approach. *Alexandria Eng. J.* 61 (2): 1353–1367.

16 Dharmapala, K.D., Rajapakse, A., Narendra, K., and Zhang, Y. (2020). Machine learning based real-time monitoring of long-term voltage stability using voltage stability indices. *IEEE Access* 8: 222544–222555.

17 Rizvi, S.M.H., Sadanandan, S.K., and Srivastava, A.K. (2021). Data-driven short-term voltage stability assessment using convolutional neural networks considering data anomalies and localization. *IEEE Access* 9: 128345–128358.

18 Bai, J. and Tan, X. (2020). Contingency-based voltage stability monitoring via neural network with multi-level feature fusion. *IFAC-PapersOnLine* 2: 13483–13488, 53AD.

19 Breiman, L., Friedman, J.H., Olshen, R.A., and Stone, C.J. (2017). *Classification and Regression Trees*. Routledge.

20 Alpaydin, E. (2005). *Introduction to Machine Learning*. London: MIT Press.

21 Bifet, A., et al. (2018). Machine Learning Software in Java. http://www.cs.waikato.ac.nz/ml/weka/.

22 Diao, R., Sun, K., Vittal, V. et al. (2009). Decision tree-based online voltage security assessment using PMU measurements. *IEEE Trans. Power Syst.* 24 (2): 832–839. https://doi.org/10.1109/TPWRS.2009.2016528.

23 Nuqui, R.F., Phadke, A.G., Schulz, R.P., and Bhatt, N. (2001). Fast on-line voltage security monitoring using synchronized phasor measurements and decision trees. *Proc. IEEE Power Eng. Soc. Transm. Distrib. Conf.* 3 (Winter Meeting): 1347–1352. https://doi.org/10.1109/pesw.2001.917282.

24 Zheng, C., Malbasa, V., and Kezunovic, M. (2012). A fast stability assessment scheme based on classification and regression tree. *IEEE International Conference on Power System Technology, POWERCON 2012*, New Zealand Auckland (30 October to 2 November 2012). https://doi.org/10.1109/PowerCon.2012.6401453.

25 Li, Z. and Wu, W. (2009). Phasor measurements-aided decision trees for power system security assessment. *2nd International Conference on Information and*

Computing Science, ICIC 2009, England UK Manchester (21–22 May 2009), vol. 1, pp. 358–361. https://doi.org/10.1109/ICIC.2009.98.

26 Beiraghi, M. and Ranjbar, A.M. (2013). Online voltage security assessment based on wide-area measurements. *IEEE Trans. Power Deliv.* 28 (2): 989–997. https://doi.org/10.1109/TPWRD.2013.2247426.

27 Mohammadi, H. and Dehghani, M. (2015). PMU based voltage security assessment of power systems exploiting principal component analysis and decision trees. *Int. J. Electr. Power Energy Syst.* 64: 655–663. https://doi.org/10.1016/j.ijepes.2014.07.077.

28 Mohammadi, H., Khademi, G., Simon, D., and Dehghani, M. (2016). Multi-objective optimization of decision trees for power system voltage security assessment. *2016 Annual IEEE Systems Conference (SysCon)*, Orlando, FL (April 18–21 2016). 2016. https://doi.org/10.1109/SYSCON.2016.7490524.

29 Khatib, A., Nuqui, R., Ingram, M., and Phadke, A. (2004). Real-time estimation of security from voltage collapse using synchronized phasor measurements. *IEEE Power Engineering Society General Meeting*, Denver, CO (6–10 June 2004), pp. 582–588.

30 Meng, X., Zhang, P., Xu, Y., and Xie, H. (2020). Construction of decision tree based on C4.5 algorithm for online voltage stability assessment. *Int. J. Electr. Power Energy Syst.* 118 (December 2019): 105793. https://doi.org/10.1016/j.ijepes.2019.105793.

31 Vanfretti, L. and Arava, V.S.N. (2020). Decision tree-based classification of multiple operating conditions for power system voltage stability assessment. *Int. J. Electr. Power Energy Syst.* 123 (June): 106251. https://doi.org/10.1016/j.ijepes.2020.106251.

32 Lashgari, M. and Shahrtash, S. (2022). Fast online decision tree-based scheme for predicting transient and short-term voltage stability status and determining driving force of instability. *Int. J. Electr. Power Energy Syst.* 137: 107738.

33 Boser, B.E., Guyon, I.M., and Vapnik, V.N. (1992). A training algorithm for optimal margin classifiers. *Proceedings of the 5th annual ACM Workshop on Computational Learning Theory*, Pittsburgh, PA (27–29 July 1992), pp. 144–152.

34 Cortes, C. and Vapnik, V. (1995). Support-vector networks. *Mach. Learn* 20 (3): 273–297.

35 Gunn, S. (1998). Support vector machines for classification and regression. *ISIS Tech. Rep.* 14 (1): 5–16.

36 Meyer, D. and Wien, F. (2015). Support vector machines. *Interface to libsvm Packag. e1071* 28: 1–8.

37 Brownlee, J. (2016). Neural Network Machine Learning Algorithm From Scratch in Python. https://dhirajkumarblog.medium.com.

38 Sajan, K.S., Kumar, V., and Tyagi, B. (2015). Genetic algorithm based support vector machine for on-line voltage stability monitoring. *Int. J. Electr. Power Energy Syst.* 73: 200–208. https://doi.org/10.1016/j.ijepes.2015.05.002.

39 Yang, H., Zhang, W., Chen, J., and Wang, L. (2018). PMU-based voltage stability prediction using least square support vector machine with online learning. *Electr. Power Syst. Res.* 160: 234–242. https://doi.org/10.1016/j.epsr.2018.02.018.

40 Poursaeed, A. and Namdari, F. (2022). Real-time voltage stability monitoring using weighted least square support vector machine considering overcurrent protection. *Int. J. Electr. Power Energy Syst.* 136: 107690.

41 Suganyadevia, M.V. and Babulalb, C.K. (2014). Fast assessment of voltage stability margin of a power system. *J. Electr. Syst.* 10 (3): 305–316.

42 Suganyadevi, M.V. and Babulal, C.K. (2014). Support vector regression model for the prediction of loadability margin of a power system. *Appl. Soft Comput. J.* 24: 304–315. https://doi.org/10.1016/j.asoc.2014.07.015.

Part III

Methods of Preventing Voltage Instability

10

Preventive Control of Voltage Instability

10.1 Introduction

If voltage stability assessment indicates the risk of voltage instability in the current situation or after occurrence of possible contingencies, the system should be placed in a safe state by taking actions. All actions used to prevent voltage instability can be divided into two categories: preventive and emergency. The purpose of preventive control is to create a sufficient voltage stability margin against possible contingencies. Pre-contingency operating point is modified so that the necessary stability margin be maintained against a set of probable contingencies [1]. Therefore, preventive actions are applied when the system is stable, but there is no desired stability margin. On the other hand, the aim of emergency control is to stabilize an unstable system [2]. Relevant actions are used when the voltage instability has started and if the appropriate actions are not applied, voltage collapse will occur until next moments or minutes.

Actions that are used as preventive actions to increase the voltage stability margin can be categorized based on their cost [3]. Changing the voltage of the generators, changing the transformers taps, and changing the reactive power injected by compensators such as parallel capacitors and inductors have the lowest cost. But the range of changes in the control variables is limited while the changes increase the risk of pre-contingency overvoltages [1]. Changes in the output active powers of generators and load curtailments are other control actions that are used to increase the voltage stability margin. These actions are more costly and should be applied optimally.

Voltage stability margin, also called load margin (LM), is the distance from the base case to the maximum loadability point (MLP), which indicates the maximum load increase that the system can withstand without a sharp drop in voltage. Occurrence of various contingencies changes the LM, and control actions must

Voltage Stability in Electrical Power Systems: Concepts, Assessment, and Methods for Improvement, First Edition. Farid Karbalaei, Shahriar Abbasi, and Hamid Reza Shabani.

Published 2023 by John Wiley & Sons, Inc.

provide the minimum required amount of LM for a set of contingencies determined by the system operator. Therefore, one of the important steps in determining the actions in preventive control is calculation of LM, which is obtained with different accuracies depending on the model used for the system.

Determining the preventive actions is usually done using optimization methods. Increasing the LM can be considered as an objective function [4] or as a constraint [1, 2]. In the latter, minimizing control actions is usually chosen as the objective function. In real-world power systems, there are so many control actions that this makes it difficult to solve the optimization problem. Therefore, the number of control variables should be reduced by selecting the most effective actions. Most of these methods are based on sensitivity analysis [3, 5, 6].

10.2 Determination of LM

As mentioned, LM is the distance from the base case to the MLP, which indicates the maximum load increase at which the system does not encounter with voltage instability or collapse. Accurate determination of LM requires proper modeling of power system elements. For example, different modeling of synchronous generator controllers and limiters leads to different results that can make a significant difference in the calculated control actions [7]. Determination of LM is done in two methods, static and dynamic, which are described later.

10.2.1 Static Analysis

In static analysis, only the presence or absence of the steady-state equilibrium point is investigated. The importance of this issue is based on the fact that one of the important reasons for the occurrence of voltage instability is the lack of steady-state equilibrium. This point is obtained by solving steady-state equations. In many cases, for simplicity, the power flow equations are considered as the steady-state equations, and the MLP is determined based on the divergence of the power flow calculations. Continuation power flow and PV-curve fitting methods described in Chapters 5 and 6 are examples of these methods. The use of the power flow equations as the steady-state equations usually leads to optimistic results, because in these equations a number of factors affecting voltage stability are neglected [2]. For example, as seen in Chapter 4, the replacement of the power received from the transmission bus by distribution network with a constant power load, which is common in power flow calculations, may not indicate the occurrence of voltage instability. The PV and slack bus assumptions are other unrealistic assumptions used in power flow calculations.

The more accurate method of calculating the steady-state equilibrium point is to use all the differential and algebraic equations of the system as Eqs. (10.1) and (10.2).

$$\boldsymbol{X}^{o} = \boldsymbol{F}(\boldsymbol{X}, \boldsymbol{Y}, \boldsymbol{U}, \lambda) \tag{10.1}$$

$$0 = \boldsymbol{G}(\boldsymbol{X}, \boldsymbol{Y}, \boldsymbol{U}, \lambda) \tag{10.2}$$

In these equations, $\boldsymbol{X}$ is the vector of state variables and $\boldsymbol{Y}$ is the vector of algebraic variables. State variables include the state variables of generators, motors, dynamic loads, and other equipment that affect the occurrence of voltage instability. Algebraic variables, for example, include the magnitude and angle of the buses voltages. $\boldsymbol{U}$ is the vector of control variables and λ is the loading factor.

The steady-state equilibrium point is obtained by simultaneously solving the equilibrium equations of the differential equations (obtained by equating the derivatives to zero) and the algebraic equations, which are as follows.

$$0 = \boldsymbol{F}(\boldsymbol{X}, \boldsymbol{Y}, \boldsymbol{U}, \lambda) \tag{10.3}$$

$$0 = \boldsymbol{G}(\boldsymbol{X}, \boldsymbol{Y}, \boldsymbol{U}, \lambda) \tag{10.4}$$

Using the excitation system and the governor equations (Sections 3.2.8 and 3.2.9), the unrealistic assumptions of PV and slack buses are eliminated. Under these conditions, the active power and voltage magnitude of the generators are determined based on the characteristics of their governor and excitation system [7]. Load increment and moving toward the MLP can be done using several methods. For example, in [1, 2, 7], the Eq. (10.5) is used for this purpose.

$$\boldsymbol{P} = \boldsymbol{P}^{o} + \lambda \mathbf{K} \tag{10.5}$$

where $\boldsymbol{P}$ is the vector of active and reactive powers injected into the buses, $\boldsymbol{P}^{o}$ denotes the values of this vector in the base case, λ is the loading factor to push the system toward the MLP, and $\mathbf{K}$ is also a vector that shows the direction of increase of powers. In buses with synchronous generators connected to them, to insert the governor characteristic effect, $\boldsymbol{P}_{C}$ [Eq. (3.125)] replaces the injected active power in Eq. (10.5) [7]. λ_{max} is the highest loading factor at which the system has a steady-state equilibrium point represents MLP, which is also called saddle-node bifurcation [8]. At this point, similar to what was stated in Section 8.2, the Jacobin matrix related to Eqs. (10.3) and (10.4) is singular. By defining the vector $\mathbf{Z}$ as the Eq. (10.6), the Eqs. (10.3) and (10.4) are expressed as the Eq. (7.10). In λ_{max}, the Jacobin matrix $H_{Z}(\boldsymbol{Z}, \boldsymbol{U}, \lambda)$ becomes singular.

$$\mathbf{Z} \triangleq \begin{bmatrix} \boldsymbol{X} \\ \boldsymbol{Y} \end{bmatrix} \tag{10.6}$$

$$0 = \mathbf{H}(\boldsymbol{Z}, \boldsymbol{U}, \lambda) \tag{10.7}$$

10.2.2 Dynamic Analysis

Assessing the voltage stability based on the presence or absence of a steady-state equilibrium point, even if the equilibrium equations are selected correctly, may have error. Because the effect of the reaction time of the controllers and limiters as well as the time constant of the dynamic equipment response that can lead to the exit from the attraction region is not seen. A more accurate solution is to dynamically simulate and determine the time response of the system. The fast time-domain quasi-steady-state (QSS) simulation method can be used to analyze the long-term voltage stability [8]. In this method, only slow dynamics of system such as load dynamics are considered, and fast dynamics are replaced by their equilibrium equations.

In dynamic analysis, λ_{max} represents the maximum loading factor at which the system time response does not show an unacceptable drop in the voltage of any of the buses. Therefore, due to the mentioned reasons, λ_{max} obtained from dynamic analysis may be different from λ_{max} calculated from static analysis in which just the presence or absence of equilibrium point is checked. So, the LM obtained from the two analyzes can be named dynamic load margin (DLM) and static load margin (SLM), respectively.

To determine DLM considering a contingency, a simple binary search method is proposed in [9]. Wherein, λ_s and λ_u are determined first. The power system is stable for λ_s and unstable for λ_u. λ_s can be chosen equal to zero (base case) and λ_u equal to the desired maximum loading factor (if a contingency is stable for the desired maximum loading factor, that contingency will be eliminated from the list of harmful contingencies). The purpose of the search is to increase the value of λ_s and decrease the value of λ_u so that $\lambda_u - \lambda_s$ becomes less than a threshold value. In this case, λ_s is selected as λ_{max}. At each search step, the loading factor is placed in the middle point between λ_s and λ_u, and the system's time response is checked for it. If the system becomes stable (unstable) at this λ, the value λ is replaced by λ_s (λ_u), and the loading factor is chosen between λ_s and λ_u, again.

Figure 10.1 shows a simple representation of dynamic analysis in a two-dimensional power injection space. Without losing the generality, it is assumed that P_1 and P_2 are the active and reactive powers consumed at a bus. In this figure, P^0, P^1, and P^2 represent the power consumption vectors that are obtained according to Eq. (10.5) as λ increases. The directed curve also shows the transient changes of powers obtained during dynamic simulation. Also, Σ shows the maximum loading surface (MLS) based on the presence or absence of steady-state equilibrium point. The MLS is generally a hypersurface consisting of infinite number of points that depending on the direction of the power increment, the system's response trajectory in steady-state reaches to one of them. All points on Σ lead to singularity of the Jacobin matrix. The feasible and infeasible regions separated by this surface are actually separated based on the presence or absence of the equilibrium point.

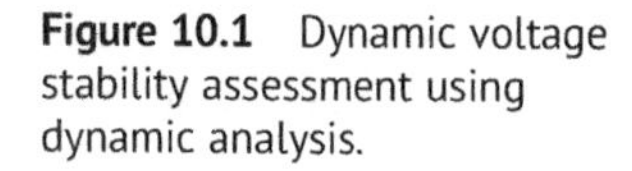
Figure 10.1 Dynamic voltage stability assessment using dynamic analysis.

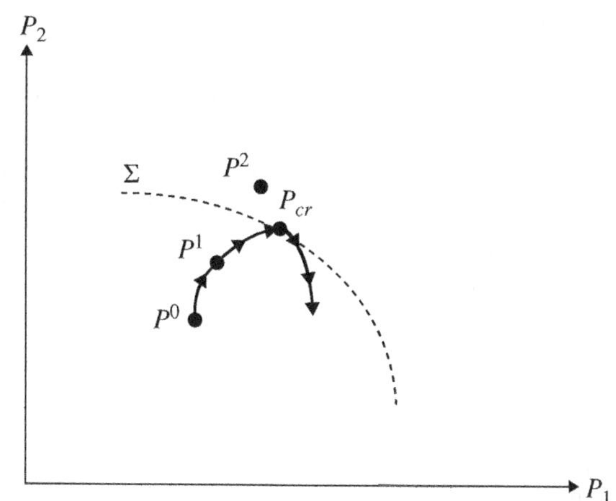

Therefore, the distance of the base case to this level indicates SLM. It is observed that since P^2 is outside the feasible region, increase of demand to P^2 leads to a voltage collapse and consequently a collapse in power consumption. P_{cr} is the point at which the trajectory of system response collides with the MLS. Instability may also occur with demands within the feasible region that can only be detected using dynamic analysis. But the system response trajectory will definitely collide with the MLS, which the information of collision point are used to determine the actions that have the greatest impact on increasing LM. Although these actions directly increase SLM, they also increase DLM.

The point P_{cr} can be detected using different voltage stability indices. In Refs. [1, 9, 10], the index of sensitivity of total generated reactive power to reactive power consumed by a load is used. As seen in Chapter 8, this sensitivity value reaches a big positive value in the vicinity of the MLS and jumps to a big negative value after crossing MLS. Also, when this level is reached, the Jacobin matrix related to the equilibrium equations becomes singular, and its minimum eigenvalue becomes zero. This eigenvalue changes sign and increases in size after passing through this surface. Accurate determination of P_{cr} requires continuous changes of the variables, but operation of excitation current limiters and other system protections may cause jumps in the variables' values. In this case, the mentioned sensitivity, without reaching very big values, changes the sign. The eigenvalue of the Jacobin matrix also changes the sign without reaching zero. To cross the MLS slowly and accurate determination of P_{cr}, as much as possible, it is better to occur instability at a load level close to the MLS for each contingency. A quick way to determine P_{cr} is to obtain it from studying the system's time response for each contingency for $P^d = P^0 + \lambda_{\mathrm{d}}^M K$ that $\lambda_{\mathrm{d}}^{\mathrm{M}}$ represents the desired LM. Contingencies that become unstable at the power level P^d are considered as harmful contingencies in preventive control. In this cases, due to the binary search methods are not used, the process of making λ_s and λ_s closer to each other does not occur and consequently LM is not calculated. In [1], a method based on the vector perpendicular to Σ is presented that calculates control actions without calculating LM. Of course, this method can only be implemented to change the active power of generators and load curtailment. Also, since LM is not determined, the probability of accurate determination of P_{cr} is also low.

Figures 10.2a and b show the probable impact of a contingency on MLS. The contingency can be outage of a line or a generator. In these figures, Σ' denotes the post-contingency situation. It is observed that the contingency causes the feasible region

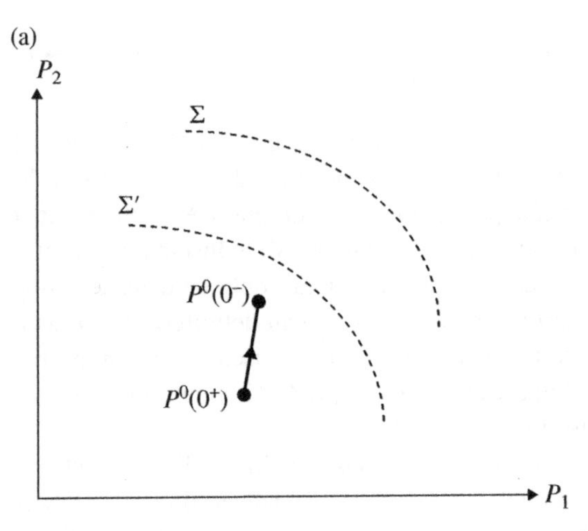

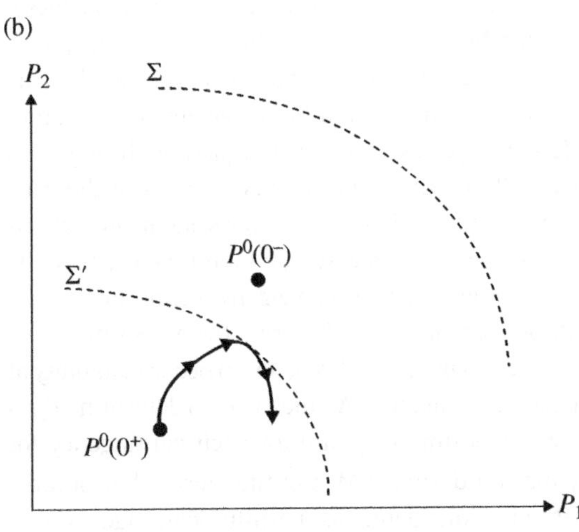

Figure 10.2 Impact of contingency on MLS. (a) Reduction of LM, (b) base case instability.

to shrink. Due to the instantaneous drop of voltages after occurrence of contingency, the consumed powers also immediately decrease. In these figures, $P^0(0^-)$ and $P^0(0^+)$, respectively, are the power vector before and after contingency occurrence. Assuming the constant power characteristic for loads, the powers are expected to recover to their base values after moments (Figure 10.2a). The contingency may be so severe that it is not even possible to provide the demanded base case power (Figure 10.2b). In this case, the system requires emergency actions.

10.3 Determination of the Optimal Value of Control Actions

As mentioned, control actions mainly include changing the voltage and active power of generators, changing the transformers taps, and changing the injected reactive power of compensators and load curtailment. The goal is to provide the desired LM for a range of probable and harmful contingencies by the least actions. Priority is given to actions that have minimum cost. For the sake of better comparison of actions, control variables are usually normalized so that their current value is zero and their maximum value is one. It is obvious that values less than their current values get negative sign. To provide the desired margin with the least cost, an optimization problem can be formulated with the following objective function [1, 2].

$$\min \sum_{j=1}^{n} W_j |\Delta u_j| \tag{10.8}$$

where $|\Delta u_j|$ is the magnitude of the change of the jth control variable, in other words, the magnitude of the jth control action. W_j is the weighting factor related to jth control action, which indicates its cost. n is the number of used control actions that are selected based on their effectiveness. To achieve the desired LM for NCT of the harmful contingencies, the inequality constraint (10.9) can be used.

$$\sum_{j=1}^{n} S_{ij} \Delta u_j \geq \lambda_{\mathrm{d}}^{\mathrm{M}} - \lambda_i^{\mathrm{M}}(U^0) \quad i = 1,\ldots,\mathrm{NCT} \tag{10.9}$$

$$u_j^{\min} \leq u_j \leq u_j^{\max} \tag{10.10}$$

where U^0 is the vector of control variables in the current state, λ_i^{M} is the LM value for these actions, $\lambda_{\mathrm{d}}^{\mathrm{M}}$ is also the desired value of LM, and S_{ij} is the sensitivity of LM to the jth control variable in the ith contingency, that its calculation will be explained later. Equation (10.9) with a linear approximation determine appropriate control actions for desired increase in LM. These actions are implemented in the current system, and it must be ensured that they do not cause exit of variables from the allowable limits. Meanwhile, if the change of the generated active power by generators and also load curtailment are used as control actions, maintaining equality between power generation and consumption in the system should be considered. To this purpose, each of these actions must be compensated by changing the power of some generators that accurate computation of compensation level is possible only when the change in the system losses is known. According to what mentioned above, it is necessary to add nonlinear equilibrium equations of the system as equality constraints to the optimization problem to calculate the amount of losses as well as the change of the variables values.

In addition to the equilibrium equations, which are in the form of Eqs. (10.3) and (10.4), the inequality constraints (10.11) must be added to ensure the allowable changes of the variables. In which, x_i denotes the ith variable, which can be, for example, a bus voltage magnitude or the amount of the power passing through a line. K is the number of these variables.

$$x_i^{\min} \leq x_i \leq x_i^{\max} \quad i = 1,\ldots, K \tag{10.11}$$

The objective function (10.8), which is a L_1 – norm, causes that among the control actions with the same cost those that have more sensitivity be selected, even if difference between these sensitivities is small. By doing so, the number of used control actions will be reduced and consequently their magnitudes will increase. This increases the error of using the sensitivities and used approximations. A solution is to limit the magnitude of control actions. Another solution is to use L_2 – norm objective as Eq. (10.12) [1].

$$\min \sum_{j=1}^{n} W_j (\Delta u_j)^2 \tag{10.12}$$

This equation leads to more control actions with less magnitudes.

When, as Eq. (10.9), the magnitude of control actions is obtained based on linearization solving the optimization problem may need to be repeated due to two reasons [1].

1) Since the calculated actions are determined based on linearization and sensitivity analysis, in comparison with true nonlinear equations, they may lead to LM less than $\lambda_{\mathrm{d}}^{\mathrm{M}}$ for a number of contingencies or more than $\lambda_{\mathrm{d}}^{\mathrm{M}}$ for all contingencies. To alleviate this problem, all sensitivities related to a contingency will be modified using the following coefficient.

$$\frac{\lambda_i^{\mathrm{M}}(U^0 + \Delta U) - \lambda_i^{\mathrm{M}}(U^0)}{\sum_{j=1}^{n} S_{ij} \Delta u_j} \tag{10.13}$$

where $\lambda_i^{\mathrm{M}}(U^0)$ and $\lambda_i^{\mathrm{M}}(U^0 + \Delta U)$ are the LM values related to ith contingency before and after implementation of actions calculated based on nonlinear equations.

2) In the optimization problem, only harmful contingencies (contingencies with LM less than $\lambda_{\mathrm{d}}^{\mathrm{M}}$) are selected. However, after applying actions, some of harmless contingencies may become harmful. A solution to alleviate this problem is taking contingencies with LM between $\left[\lambda_{\mathrm{d}}^{\mathrm{M}}\ \lambda_{\mathrm{d}}^{\mathrm{M}'}\right]$ into the account. In fact, it is assumed that after applying the actions, contingencies with margin more than $\lambda_{\mathrm{d}}^{\mathrm{M}'}$ (much more than $\lambda_{\mathrm{d}}^{\mathrm{M}}$) will not have margins less than $\lambda_{\mathrm{d}}^{\mathrm{M}}$. Another way is to select the contingencies initially based on $\lambda_{\mathrm{d}}^{\mathrm{M}}$. If after solving the optimization problem and calculating the margin related to contingencies using nonlinear equations, some of contingencies get margins less than $\lambda_{\mathrm{d}}^{\mathrm{M}}$, they will also be added to the set of harmful contingencies and the problem is solved again.

The criterion to finish the calculations is providing desired margin for all contingencies, while the margin be approximately equal to $\lambda_{\mathrm{d}}^{\mathrm{M}}$ for at least one of the contingencies.

10.4 Computation of Sensitivities

10.4.1 Computation of Sensitivities Based on the Computation of MLP

Owing to singularity of the Jacobin matrix of equilibrium equations at MLS, there is the following equation [11],

$$W(\boldsymbol{Z}_*, \boldsymbol{U}_*, \boldsymbol{P}_*)\boldsymbol{H}_{\boldsymbol{Z}}(\boldsymbol{Z}_*, \boldsymbol{U}_*, \boldsymbol{P}_*) = 0 \tag{10.14}$$

where W is the left eigenvector related to the zero eigenvalue of matrix $\boldsymbol{H}_{\boldsymbol{Z}}$. $\boldsymbol{P}$ is the vector of injected powers that due to its relationship with λ [Eq. (10.5)] has been replaced in Eq. (10.7). The sign $*$ indicates the values of the vectors in MLS. By linearization of equilibrium equations in MLP, the following equation is produced [11]:

$$\boldsymbol{H}_{\boldsymbol{Z}}|_*\Delta\boldsymbol{Z}_* + \boldsymbol{H}_{\boldsymbol{U}}|_*\Delta\boldsymbol{U}_* + \boldsymbol{H}_{\boldsymbol{P}}|_*\Delta\boldsymbol{P}_* = 0 \tag{10.15}$$

where $\boldsymbol{H}_{\boldsymbol{U}}$ and $\boldsymbol{H}_{\boldsymbol{P}}$, respectively, are the derivatives of the matrix $\boldsymbol{H}$ with respect to the vectors $\boldsymbol{U}$ and $\boldsymbol{P}$. Furthermore, " $|_*$ " means that these derivatives are calculated at $(\boldsymbol{Z}_*, \boldsymbol{U}_*, \boldsymbol{P}_*)$. By multiplying the both sides of the Eq. (10.15) by $W(\boldsymbol{Z}_*, \boldsymbol{U}_*, \boldsymbol{P}_*)$ and using the Eq. (10.14),

$$\boldsymbol{W}\boldsymbol{H}_{\boldsymbol{U}}|_*\Delta\boldsymbol{U}_* + \boldsymbol{W}\boldsymbol{H}_{\boldsymbol{P}}|_*\Delta\boldsymbol{P}_* = 0 \tag{10.16}$$

On the other hand, from the Eq. (10.5),

$$\Delta\boldsymbol{P}_* = \boldsymbol{K}\Delta\lambda_* \tag{10.17}$$

By substituting in (10.16),

$$\boldsymbol{WH_U}|_* \Delta \boldsymbol{U}_* + \boldsymbol{WH_P}|_* \boldsymbol{K} \Delta \lambda_* = 0 \tag{10.18}$$

From the earlier equation, the vector of sensitivity of LM to control actions is obtained as Eq. (10.19).

$$\boldsymbol{S} = \frac{-\boldsymbol{WH_U}|_*}{\boldsymbol{WH_P}|_* \boldsymbol{K}} \tag{10.19}$$

10.4.2 Computation of Sensitivities Without the Computation of MLP

Ref. [5] provides an approximate method for calculating LM sensitivity to control actions that determines this sensitivity without accurate calculation of MLP. Of course, due to the inaccurate amount of this sensitivity, it is used only to determine effective actions (not to determine their optimal amount). Another used assumption is that the conventional power flow equations are used as the equilibrium equations. The look-ahead method is used to estimate MLP [12]. This method is described in Chapter 6 and is briefly described here as a reminder. In this method, estimation of MLP is done using approximation of the PV curve as a quadratic function. To approximate the PV curve, two points obtained from the power flow solutions are used. This approximation is done at the most sensitive bus of system that is called the pilot bus. In Ref. [5], the Eq. (10.20) is used to select the most sensitive bus. This equation shows the relative variations in voltage magnitude at the bus i in two power flow solutions. The bus with the most voltage variation is chosen as the pilot bus.

$$\Delta V_i = \frac{V_i|\lambda_1 - V_i|\lambda_2}{V_i|_{\lambda_1}} \tag{10.20}$$

The following quadratic function is used to approximate the PV curve:

$$\lambda = \alpha + \beta V_{\mathrm{P}} + \gamma V_{\mathrm{P}}^2 \tag{10.21}$$

where V_{P} is the voltage magnitude at the pilot bus and α, β, and γ are the parameters used to approximate the PV curve. By solving power flow equations with the loading factors λ_1 and λ_2 and calculating the pilot bus voltage at these factors, two equations of (10.22) and (10.23) are obtained.

$$\lambda_1 = \alpha + \beta V_{\mathrm{P},1} + \gamma V_{\mathrm{P},1}^2 \tag{10.22}$$

$$\lambda_2 = \alpha + \beta V_{\mathrm{P},2} + \gamma V_{\mathrm{P},2}^2 \tag{10.23}$$

To find the parameters' values, three equations are required, that the third one is obtained from derivative of the second equation with respect to λ. The matrix representation of these three equations is as Eq. (10.24).

$$\begin{bmatrix} \lambda_1 \\ \lambda_2 \\ 1 \end{bmatrix} = \begin{bmatrix} 1 & V_{P,1} & V_{P,1}^2 \\ 1 & V_{P,2} & V_{P,2}^2 \\ 0 & \frac{dV_{P,2}}{d\lambda_2} & 2\frac{dV_{P,2}}{d\lambda_2}V_{P,2} \end{bmatrix} \begin{bmatrix} \alpha \\ \beta \\ \gamma \end{bmatrix} \tag{10.24}$$

How to calculate $\frac{dV_{P,2}}{d\lambda_2}$ has been described in Section 6.3. The estimated value of MLP is calculated from derivative of λ with respect to V_P.

$$\frac{d\lambda}{dV_P} = \beta + 2\gamma V_P^* = 0 \Rightarrow V_P^* = \frac{-\beta}{2\gamma} \tag{10.25}$$

where V_P^* is the magnitude of the pilot bus voltage at MLP. Now, by substituting V_P^* in Eq. (10.21), the estimated value of MLP is obtained.

$$\lambda_{max}^{estimated} = \alpha - \frac{\beta^2}{4\gamma} \tag{10.26}$$

The sensitivity of the estimated value of MLP to the control variable u_C can be obtained from derivative of the Eq. (10.26).

$$\frac{d\lambda_{max}^{estimated}}{du_C} = \frac{\partial\alpha}{\partial u_C} - \frac{\beta}{2\gamma}\frac{\partial\beta}{\partial u_C} + \frac{\beta^2}{4\gamma^2}\frac{\partial\gamma}{\partial u_C} \tag{10.27}$$

Calculation of this sensitivity requires determining the derivatives of α, β, and γ with respect to u_C. To do this, at first, from Eq. (10.24)

$$\alpha = f_\alpha\left(V_{P,1}, V_{P,2}, V_2^0, \lambda_1, \lambda_2\right) \tag{10.28}$$

$$\beta = f_\beta\left(V_{P,1}, V_{P,2}, V_2^0, \lambda_1, \lambda_2\right) \tag{10.29}$$

$$\gamma = f_\gamma\left(V_{P,1}, V_{P,2}, V_2^0, \lambda_1, \lambda_2\right) \tag{10.30}$$

where $\frac{\partial V_{P,2}}{\lambda}$ has been indicated by V_2^0. By maintaining λ_1 and λ_2 constant,

$$\frac{d\alpha}{du_C} = \frac{\partial f_\alpha}{\partial V_{P,1}}\frac{\partial V_{P,1}}{\partial u_C} + \frac{\partial f_\alpha}{\partial V_{P,2}}\frac{\partial V_{P,2}}{\partial u_C} + \frac{\partial f_\alpha}{\partial V_2^0}\frac{\partial V_2^0}{\partial u_C} \tag{10.31}$$

$$\frac{d\beta}{du_C} = \frac{\partial f_\beta}{\partial V_{P,1}}\frac{\partial V_{P,1}}{\partial u_C} + \frac{\partial f_\beta}{\partial V_{P,2}}\frac{\partial V_{P,2}}{\partial u_C} + \frac{\partial f_\beta}{\partial V_2^0}\frac{\partial V_2^0}{\partial u_C} \tag{10.32}$$

$$\frac{d\gamma}{du_C} = \frac{\partial f_\gamma}{\partial V_{P,1}}\frac{\partial V_{P,1}}{\partial u_C} + \frac{\partial f_\gamma}{\partial V_{P,2}}\frac{\partial V_{P,2}}{\partial u_C} + \frac{\partial f_\gamma}{\partial V_2^0}\frac{\partial V_2^0}{\partial u_C} \tag{10.33}$$

In these equations, $\frac{\partial V_{P,1}}{\partial u_C}$, $\frac{\partial V_{P,2}}{\partial u_C}$, and $\frac{\partial V_2^0}{\partial u_C}$ are unknowns that must be calculated. V_P is a power flow variable. Let us assume that the vector of power flow variables is represented as $\boldsymbol{X}$, and the power flow equations are compactly as.

$$\boldsymbol{G}(\boldsymbol{X}, \boldsymbol{u}_C, \lambda) = 0 \tag{10.34}$$

By differentiating from the earlier equation,

$$\frac{\partial \boldsymbol{G}}{\partial \boldsymbol{X}} \Delta \boldsymbol{X} + \frac{\partial \boldsymbol{G}}{\partial \boldsymbol{u}_C} \Delta \boldsymbol{u}_C + \frac{\partial \boldsymbol{G}}{\partial \lambda} \Delta \lambda = 0 \tag{10.35}$$

Assuming that λ is constant,

$$\frac{\partial \boldsymbol{G}}{\partial \boldsymbol{X}} \frac{\Delta \boldsymbol{X}}{\Delta \boldsymbol{u}_C} = -\frac{\partial \boldsymbol{G}}{\partial \boldsymbol{u}_C} \tag{10.36}$$

where $\frac{\partial \boldsymbol{G}}{\partial \boldsymbol{X}}$ is the Jacobian matrix of the power flow equations and $\frac{\partial \boldsymbol{G}}{\partial \boldsymbol{u}_C}$ is a known vector. Solving Eq. (10.36) for λ_1 and λ_2, an approximation of $\frac{\partial V_{P,1}}{\partial u_C}$ and $\frac{\partial V_{P,2}}{\partial u_C}$ is obtained. To calculate $\frac{\partial V_2^0}{\partial u_C}$, initially from Eq. (10.35) under the assumption that u_C is constant,

$$\frac{\partial \boldsymbol{G}}{\partial \boldsymbol{X}} \frac{\Delta \boldsymbol{X}}{\Delta \lambda} + \frac{\partial \boldsymbol{G}}{\partial \lambda} = 0 \tag{10.37}$$

The method of calculation $\Delta \boldsymbol{X}^0$ (i.e. $\frac{\Delta \boldsymbol{X}}{\Delta \lambda}$) has been described in Section 6.3. Now, using differentiation $\frac{\Delta \boldsymbol{X}}{\Delta \lambda}$ with respect to u_C,

$$\frac{\partial \boldsymbol{G}}{\partial \boldsymbol{X}} \frac{\Delta \boldsymbol{X}^0}{\Delta \boldsymbol{u}_C} = -\frac{\partial^2 \boldsymbol{G}}{\partial \boldsymbol{u}_C \partial \boldsymbol{X}} \frac{\Delta \boldsymbol{X}}{\Delta \lambda} - \frac{\partial^2 \boldsymbol{G}}{\partial \boldsymbol{u}_C \partial \lambda} \tag{10.38}$$

From the earlier equation, $\frac{\Delta \boldsymbol{X}^0}{\Delta \boldsymbol{u}_C}$ is obtained. With calculation of this derivative and derivatives of $\frac{\partial V_{P,1}}{\partial u_C}$ and $\frac{\partial V_{P,2}}{\partial u_C}$, the values of sensitivities in the Eqs. (10.31)–(10.33) and then $\frac{d\lambda_{max}^{estimated}}{du_C}$ are obtained.

There are two points about calculating the sensitivity of LM to the control actions. The first one, reducing the power consumption of a load leads to a reduction in both active and reactive powers. Therefore, if only reduction of the load active power is considered as a control action, the effect of the reactive power reduction must also be considered. Assuming that, before load reduction, the ratio of active power to reactive power at the bus j is α and the reduction of active and reactive powers is done with the same ratio, the sensitivity of the active power of the bus j must be modified as follow [1]:

$$S_{ij} = \frac{\partial \lambda_i^{\mathrm{M}}}{\partial P_j} + \alpha_j \frac{\partial \lambda_i^{\mathrm{M}}}{\partial Q_j} \tag{10.39}$$

Changing the active power of generators also changes their reactive power. This happens even when the generator has reached its reactive power limit, due to the dependence of the reactive power limit on the generated active power (Section 3.2.7). The change in reactive power has no effect on the LM until the generator does not reach its reactive power limit and its voltage is controlled. But, after encountering the reactive power limit, the reactive power change is effective, and it must be taken into the sensitivity calculation account. Therefore, in these conditions, the sensitivity related to the generator active power must be modified as Eq. (10.39). In this case, the coefficient α becomes a negative number obtained from the slope of the generator power capability curve (Figure 3.6) [1].

The second point is that, as mentioned, if the aim is change in a generator active power or the load power consumption reduction, these actions must be compensated by changing the power of some other generators. Therefore, the LM sensitivity to the mentioned actions must be modified by its sensitivity to compensating actions. For example, if a generator power increment is compensated by decrease in another generator production, the difference between the two generators' sensitivities should be inserted in Eq. (10.9).

10.5 Determination of the Most Effective Actions

In [5], the sensitivity of LM to control actions is used to determine the most effective actions. By normalizing the control variables, these sensitivities show how much the LM changes as a control variable changes. Hence, this sensitivity shows the availability and effectiveness of a control action [6]. It is possible that a control action has significant effect on increasing LM for a number of contingencies, and its effect on other contingencies be small or even destructive. In Ref. [6], a method is proposed in which selection of effective actions is based on their impact on all considered contingencies, while the cost of each action is also taken into account. To this aim, the normalized sensitivity vector is first calculated for different contingencies. Eq. (10.40) shows this vector for the ith contingency.

$$S_i^n = \frac{S_i}{\max\left(|S_i|\right)} \tag{10.40}$$

The elements of vector S_i are sensitivities of LM to various actions when the ith contingency occurs, and $\max(|S_i|)$ represent the maximum magnitude of these sensitivities. If C_C be the cost of increasing the control variable u_C from the current value to the maximum value, to insert the control action cost, the normalized value of each sensitivity is divided by its cost.

$$S_{iC}^{cost} = \frac{S_{iC}^{n}}{C_C} \tag{10.41}$$

where S_{iC}^{n} is the normalized value of the LM sensitivity to u_C when the ith contingency occurs. The index $\mathrm{CEI_C}$ is now defined as Eq. (10.42), which indicates the effect of the u_C change to eliminate the criticality of all contingencies, taking into account its availability, effectiveness, and cost.

$$\mathrm{CEI_C} = \sum_{i=1}^{\mathrm{NCT}} S_{iC}^{cost} \tag{10.42}$$

where NCT is the number of considered contingencies. It is assumed that Ω represents the vector of MLPs for all contingencies.

$$\Omega = [\, \lambda_{\max_1}, \quad \lambda_{\max_2}, \quad \cdots \quad \lambda_{\max_{\mathrm{NCT}}} \,] \tag{40.43}$$

The goal is to provide the minimum desired value of MLP (λ_d^M) for all contingencies at the lowest cost. Determination of actions is done in two phases. The number of actions is assumed to be NC.

- Top-Down Phase

In this phase, the controls are first sorted based on the CEI value in a descending order. Then, the control actions with the highest CEI value are implemented one by one and after implementation of each control action, the change in Ω is determined approximately as follows:

$$\Omega_i \leftarrow \Omega_i + S_{iC} \quad i = 1, \ldots, \mathrm{NCT} \tag{10.44}$$

If $\min(\Omega) \le \lambda_d^M$ and $C \le NC$, the next control action is applied based on the CEI value. This continues until $\min(\Omega) \ge \lambda_d^M$.

- Bottom-Up Phase

A number of low-effectiveness control actions may be prioritized due to low cost. But, more effective actions such as load curtailment due to the high cost will be transferred to the bottom of the list. Accordingly, the control actions obtained in the top–down phase may be more than necessary. To alleviate this problem in bottom–up phase, from the bottom of the list obtained in top–down phase, the actions are removed one by one, and the following algorithm is persuaded.

$$\Omega_i \leftarrow \Omega_i - S_{iC} \quad i = 1,\ldots,\mathrm{NCT}$$

$$\text{If } \min(\Omega) \geq \lambda_{\mathrm{d}}^{\mathrm{M}} \quad \text{then}$$

$$G \leftarrow G - u_{\mathrm{C}}$$

$$\Omega \leftarrow \Omega$$

In this algorithm, G is the set of actions obtained in the top–down phase.

In Ref. [3], the sensitivity of a voltage stability index to control actions is used to determine appropriate actions. This index, called the load impedance modulus margin (LIMM), is based on the distance between the load impedance magnitude and the Thevenin impedance magnitude, as shown in Eq. (10.45). The bus with the lowest index value is selected as the pilot bus, and the sensitivity of the index at that bus to control actions will be investigated. A full description of how this sensitivity is calculated is given in [3].

$$\mathrm{LIMM} = \frac{|Z_{\mathrm{L}}| - |Z_{\mathrm{th}}|}{|Z_{\mathrm{L}}|} \tag{10.45}$$

10.6 Summary

This chapter was about description of preventive control. The purpose of preventive control is to determine the proper control actions to provide a desired load margin for a set of probable contingencies. Achieving this goal requires solving an optimization problem. To simplify control actions, instead of using true nonlinear equations, linear equations obtained from calculating the LM sensitivity to control measures are used. Obtained actions using linear equations may not provide the expected LM for an actual system consisting of nonlinear equations. This issue should be investigated and the calculations repeated if necessary. In this chapter, methods for determining more effective control actions were also presented.

References

1 Capitanescu, F. and Van Cutsem, T. (2002). Preventive control of voltage security margins: a multicontingency sensitivity-based approach. *IEEE Trans. Power Syst.* 17 (2): 358–364.

2 Feng, Z., Ajjarapu, V., and Maratukulam, D.J. (2000). A comprehensive approach for preventive and corrective control to mitigate voltage collapse. *IEEE Trans. Power Syst.* 15 (2): 791–797.

3 Li, S., Tan, Y., Li, C. et al. (2017). A fast sensitivity-based preventive control selection method for online voltage stability assessment. *IEEE Trans. Power Syst.* 33 (4): 4189–4196.
4 Canizares, C.A. (1998). Calculating optimal system parameters to maximize the distance to saddle-node bifurcations. *IEEE Trans. Circuits Syst. I Fundam. Theory Appl.* 45 (3): 225–237.
5 Mansour, M.R., Geraldi, E.L., Alberto, L.F.C., and Ramos, R.A. (2013). A new and fast method for preventive control selection in voltage stability analysis. *IEEE Trans. Power Syst.* 28 (4): 4448–4455.
6 Mansour, M.R., Fernando, L., Alberto, C. et al. (2015). Preventive control design for voltage stability considering multiple critical contingencies. *IEEE Trans. Power Syst.* 31 (2): 1517–1525.
7 Feng, Z., Ajjarapu, V., and Long, B. (2000). Identification of voltage collapse through direct equilibrium tracing. *IEEE Trans. Power Syst.* 15 (1): 342–349.
8 Van Cutsem, T. and Vournas, C. (1988). *Voltage Stability of Electric Power Systems.* Norwell, MA: Kluwer.
9 Van Cutsem, T., Capitanescu, F., Moors, C. et al. (2000). An advanced tool for preventive voltage security assessment. *Proceedings of the 7th Symposium of Specialists in Electric Operational and Expansion Planning (SEPOPE)*, Curitiba, Brazil, May 2000.
10 Van Cutsem, T. (1995). An approach to corrective control of voltage instability using simulation and sensitivity. *IEEE Trans. Power Syst.* 10 (2): 616–622.
11 Greene, S., Dobson, I., and Alvarado, F.L. (1997). Sensitivity of the loading margin to voltage collapse with respect to arbitrary parameters. *IEEE Power Eng. Rev.* 12 (1): 262–272.
12 Chiang, H.D., Wang, C.S., and Flueck, A.J. (1997). Look-ahead voltage and load margin contingency selection: functions for large-scale power systems. *IEEE Power Eng. Rev.* 12 (1): 173–180.

11

Emergency Control of Voltage Instability

11.1 Introduction

As mentioned in Chapter 10, in preventive control, the system situation when probable contingencies occur is assessed. In which, by applying appropriate control actions, the desired stability margin against these contingencies is obtained. However, the considered contingencies usually include just an element outage, and the simultaneous outages of two or more elements are not considered. On the other hand, it may not be economically viable to take preventive actions against a number of major contingencies due to their low probability of occurrence [1]. Hence, the power system may always encounter contingencies for which appropriate actions have not already been taken. In this situation, the system needs emergency actions. Therefore, the purpose of emergency actions, unlike preventive actions, is not to increase the stability margin, but the goal is to stabilize an unstable system, which if not done, the voltage collapse will occur in a few moments or minutes.

Emergency actions against voltage instability and collapse are from measures called System Protection Schemes (SPS) that aim to detect abnormal system situation and take the necessary control actions. In response-based SPS, the emergency situation is determined based on the measured variables, but in the event-based type, the occurrence of predetermined critical contingencies indicates the emergence of an emergency situation [1]. Of course, in this type of SPS, it is necessary for the control center to determine the impact of critical contingencies based on the current operating point [2]. Event-based SPS is suitable for cases where there is very little opportunity to take actions, which usually does not involve long-term voltage instability [3]. Response-based SPSs have the ability to determine emergency actions based on the contingency severity.

Actions commonly used as emergency actions include changing generators' voltage, shunt capacitor, and inductor switching, changing the transformers'

Voltage Stability in Electrical Power Systems: Concepts, Assessment, and Methods for Improvement, First Edition. Farid Karbalaei, Shahriar Abbasi, and Hamid Reza Shabani.

Published 2023 by John Wiley & Sons, Inc.

tap, changing the output reactive power of distributed generation sources, and load shedding. Appropriate actions must be applied immediately after an emergency situation is diagnosed. Due to the problem of exiting from the attraction region, the speed of applying actions is very important because the longer the actions are delayed, the more actions will be needed to prevent the voltage collapse.

In addition to exiting from the attraction region, delays in applying the actions can cause problems such as stalling of motors and loss of synchronization, sudden voltage drop in generators that have reached their excitation current, and consequently outage of them due to undervoltage protection as well as line outage due to their protection system activation. Meanwhile, the durability of the voltage drop at the terminals of consumers can cause problems in their performance [1]. Traditional SPSs are mostly rule-based, meaning that emergency actions are calculated and applied based on a set of predetermined rules [4]. The advantage of this method is its simplicity and fast applying speed [5]. But, it is very difficult to coordinate actions [6]. Under voltage load shedding (UVLS) is an example of this type of SPS.

It is desirable that in emergency situations, among the various available controls, the optimal actions are determined and applied in real-time. In real-time SPS, emergency actions are calculated using the real-time system model. This method, which usually uses the model predictive control (MPC), allows coordination among different control actions, but requires solving a large optimization problem with a combination of continuous and discrete variables. In addition, its implementation also requires global state estimation and appropriate telecommunication infrastructures. These problems, in addition to the very limited time available to calculate and apply actions for voltage emergency control, have led many of the proposed methods to focus only on how to apply load shedding as the most effective action to prevent voltage collapse, rather than coordinating and optimally determining different actions. Determining the time, place, and amount of load shedding is done either with the help of if-then rules, which are mainly based on off-line calculations [1, 3, 7–9] or based on limited real-time calculations [10].

11.2 Load Shedding

Load shedding as an emergency action has always been of interest to researchers. In whole studies, the main goal is to recover voltage with the minimum shed load. In many of the proposed methods, detection of emergency situations and need to shed load is done by measuring voltages, which is called under voltage load shedding (UVLS) [1, 3, 7–9]. Of course, other indices such as change in produced reactive power by generators [6] and sensitivity of total generated reactive power to loads' reactive power [10] have also been used.

In UVLS, each controller may monitor a transmission bus voltage [1, 7, 8] or the average value of voltages at several transmission buses is used [3, 9]. In the first case, the outage command will be send to the distribution-level feeders associated with that transmission bus. Distribution feeders may be fed from the secondary side of transformer connected to the transmission bus [7, 8], or may be indirectly connected to the bus via a subtransmission network [1]. In the second case, ranking of load buses for load shedding is done based on the information contained in the eigenvectors of the Jacobin matrix. In addition to the magnitude of the voltages, other variables such as incremental rate of the generators' excitation current may also be used to assess the system state [7, 8]. According to what mentioned above, load shedding may be based on local information [1] or a wide-area view of system [6–8, 10], called local and wide-area load shedding, respectively.

11.2.1 UVLS against Long-term Voltage Instability

11.2.1.1 Centralized Rule-based Controller

UVLS methods are commonly based on a number of if-then rules [3, 9]:

$$\text{if } V < V_i^{\text{th}} \text{ during } \tau_i \text{ seconds, shed } \Delta P_i^{\text{sh}} \text{ MW}$$

where it states that if the voltage remains less than the threshold value V_i^{th} for τ_i seconds, load shedding will be applied with the amount of ΔP_i^{sh} MW. Here, V is the average value of voltages at several transmission buses in an area prone to voltage instability. The system weak areas can be found with different voltage stability indices. The delay τ_i is intended to ensure that, first, the voltage drop is due to voltage instability and is not related to transient faults, and second, there is enough time to take other actions that may automatically be activated to help voltage recovery when the voltage drops (shunt reactor tripping and capacitor switching are from these actions).

Voltages are continuously monitored, and the load shedding is continued until the voltages are recovered to the desired values. Hence, the rule-based controller creates a closed-loop controller. If the delay and step size of the load shedding be constant, the corresponding controller is called fixed-steps fixed delays (FSFD) [9]. In these conditions, two or three rules are usually considered. Optimal performance is achieved when the delay in load shedding, and the amount of shed load is adjusted according to the rate of voltage drop. The higher the rate, the more load must be shed to return the system's demanded load to the loadability region. Accordingly, in reference [9], three other types of rule-based controllers are proposed as follows:

- Variable-step variable delay controller (VSFD) in which both delay and load shedding step size are variable.

- Variable-step fixed delay (VSFD) controller in which delay is fixed and load shedding step size is variable.
- Fixed-step variable delay (FSVD) controller in which delay is variable and load shedding step size is fixed.

The delay proportional to the rate of voltage drop is obtained according to the following equation:

$$\int_{t_0}^{t_0+\tau} \left(V^{\text{th}} - V\right)dt = C, \quad \tau_{\min} \le \tau \le \tau_{\max} \tag{11.1}$$

where t_0 is the time at which V becomes less than V^{th}. C and V^{th} are parameters that must be specified in designing of controller. Given Eq. (11.1), it is clear that the faster the voltage drops, the less delay there will be for load shedding.

Also, the load shedding step size in proportion to the rate of voltage drop is determined by Eq. (11.2).

$$\Delta P^{\text{sh}} = K\Delta V_{\text{avg}}, \quad \Delta P_{\min} \le \Delta P^{\text{sh}} \le \Delta P_{\max} \tag{11.2}$$

where ΔV_{avg} is the average voltage drop during $[t_0 \ t_0 + \tau]$, and K is another controller parameter.

$$\Delta V_{\text{avg}} = \frac{1}{\tau}\int_{t_0}^{t_0+\tau} \left(V^{\text{th}} - V\right)dt \tag{11.3}$$

Here, it is also observed that a faster voltage drop causes increase in ΔV_{avg} and consequently increase in load shedding. Since load shedding is done by disconnecting the feeders, the minimum load shedding can be found based on the load of feeders. The maximum allowable amount of load shedding is also considered to prevent unacceptable transients [1]. Based on what has been mentioned, the vector of parameters that must be determined in designing of controllers is as Eqs. (11.4)–(11.7). In the FSFD controller, the vector of parameters is K rules.

$$\boldsymbol{X}_{\text{FSFD}} = \left[V_1^{\text{th}}, \tau_1, \Delta P_1^{\text{sh}}, \ldots, V_k^{\text{th}}, \tau_k, \Delta P_k^{\text{sh}}\right] \tag{11.4}$$

$$\boldsymbol{X}_{\text{VSVD}} = \left[V^{\text{th}}, C, \tau_{\min}, \tau_{\max}, K, \Delta P_{\min}, \Delta P_{\max}\right] \tag{11.5}$$

$$\boldsymbol{X}_{\text{FSVD}} = \left[V^{\text{th}}, C, \tau_{\min}, \tau_{\max}, \Delta P^{\text{sh}}\right] \tag{11.6}$$

$$\boldsymbol{X}_{\text{VSFD}} = \left[V^{\text{th}}, \tau, K, \Delta P_{\min}, \Delta P_{\max}\right] \tag{11.7}$$

In addition to rules related to load shedding, some rules for coordinating with other controllers can be considered. For example, shunt compensation switching is one of these actions that can be applied by the following rule [9].

$$\text{if } V < V_{\text{Com}}^{\text{th}} \text{ during } \delta \text{ (or } \delta'\text{) seconds,switch one more shunt}$$

Usually, $V_{\text{Com}}^{\text{th}}$ is chosen larger than V^{th} related to the load shedding. δ is related to the delay in the first action, which should be large enough to do not react to temporary voltage drops. δ' is the delay between two successive switching, too.

- Design of Controllers [3, 9]

The design will be done in two stages. In the first stage, a number of training scenarios are produced, which can include outage of one or more elements at different operating points. For each unstable scenario, the minimum load shedding to stabilize that scenario is determined. This can be done using different search methods such as binary search. The exact effect of load shedding can be determined by time-domain simulations of system behavior for different amounts of load shedding. Time simulation of an unstable scenario is used to find proper locations for load shedding. When the system response trajectory collides with the loadability limit, which is when the eigen or singular value of the Jacobin matrix becomes zero, the right singular vector elements indicate the system's weak buses and ranking of load buses for load shedding. Given that the controllers of different areas are designed independently, it is assumed that for all unstable scenarios related to an area, there are the same suitable locations for load shedding. When the load shedding is required, initially all the interruptible load at the bus with the highest priority is shed. Then if necessary, the load of the other buses that have the next priority will be shed, one by one.

In the second stage, the parameters are determined in such a way that in all scenarios, the shed load is as close as possible to the values obtained in the first stage. In addition, the parameters must be set in such a way that all unstable scenarios become stable (dependability) and in stable scenarios no load shedding occurs (security). Another goal may be that the voltages should not be less than a certain value for a certain time. A complete description of objective functions, constraints, and optimization methods for design of load shedding is given in [9].

11.2.1.2 Distributed Rule-based Controller

Rule-based controllers proposed in [3, 9] require a central controller, because they detect the need to load shedding based on measuring the average value of voltage at several buses and also they send the load shedding command to different buses in a coordinated manner. In Ref. [1], a distributed rule-based controller with variable load shedding step size and variable delay has been presented. In this method, each controller monitors only one transmission bus voltage and send the load shedding command to a number of distribution network feeders connected to that transmission bus (Figure 11.1).

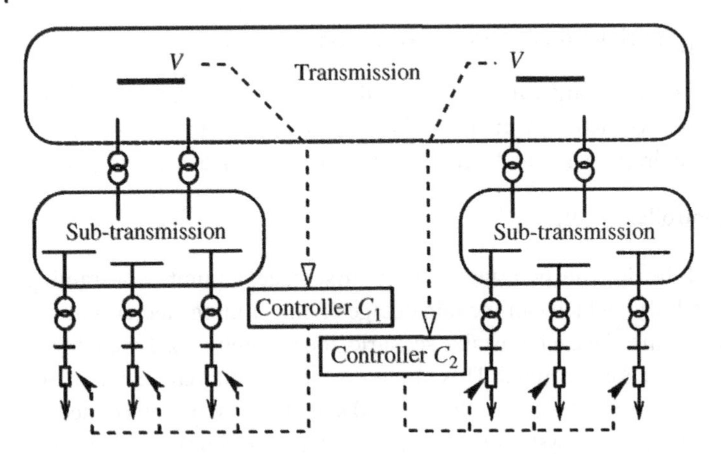

Figure 11.1 The structure of the distributed rule-based controller [1].

The controllers are distributed at points prone to voltage instability. Although they seem to operate independently and without any coordination with each other, when a controller tries to shed load, not only voltage of the related bus but also voltage at other nearby buses start to increase. This issue due to the Eqs. (11.1) and (11.2) causes more delay in the adjacent controllers and also reduction of the shed load applied by them. Therefore, without transmitting information through a communication network, a cooperation is made in the controllers of an area that prevent additional load shedding.

- Design of Controllers

The parameters of method including V^{th}, K, and C [Eqs. (11.1)–(11.3)] are determined based on a set of different scenarios that are generated using different contingencies in different system conditions. The parameters are selected so that the following goals be achieved in the all scenarios:

1) The controllers do not react to the stable scenarios.
2) The controllers recover the voltage in facing with each unstable scenarios, with the minimum shed load.

The value of Vth should be selected large enough to prevent excess delay that may lead to additional load shedding. It should also be chosen small enough not to react to stable scenarios. Therefore, in reference [1], it is suggested that V^{th} be a slight less than the minimum voltage appeared in the stable scenarios. After selecting V^{th}, to determine C and K, different scenarios in which load shedding is required will be simulated with different combinations of C and K. Then, a combination at which in all scenarios, the stability is ensured with the minimum shed load will be selected. Of course, in this selection, some considerations must also be respected. For example, a very small amount of K is not practical because, as mentioned, load shedding is applied by disconnecting the feeders, so the shed load cannot be less than the minimum load supplied by feeders.

11.2.1.3 Two-level Rule-based Controller

In the distributed controller described in the previous section, V^{th} is determined offline and based on different contingencies in different conditions. In reference [11], the distributed controller, which is a local controller, is called the lower-level controller, whose value V^{th} is determined real-time using an upper-level controller. The purpose of the upper-level controller, which is a central controller, is early detection of long-term voltage instability. For this purpose, whether or not the system trajectory reaches the maximum loadability limit is determined. This can be done using sensitivity analysis (Section 11-2-4) or the method used in section 7-4. The transmission buses voltage when the system trajectory reaches the maximum loadability level is chosen as V^{th} and sent to the lower-level controller.

Another emergency action suggested in [11] is to modify LTC control in transmission transformers. As long as the transmission side voltage is higher than $V^{th} + \delta$, the LTC operation is normal and the distribution side voltage is controlled. But when the transmission side voltage becomes less than $V^{th} - \delta$, the LTC preserves the transmission side voltage instead of regulating the distribution side voltage. In distance between $[V^{th} - \delta, V^{th} + \delta]$ the tap is kept constant. The value of 2δ is considered equal to the dead-band of usual LTCs.

11.2.2 UVLS Against Both Short- and Long-term Voltage Instability

In references [7, 8], the distributed rule-based controller is developed in such a way that it can be used to prevent rapid voltage collapse due to short-term voltage instability. This mode of instability happens due to stalling of the induction motors and causes rapid voltage drop, and therefore, delay in load shedding must be reduced. Of course, due to the security of the protection, the delay cannot be less than a threshold. To relieve this problem, controllers are proposed in [7, 8]. The OEL activation signals of the generators near the controller place are also used. OEL activation, in addition to being a sign of high reactive power consumption due to reduction of motor speeds, causes outage of the generator from the synchronous mode and acceleration of voltage collapse. With receiving the OEL activation signal from at least one of the nearby generators, the controller tries to shed load with less delay. This signal is sent when the generator excitation current be greater than the excitation current limit ($i_f^{\lim}$) for a threshold minimum time. The reason for the delay in sending signal by the generators is that the excitation current may increase temporarily due to a fault occurred near of the generator and it had

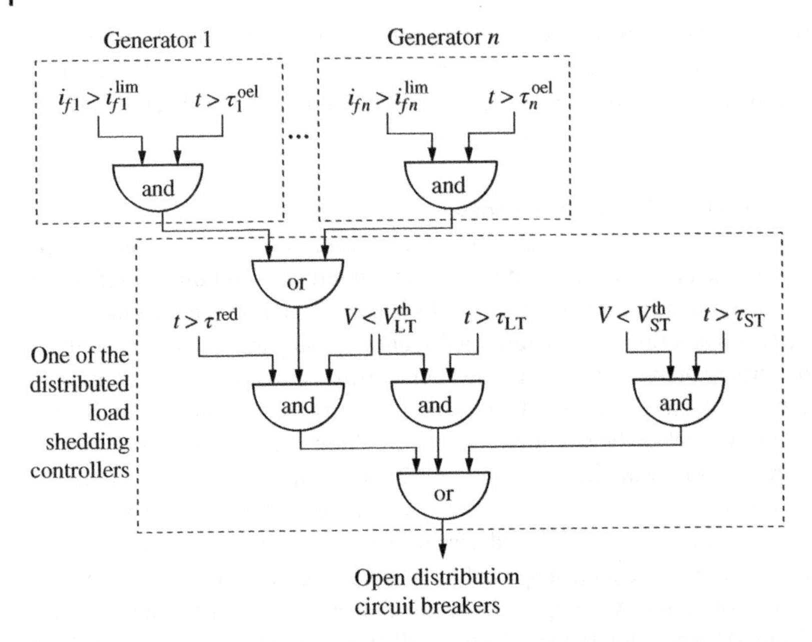

Figure 11.2 The logic of wide-area load shedding against short- and long-term voltage instabilities [8].

nothing to do with the generator voltage control. Due to the fact that the controller, in addition to a magnitude of load buses' voltage, depends on the generators' excitation current, it falls into the category of wide-area protection. In Ref. [7], V^{th} is taken the same in both short- and long-term voltage instabilities, and the difference is only the delay that it is determined based on the generators' excitation current increment. This approach may not prevent voltage instability when the percentage of motor loads is high and voltage collapse rapidly occurs. Hence, the delay must be decreased. In order to the delay decrement does not affect protection and security, [8] has suggested that V^{th} related to the short-term voltage instability is also reduced. Figure 11.2 shows the wide-area load shedding logic against short- and long-term voltage instabilities. In this figure, $V_{\text{ST}}^{\text{th}}$ and $V_{\text{LT}}^{\text{th}}$ are the threshold voltages related to short- and long-term instabilities, and τ_{ST} and τ_{LT} are the corresponding delays, respectively. The load shedding value is computed according to Eq. (11.2), and the values of K and delays are specified using simulation of various scenarios, similar to the method of [1].

To simulate short-term voltage instability, a portion of the load must be substituted by the induction motor load type. When the load shedding is applied to the motors, the resistance and inductance of the motor which are in p.u. based on the motor MVA are kept constant, but the base MVA is reduced. Also, the mechanical torque based on the new base MVA is remained constant [8].

11.2.3 Load Shedding Based on Incremental Value of Generator Reactive Power

In Ref. [6], the change in reactive power produced by generators is considered as a criterion of severity a disturbance and is used to determine the amount of required load shedding. To this aim, the load buses are first divided into several voltage control areas (VCAs) with the same voltage instability problems. Furthermore, the reactive power sources, based on their role in controlling voltage of respective VCAs, are also classified in groups called reactive power reserve basins (RPRBs). Given that voltage instability, especially at the beginning of its occurrence, is a local phenomenon, dividing the power system into different areas can help to quickly determine effective actions in real-time voltage control [12]. In Ref. [6], a curve fitting model is used to determine the appropriate relationship between the generators' reactive power increment and the amount of reactive power that must be reduced in load shedding. Load shedding is applied sequentially until the voltages are recovered to the desired values. Here, similar to under voltage load shedding, a deliberate delay is considered in implementing the first load shedding as well as between two consecutive load shedding, to ensure the reaction of the other voltage control equipment.

At each step of load shedding, the reactive power to be shed based on the curve fitting model is distributed among the areas according to the following equation.

$$Q_{\text{shed,VCA}_j} = \left(\frac{100 - \text{vcaRVSA}_j}{100n - \sum_{j=1}^{n} \text{vcaRVSA}_j} \right) \times Q_{\text{shed}} \tag{11.8}$$

where Q_{shed} is the total reactive power to be shed and $Q_{\text{shed,VCA}_j}$ is the amount of reactive power to be shed in the jth area. n is the number of areas and vcaRVSA_j is an index that determines the criticality of each area. This index is based on the distance between the reactive power produced by the generators in the area and their maximum reactive power.

11.2.4 Adaptive Load Shedding Based on Early Detection of Voltage Instability

In reference [13], a method based on wide-area system monitoring for early detection of voltage instability is presented. In this method, instead of simulating the system response to various disturbances, the trajectory of voltage changes is monitored using phasor measurement units (PMUs) as well as state estimators. Having voltage phasors and using the system model, other variables such as injection current to buses and generator excitation currents are obtained. Having the trajectory of changes of variables and using the system model, it is possible to detect the collision of the system trajectory with the maximum loadability level which is a sign of voltage instability. The advantage of measuring variables instead of simulating a system response is that it reduces the required model dimensions. For example, since the goal is to detect the loadability limit, there is no need to model the load, which usually leads to a large error due to the complexity of the downstream network.

The trajectory of long-term changes of variables is assumed to be as the following algebraic equations, in which the vector $\boldsymbol{S}$ represents the system loads and the vector $\boldsymbol{Z}$ represents the system variables which include the algebraic and state variables.

$$\boldsymbol{\varphi}(\boldsymbol{Z}, \boldsymbol{S}) = 0 \tag{11.9}$$

φ consists of the system algebraic equations plus equilibrium equations of the system fast dynamics. The performance of equipment such as OELs and LTCs that have slow dynamics is reflected at any time in the measured voltages. For example, by calculating the generators excitation current and taking the OEL time delay into account, it is determined that the generator is in the voltage control state or field current limit.

In [10], sensitivity-based indices are used to diagnose the voltage instability occurrence and apply load shedding. For this purpose, the sensitivity of the total generated reactive power to the consumed reactive power is used to detect the collision of the system trajectory with the loadability limit. This sensitivity value reaches a large positive number in the vicinity of the maximum loadability point and then jumps to a large negative value after passing it. Assuming that the vector of reactive power consumed by loads is $\boldsymbol{q} = \left[Q_1, \ldots, Q_{N_L}\right]^T$, the vector of sensitivity of total generated reactive power to the reactive power consumed by loads is as follows [10]:

$$S_{Q,\boldsymbol{q}} = -\boldsymbol{\varphi}_{\boldsymbol{q}}^T \left(\boldsymbol{\varphi}_{\boldsymbol{z}}^T\right)^{-1} \nabla_{\boldsymbol{z}} Q_g \tag{11.10}$$

where $\nabla_{\boldsymbol{z}} Q_g$ is the gradient of total generated reactive power Q_g with respect to $\boldsymbol{z}$. $\boldsymbol{\varphi}_{\boldsymbol{z}}$ and $\boldsymbol{\varphi}_{\boldsymbol{q}}$ are also gradients of $\boldsymbol{\varphi}$ with respect to $\boldsymbol{z}$ and $\boldsymbol{q}$, respectively. When the system trajectory reaches the loadability limit, which is detected from the sensitivity sign changes, each transmission bus voltage in the desired area is considered as V^{th} of that bus, and load shedding is implemented to recover all voltages to values higher than respective V^{th}.

To determine the locations of load shedding, when the system trajectory reaches the loadability limit, the bus voltage sensitivity to the loads active power is calculated as follows:

$$\boldsymbol{S}_{\mathrm{V_{LP}}} = -\boldsymbol{\varphi}_{\boldsymbol{p}}^{T}\left(\boldsymbol{\varphi}_{\boldsymbol{z}}^{T}\right)^{-1}\nabla_{\boldsymbol{z}}V_{\mathrm{L}} \tag{11.11}$$

where $\boldsymbol{P} = [P_1, \ldots, P_{N_\mathrm{L}}]^T$ is the loads active power vector. The high sensitive buses are chosen and initially the maximum possible load shedding is done at the most sensitive bus. If the voltage does not recover, load shedding will be implemented to the next bus, and so on.

According to Eq. (11.12), the load shedding amount is calculated from the difference of the buses power consumption when the system trajectory collides with the maximum loadability with their predisturbance powers. This difference is called unrestored power.

$$\Delta P_{\mathrm{un}} = \sum_{i \in I}\left(P_{L_i}^{\mathrm{pre}} - P_{L_i}^{\mathrm{cri}}\right) \tag{11.12}$$

where I denotes the set of load buses in the considered area.

Given that ΔP_{un} is a rough estimate of the load to be shed, load shedding in different steps in the size of Eq. (11.13) will be done, sequentially.

$$\Delta P = \alpha \Delta P_{\mathrm{un}}, \quad 0 < \alpha \leq 1 \tag{11.13}$$

α should be large enough to prevent long delays in load shedding that may lead to no voltage recovery or need to more load shedding. It should also be small enough to prevent unnecessary load shedding and overvoltage. In [10], the delay between the sequential load shedding is three seconds.

11.3 Decentralized Voltage Control

As mentioned earlier, voltage instability is initially a local phenomenon. Therefore, it is possible to prevent its spread by taking appropriate and timely actions in the unstable area. The unstable area is a part of the power system in which the voltage collapse occurs at its load buses earlier than elsewhere. Load buses in an unstable area are electrically close to each other. On the other hand, the most effective actions to prevent voltage collapse should be actions that are applied in unstable area load buses or buses close to them.

In order to take advantage of the local nature of voltage instability and to quickly determine appropriate action in an emergency, the power system can be divided into local zones. The concept of electrical distance is used for this purpose. There are several definitions of electrical distance. One of the most widely used is based on the inverse of the admittance matrix, which according to equation 11-14 the absolute value of each element of the resulting matrix is considered as the electrical distance of different buses [15][16]. For example, D_{ij} indicates the electrical distance between the buses i and j.

$$[D] = \left|[\mathbf{Y}_{bus}]^{-1}\right| \tag{11.14}$$

There are some other measures for electrical distance. For example in [17], the attenuation of voltage variations between two buses is used as the electrical distance. This is calculated for buses i and j by,

$$\alpha_{ij} = \frac{\partial V_i}{\partial V_j} \tag{11.15}$$

Determining the zones can be done off-line [15] or real-time [16]. In the method used in [15][16], first the number of zones is determined by determining the generators close to each other. In other words, zones are formed in such a way that the generators close to each other are located in a zone. For example, the generators G_i and G_j are in the same zone if,

$$D_{ij} = min\,(D_{ik}), k = 1,2,\ldots,N_G \tag{11.16}$$

where, N_G is the number of generators. Each generator is located in a zone if the generator is closest to at least one of the generators in that zone. After identifying the generators in each zone, the load buses are assigned to different zones one by one based on their electrical distance to the generators. Figure 11.3 shows an example of the zoning in the IEEE 39 bus system.

In reference [15], a Performance Index (PI) according to equation 11.17 is used to identify zones prone to instability.

$$\mathrm{PI} = W_{V_i}\sum\nolimits_{i=1}^{N_{LD}}(V_{L,min} - V_i) + W_{g_i}\sum\nolimits_{i=1}^{N_{GD}}(Q_i - Q_{max,i}) \tag{11.17}$$

where, N_{LD} is the number of load buses whose voltage is less than $V_{L,\,min}$. N_{GD} is the number of generators whose output reactive power has exceeded their maximum limit. The maximum output reactive power, as described in the chapter 3, depends on the generator output active power and voltage. Weight coefficients W_{V_i} and W_{g_i} show the relative importance of decreasing voltage and increasing reactive power in assessing the voltage stability of the zone. For strong systems where the distance between the generators and the loads is small, instability usually occurs when one or more generators reach their reactive power limits and lose voltage control. In these systems, the value W_{g_i} must be greater than W_{V_i}. On the

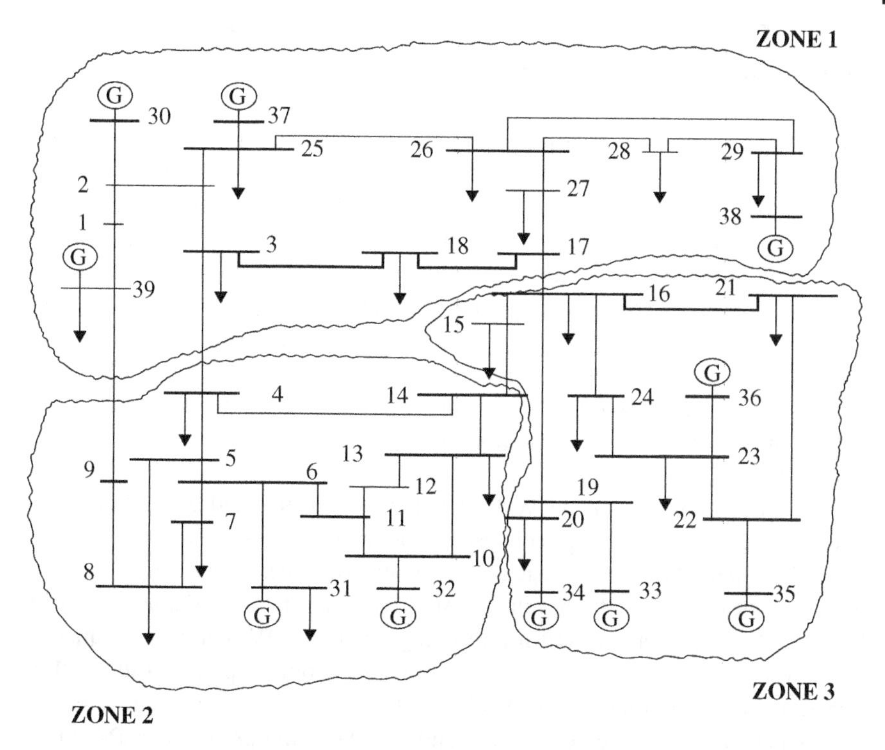

Figure 11.3 An example of zoning in the IEEE 39-bus system [15].

other hand, in systems with a large distance between generators and loads, instability usually occurs before the generators reach their reactive power generation limits. In these systems, choosing W_{V_i} larger than W_{g_i} is more appropriate.

Control actions are applied in each zone when the integral of PI integral in that zone exceeds a threshold value. Implementing of actions continues until the PI is reduced to zero and remains at that value for the time period t_{stop}. Actions include capacitor switches and, if necessary, load shedding, which are applied at certain steps and at specific time delays. In reference [16], load shedding and increase of the reference voltage value of the generators are used as control actions. Load shedding is done only in the unstable zone, but it may be possible to increase the generator voltage in other zones as well. Identification of generators effective in voltage recovery as well as suitable locations for load shedding is performed using sensitivity analysis. For this purpose, the voltage sensitivity of the bus that has the highest voltage drop to the generators voltage change and also to change in loads is used. A complete description of the calculation of these sensitivities is given in [16].

11.4 The use of Active Distribution Networks in Emergency Voltage Control

As mentioned earlier, if the exiting from the attraction region is not the problem, the inability in regulating voltage by the LTC transformer occurs when it is not possible to transfer the requested power to the network fed by the transformer. This inability leads to instability and long-term voltage collapse. A way to prevent occurrence of voltage instability is to reduce the requested power, which can be done by load shedding or reducing the LTC setpoint voltage. Of course, it should be noted that reducing the voltage only leads to a reduction in power consumption when the loads have voltage-dependent characteristic. For example, in areas with industrial loads, because the main consumers are induction motors, reducing the voltage increases the power consumption and increases the probability of voltage instability.

In an active distribution network that there are distributed generation units (DGUs), occurrence of voltage collapse can be prevented with less load shedding using the capability of the voltage and the power of DGUs. For example, if these units are located in medium voltage (MV) distribution networks as shown in figure 11.4, the power received from the high voltage (HV) network is equal to the power consumed in the MV network minus the output power of the DGUs. Therefore, it seems that a simple solution to reduce the power demanded by the HV network is to increase the active and reactive power produced by DG units. But increasing the active power is usually not possible because the main purpose of DGUs is to produce active power. Hence, these units are operated at any time depending on the availability of intermittent energy sources at the maximum possible amount of active power and the increase of it is no longer possible [18]. Also since the increase in reactive power generated by DGUs increases the loads voltage and consequently their active and reactive powers consumption, it may push the total power required by the MV network out of the feasible area. Due to this issue,

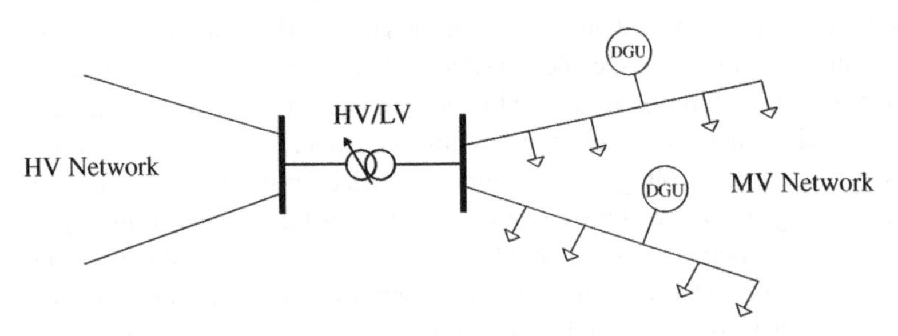

Figure 11.4 Presence of DGUs in the MV network.

reducing the reactive power of DGUs has been proposed as another solution [18]. Since reducing the reactive power output of DGUs reduces the loads voltage, depending on the characteristics of the loads, it may decrease the demanded power from the HV network.

As mentioned, reducing the LTC setpoint voltage can also be one of the ways to reduce the voltage of the MV network and consequently its power consumption. The presence of DGUs makes it possible to reduce the voltage uniformly along the MV network feeders. This allows a further reduction in the LTC reference voltage. However, there is a risk that the voltage drop causes the DGUs outage and to worse the unstable voltage situation.

In reference [18], based on what mentioned above, a method for voltage control in emergency situations is presented; to better understand it, initially, the process of appearing long-term voltage instability, which was described in detail in the first chapter, is reviewed. Figure 11.5 shows PV curves seen from the HV bus of the HV / MV transformer in both pre- and post-disturbance modes. For the sake of simplicity, it is assumed that the transformer losses are negligible compared to its passing power. P_0 is the power that is demanded from the HV network at the MV bus

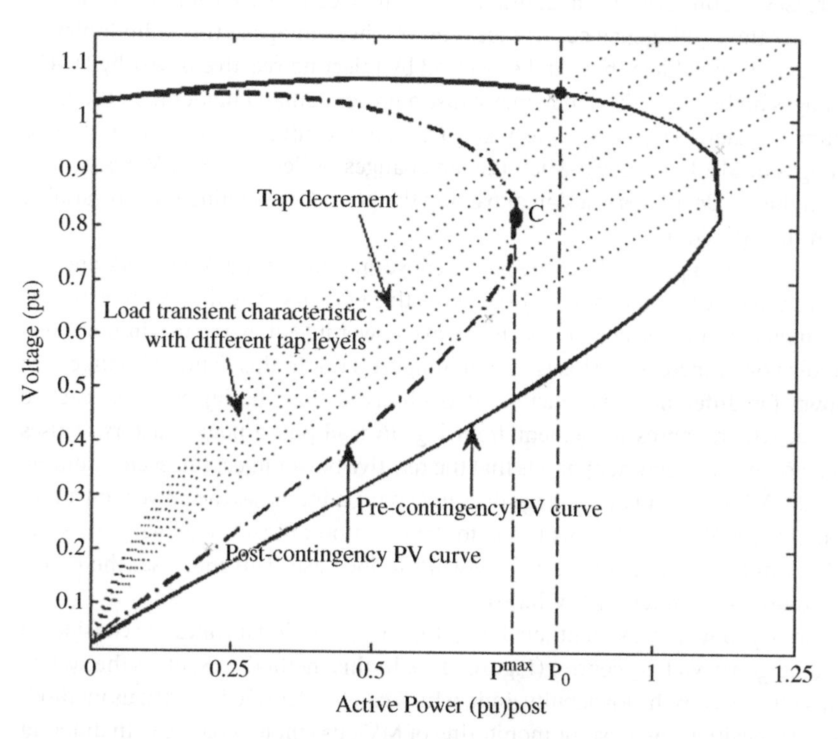

Figure 11.5 The process of long-term voltage instability occurrence.

regulated voltage. In this figure, the load characteristics seen from the transformer HV side with different tap positions are also shown. At any given moment, the intersection point of this characteristic with the PV curve shows the HV bus voltage and the power received from the HV bus. Its assumed that after a contingency occurrence, the maximum receivable power from the HV network (P_{cost}^{max}) becomes less than P_0. Therefore, it is not possible to regulate the MV bus voltage in the pre-contingency value. In this case, if the LTC setpoint voltage is still set to the pre-contingency value, the LTC operation will eventually lead to a voltage drop. The LTC reduces the tap value to recover the voltage (assuming that the variable tap is on the HV side as shown in the figure), which causes the load characteristic to move downwards.
As can be seen from figure 11.5, the characteristic downward movement leads to an increase in power received from the HV network, in other words, leads to an increase in power consumption in the MV network, which indicates an increase in the MV bus voltage. This increase in voltage and power continues until the characteristic reaches the nose of the PV curve (point C). Then, as the tap value decreases, the MV bus voltage and power received from the HV network decreases. Point C is the maximum power that can be transferred to the HV bus after the contingency occurrence. The MV bus voltage at this point indicates the maximum voltage that can be reached by injecting reactive power by DGUs. Voltages higher than this value may cause a power demand higher than the maximum transferable power. One way to detect reaching point C is to observe changes in MV bus voltage when the tap changes. A decrease in MV bus voltage when the tap decreases is a sign of passing the point C (assuming the tap variable is on the HV side).

There are two points about the power consumption of the MV network and the maximum power that can be transferred to this network. The first point is that, as mentioned in the first chapter, with reactive power injection, the maximum transferable power increases. This is shown in figure 11.6. In this figure, PV curves are shown for different power factors. It is observed that the injection of reactive power, which causes movement from lag- to lead-phase power factors, causes increase in P_{max}. Another point is that the reactive power injection, even assuming that the MV bus regulated voltage does not change, increases the power consumption of the MV network. This is due to the injection of reactive power causes the voltage different buses in the MV network to increase. This increases the power consumption of different MV buses.
According to what was mentioned, in [18], four operational zones are considered for emergency voltage control (Figure 11.7). In this method, as soon as the system trajectory reaches the loadability limit, which can be identified by various methods such as sensitivity analysis or monitoring of MV bus voltage change with different tap values, the MV and HV bus voltages are respectively selected as the maximum

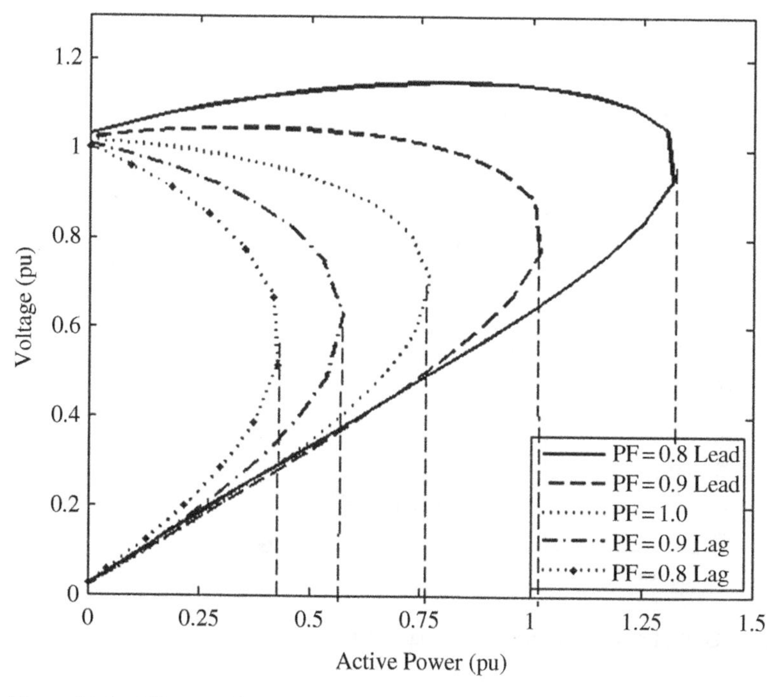

Figure 11.6 PV curves in different power factors.

allowable MV bus voltage (V_d^{max}) and the minimum allowable HV bus voltage (V_t^{min}). In zone 1, V_t is less than V_t^{min} and V_d is less than V_d^{max}. In this case, the value of tap r freezes and the reactive power of the DGUs will increase. This may cause V_t to exceed V_t^{min} before V_d passing from V_d^{max}. In this situation, operational point enters zone 4. No action is taken in this zone, i.e., the tap value and reactive power of the DGU are kept constant. In this zone, ε is selected so that V_d does not jump from one zone to another with each change in tap value. The sample value is 0.02 pu, which is equal to the usual LTC deadband. Since changes of V_t are more dependent on continuous changes of DGU reactive power, the value of less than ε is considered. In zone 4, the HV bus voltage is recovered to a value higher than V_t^{min} and there is no overvoltage in the MV bus. However, in cases where the HV bus voltage is more than $V_t^{min} + \delta$, if the MV bus voltage is less than $V_d^{max} + \varepsilon$ (zone 3), the tap value is reduced to increase the MV bus voltage. Increasing the MV bus voltage will increase the voltage of MV network and improves the performance of the network equipment. Of course, as mentioned, this voltage increment must be so that it is possible to transfer the corresponding demanded power to the MV network.

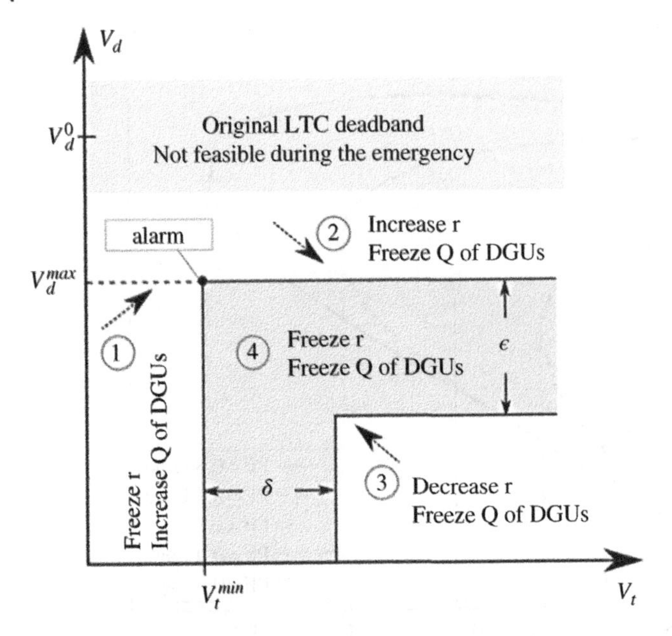

Figure 11.7 Control scheme in the (V_t, V_d) space [18].

As can be seen from figure 11.7, it is also possible to move from zone 1 to zone 2. In this zone, the MV bus voltage is higher than V_d^{max}, which is reduced to by increasing the tap value. It should be noted that according to the figure 11-7, increasing (decreasing) the tap value will decrease (increase) the MV bus voltage, if the operating point (i.e., the point where the load characteristic meets the PV curve) is in the upper region of the PV curve. In the lower part of the PV curve, opposite results are obtained; i.e., increasing (decreasing) the tap value increases (decreases) the MV bus voltage. In zones 2 and 3, since the voltage V_t is more than the voltage V_t^{min}, the operating point is expected to be above the PV curve. Regulating the MV bus voltage at V_d^{max} may also be possible by reducing the LTC voltage setpoint and tap blocking. But in this case, more load shedding is required to recover the MV bus voltage to the pre-contingency value.

It should be noted that if the voltage V_d^{max} is so low that the voltage recovery to V_d^{max} may not be able to prevent voltage instability and collapse. This happens when a significant portion of the MV network load is the induction motor type. These motors consume so much reactive power at low voltage state that DGUs may not be able to supply this demand. In this state, the MV bus voltage must be quickly recover to pre-contingency voltage, which may require load shedding.

11.5 Coordinated Voltage Control

As mentioned, it is desirable that in the event of an emergency, among the available control actions, the optimal actions are determined and implemented in real time. For this purpose, the concept of MPC is usually used, which is discussed later.

11.5.1 Model Predictive Control

This method is based on predicting the variation trajectory of system variables for different combinations of control actions. The trajectory, in addition to the system dynamics, also depends on the variables values at the implementation moment of actions, which are obtained by measuring. The goal is to determine the actions that bring the trajectory of the variables as possible as close to the desired trajectory. Due to the fact that predicting the trajectory of variables and consequently predicting the impact of control actions may be with error, the calculation and implementation of control actions are done in several steps. The prediction is made for a specific time period, called the prediction horizon, which begins from the sampling moment.

For each time horizon, instead of a set of fixed control actions, a sequence of optimal actions may be determined. For example, in emergency voltage control, many actions may be required to prevent a voltage drop at first; but these actions can lead to overvoltages in later moments which to prevent it, some of the controls values must be changed at the suitable time. Of course, determination of a sequence of actions increases computation time and may not be applicable for a large system.

After determining the sequence of optimal actions, their first step is applied. Then, at the next sampling, new actions are calculated by re-sampling the variables. The calculation and implementation of actions are continued until a favorable trajectory is created for the system variables. The time to calculate the actions should be as short as possible. It is clear that this time cannot be greater than the time interval between two consecutive samplings because with each sampling new calculations begin. Also, since the effect of measures is predicted based on the values of variables at the sampling time, if the time interval between sampling and application of actions be large, changing the variables status in this time interval may reduce the effect of actions.

A significant portion of the time required to determine appropriate actions is for the time to predict the trajectory of variables. The most accurate prediction is obtained by solving nonlinear algebraic and differential equations of the system. But the time required for this prediction does not allow it to be used for real-time applications. In the following sections, some approximate prediction methods are

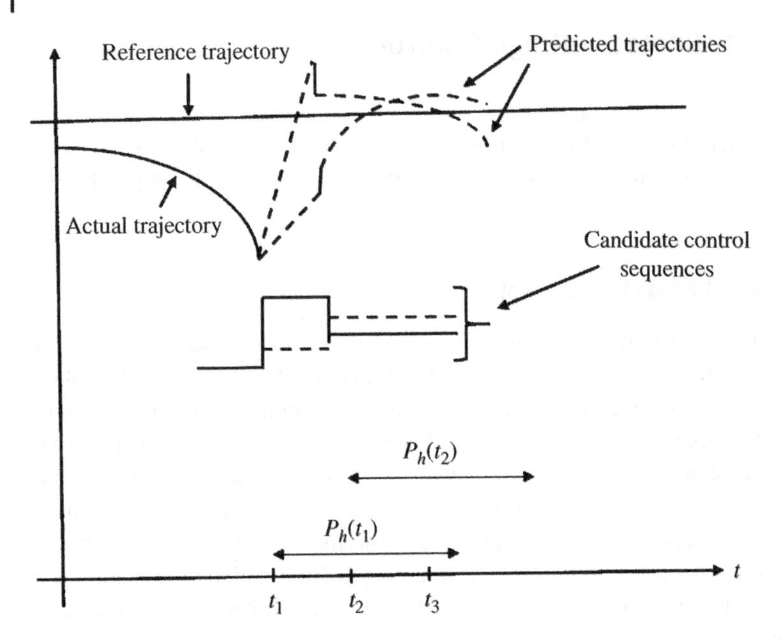

Figure 11.8 The principles of MPC.

introduced that require much less time. Given that MPC is a closed loop control method, a limited prediction error can be acceptable. Figure 11.8 conceptually illustrates the principles of the MPC method. In this figure, t_1, t_2, and t_3 represent the three sampling times. $T_{\text{Ph}}(t_1)$ and $T_{\text{Ph}}(t_2)$ are the moving prediction horizons that the times in parentheses shows when the prediction started. This figure also shows two examples of control sequences and corresponding predicted trajectories. The prediction period is selected based on the settling time of the slowest dynamics [19].

11.5.2 Prediction of Trajectory of Variables

11.5.2.1 Simplifications Required for Emergency Voltage Control in the Transmission Network

As mentioned, one of the most important part of MPC is predicting the path of variables when applying control actions. Prediction is done by solving the system equations. The power system model includes thousands of algebraic and nonlinear differential equations and that completely solving them is very time consuming and not suitable for emergency voltage control, which requires fast computing of control actions. Therefore, the equations must be reduced by applying some simplifications. Since MPC is a closed loop method, the error resulting from these simplifications can be acceptable.

Given that the application of MPC in emergency voltage control is for long-term voltage instability, only equations related to long-term dynamics can be considered and fast dynamics can be replaced with their equilibrium equations. Slow dynamics mainly include thermostatically controlled thermal loads and LTCs, which play an important role in appearing long-term voltage instability. Replacing fast dynamics with their equilibrium equations reduces the complexity of system equations but does not reduce the number of them. The number of equations depends on the number of the power system's buses, which includes the buses in distribution and transmission networks. Due to the large number of buses, it is not possible to consider both distribution and transmission networks in emergency control of voltage.

Figure 11.9 shows an example of a power system structure that includes a transmission network (TN) and some distribution networks (DNs). Transmission network is connected to distribution networks through transmission transformers that use load tap changer (LTC).

As mentioned in Chapter 4, there are LTC transformers in the distribution networks themselves, which cause voltage recovery and consequently power recovery at the buses of these networks. In power system modeling, during emergency control of voltage in the transmission network, the network connected to the transmission bus which includes distribution networks and transmission transformers is usually replaced by a dynamic load model (Figure 11.10). This model should be able to simulate as much as possible the behavior of the transmission and distribution LTC transformers in power recovery. For this purpose, exponential recovery model (Section 2.3.2) is usually used.

If the change of tap value or the change of voltage set-point of the LTC transmission transformer are considered as control actions, only the distribution

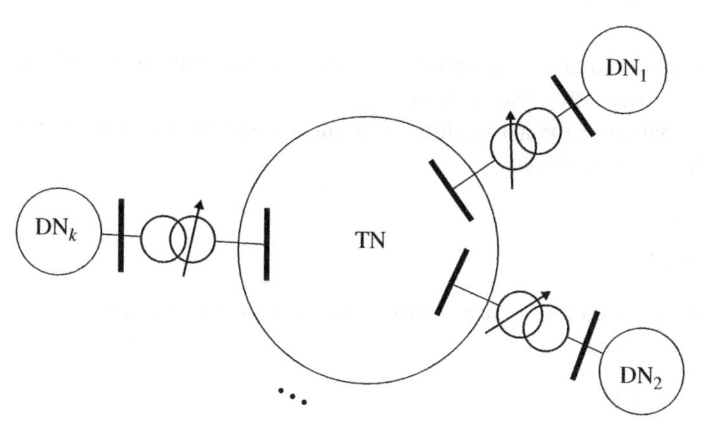

Figure 11.9 Atypical structure of power system.

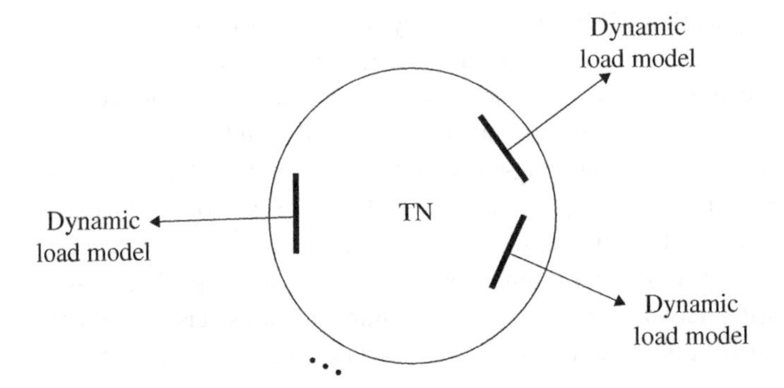

Figure 11.10 Replacement of transmission transformers and distribution networks with dynamic load models.

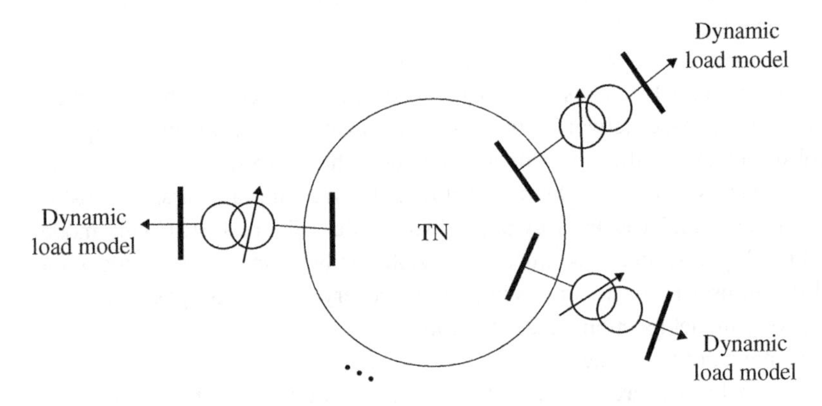

Figure 11.11 Replacement of distribution networks with dynamic load models.

networks are replaced with dynamic load model, and a discrete-time model will be used for the LTC transmission transformers (Figure 11.11).

Let us assume that, after the above simplifications, the system differential and algebraic equations are as follows:

$$\boldsymbol{x}^0 = \boldsymbol{f}(\boldsymbol{x}, \boldsymbol{y}, \boldsymbol{u}) \tag{11.18}$$

$$0 = \boldsymbol{g}(\boldsymbol{x}, \boldsymbol{y}, \boldsymbol{u}) \tag{11.19}$$

where $\boldsymbol{x} \subset R^n, \boldsymbol{y} \subset R^m$, and $\boldsymbol{u} \subset R^L$ are, respectively, the vectors of state, algebraic, and control variables, and

$$\boldsymbol{f} : R^{n+m+L} \to R^n$$

$$\boldsymbol{g} : R^{n+m+L} \to R^m$$

The exact trajectory of the variables when the vector u changes is obtained by numerically solving the nonlinear Eqs. (11.18) and (11.19). This is time consuming and is not suitable for real-time calculations. Instead, the approximate trajectory prediction methods are provided that require much less solution time. Some of these methods are described later.

11.5.2.2 Euler State Prediction (ESP)

Given that the long-term behavior of voltages is usually monotonic, the trajectory of voltages variations can be considered as a straight line [19]. Suppose that the actions are applied at the moment t_k. At this moment, the state variables do not jump, thus

$$\boldsymbol{x}(t_k^-) = \boldsymbol{x}(t_k^+) = \boldsymbol{x}(t_k) \tag{11.20}$$

where t_k^- and t_k^+ are moments immediately before and after t_k, respectively. Unlike the state variables, algebraic variables change instantly that whose jumped values are obtained by solving Eq. (11.21), in which $\boldsymbol{u}(t_k^+)$ represents the vector of control variables changed at the moment t_k.

$$0 = \boldsymbol{g}(\boldsymbol{x}(t_k), \boldsymbol{y}(t_k^+), \boldsymbol{u}(t_k^+)) \tag{11.21}$$

Assuming $\boldsymbol{f}(\boldsymbol{x}(t_k), \boldsymbol{y}(t_k^+), \boldsymbol{u}(t_k^+))$ as the slope of the state variables variations at the moment t_k, the approximate change of the state variables in the prediction horizon using Euler prediction becomes [19]:

$$\Delta\boldsymbol{x}(t_k + T_{\text{ph}}(t_k)) = T_{\text{ph}}(t_k)\boldsymbol{f}(\boldsymbol{x}(t_k), \boldsymbol{y}(t_k^+), \boldsymbol{u}(t_k^+)) \tag{11.22}$$

Using (11.22), the vector of predicted algebraic variables at the end of the prediction horizon is obtained from Eq. (11.23).

$$0 = \boldsymbol{g}(\boldsymbol{x}(t_k) + \Delta\boldsymbol{x}(t_k + T_{\text{ph}}(t_k)), \boldsymbol{y}(t_k + T_{\text{ph}}(t_k)), \boldsymbol{u}(t_k^+)) \tag{11.23}$$

Now, the trajectory of the algebraic variables including the voltages magnitudes is obtained by assuming that they move on a straight line as follows.

$$\boldsymbol{y}(t) = \boldsymbol{y}(t_k^+) + \frac{t - t_k}{T_{\text{ph}}(t_k)}(\boldsymbol{y}(t_k + T_{\text{ph}}(t_k)) - \boldsymbol{y}(t_k^+)), \quad t \in [t_k^+, t_k + T_{\text{ph}}(t_k)] \tag{11.24}$$

11.5.2.3 Two-Point Prediction Method

Similar to the ESP method, this method is also based on predicting two points of the trajectory and connecting them with a straight line [20]. The first point is the value of considered variable immediately after applying the actions (similar to ESP method). The second point is a steady-state point that assuming it does not leave the attraction region the trajectory of the variables will reach that point. This point will be found by solving the following equations.

$$0 = \boldsymbol{f}(\boldsymbol{x}, \boldsymbol{y}, \boldsymbol{u}(t_k^+)) \tag{11.25}$$

$$0 = \boldsymbol{g}(\boldsymbol{x}, \boldsymbol{y}, \boldsymbol{u}(t_k^+)) \tag{11.26}$$

If the slowest variable time constant be τ, it can be assumed that the steady-state conditions appear in 4τ after t_k. Therefore, the path of algebraic variables is predicted as follows [20].

$$\boldsymbol{y}(t) = \boldsymbol{y}(t_k^+) + \frac{t - t_k}{4\tau}(\boldsymbol{y}(t_k + 4\tau) - \boldsymbol{y}(t_k^+)), \quad t \in [t_k^+, t_k + 4\tau] \tag{11.27}$$

11.5.2.4 Prediction Using Trajectory Sensitivity

In this method, the trajectory of the variables is first determined when the current control variables are used. Then, by calculating the sensitivity of the variables at different moments of the prediction horizon to the control actions applied at the beginning of this period, the change of variables trajectory when the control variables change is predicted. For example, if u_0 be the vector of control variables at t_0, the variables trajectory for u_0 is obtained by numerically solving of Eqs. (11.28) and (11.29).

$$\boldsymbol{x}^0 = \boldsymbol{f}(\boldsymbol{x}, \boldsymbol{y}, \boldsymbol{u}_0), \quad \boldsymbol{x}(t_0) = \boldsymbol{x}_0 \tag{11.28}$$

$$0 = \boldsymbol{g}(\boldsymbol{x}, \boldsymbol{y}, \boldsymbol{u}_0) \tag{11.29}$$

Solving Eqs. (11.28) and (11.29) is possible with different numerical methods and that one of the most widely used of them is trapezoidal numerical integration, which is described later. Having the path of variables for $\boldsymbol{u}_0$ as well as trajectory sensitivities, the response of variables to changes in control variables is estimated according to Eqs. (11.30) and (11.31) [4].

$$\boldsymbol{x}(t, \boldsymbol{x}_0, \boldsymbol{u}) = \boldsymbol{x}(t, \boldsymbol{x}_0, \boldsymbol{u}_0) + \boldsymbol{x}_{\boldsymbol{u}}(t)(\boldsymbol{u} - \boldsymbol{u}_0) \tag{11.30}$$

$$\boldsymbol{y}(t, \boldsymbol{x}_0, \boldsymbol{u}) = \boldsymbol{y}(t, \boldsymbol{x}_0, \boldsymbol{u}_0) + \boldsymbol{y}_{\boldsymbol{u}}(t)(\boldsymbol{u} - \boldsymbol{u}_0) \tag{11.31}$$

where $\boldsymbol{x}_{\boldsymbol{u}}(t)$ and $\boldsymbol{y}_{\boldsymbol{u}}(t)$ are, respectively, trajectory sensitivity matrices of state and algebraic variables with respect to the vector u that are calculated for $\boldsymbol{u} = \boldsymbol{u}_0$.

- Trapezoidal numerical integration

One of the numerical integration methods is the trapezoidal method, which is widely used in dynamic analysis of power systems. To describe this method, initially, a first-order nonlinear differential equation as Eq. (11.32) is considered assuming that the value of x at the moment t_0 is equal to x_0. We want to get the time response of $x(t)$ for $t > t_0$.

$$x^0 = f(x, t), \quad x(t_0) = x_0 \tag{11.32}$$

From Eq. (11.32):

$$\mathrm{d}x = f(x, t)\mathrm{d}t \tag{11.33}$$

By integrating from both sides of Eq. (11.33) from t_0 to $t_1 = t_0 + \Delta t$:

$$\int_{x(t_0)}^{x(t_1)} \mathrm{d}x = \int_{t_0}^{t_1} f(x, t)\mathrm{d}t \tag{11.34}$$

$$x(t_1) = x(t_0) + \int_{t_0}^{t_1} f(x, t)\mathrm{d}t \tag{11.35}$$

where Δt is the integration time step. To calculate the right side integration of Eq. (11.35), the area created by it is approximated with a trapezoid (Figure 11.12). Accordingly, Eq. (11.35) becomes:

$$x(t_1) = x(t_0) + \frac{\Delta t}{2}[f(x(t_0), \ t_0) + f(x(t_1), \ t_1)] \tag{11.36}$$

It is observed that the differential equation has been converted into a nonlinear algebraic equation whose unknown is $x(t_1)$. The general formula for calculation of $x(t_{k+1})$ with knowing $x(t_k)$ is as given in Eq. (11.37).

$$x(t_{k+1}) = x(t_k) + \frac{\Delta t}{2}[f(x(t_k), \ t_k) + f(x(t_{k+1}), \ t_{k+1})] \tag{11.37}$$

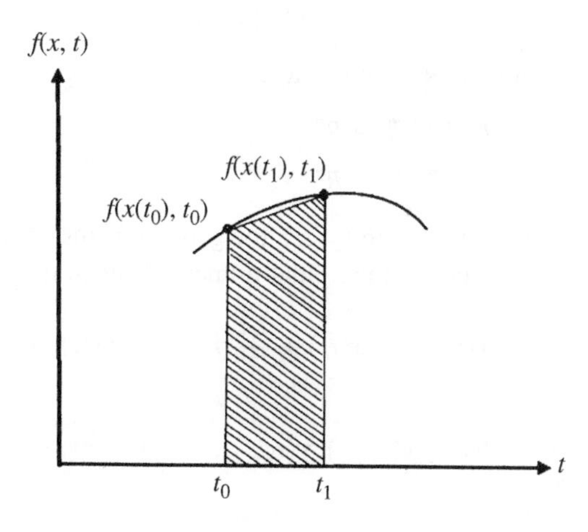

Figure 11.12 The approximation of an area with a trapezoid [21].

Now, assuming that in addition to the differential equation, there is also an algebraic equation according to Eqs. (11.38) and (11.39).

$$x^0 = f(x, y, t), \quad x(t_0) = x_0 \tag{11.38}$$

$$0 = g(x, y, t), \quad y(t_0) = y_0 \tag{11.39}$$

where y is the algebraic variable. In this case, Eq. (11.38) similar to Eq. (11.37) is converted into the algebraic equation (11.40) by applying the trapezoidal approximation for the area below the curve.

$$x(t_1) = x(t_0) + \frac{\Delta t}{2}[f(x(t_0),\ y(t_0),\ t_0) + f(x(t_1),\ y(t_1),\ t_1)] \tag{11.40}$$

The algebraic equation (11.39) remains for all times. Therefore, to calculate $x(t_1)$ and $y(t_1)$, the Eq. (11.40) with the Eq. (11.41) is used.

$$0 = g(x(t_1), y(t_1), t_1) \tag{11.41}$$

Now, having $x(t_k)$ and $y(t_k)$, general formulas for calculation of $x(t_{k+1})$ and $y(t_{k+1})$ are

$$x(t_{k+1}) = x(t_k) + \frac{\Delta t}{2}[f(x(t_k),\ y(t_k),\ t_k) + f(x(t_{k+1}),\ y(t_{k+1}),\ t_{k+1})] \tag{11.42}$$

$$0 = g(x(t_{k+1}), y(t_{k+1}), t_{k+1}) \tag{11.43}$$

If there is a differential-algebraic (DA) dynamic system with n state variables and m algebraic variables as (11.44) and (11.45),

$$\boldsymbol{x}^0 = \boldsymbol{f}(\boldsymbol{x}, \boldsymbol{y}, t), \quad \boldsymbol{x}(t_0) = \boldsymbol{x}_0 \tag{11.44}$$

$$0 = \boldsymbol{g}(\boldsymbol{x}, \boldsymbol{y}, t), \quad \boldsymbol{y}(t_0) = \boldsymbol{y}_0 \tag{11.45}$$

where $x \subset R^n$ and $y \subset R^m$,

$$\boldsymbol{f} : R^{n+m} \rightarrow R^n$$

$$\boldsymbol{g} : R^{n+m} \rightarrow R^m$$

Having $\boldsymbol{x}(t_k)$ and $\boldsymbol{y}(t_k)$, the general formulas for the calculation of $\boldsymbol{x}(t_{k+1})$ and $\boldsymbol{y}(t_{k+1})$ using trapezoidal numerical integration are

$$\boldsymbol{x}(t_{k+1}) = \boldsymbol{x}(t_k) + \frac{\Delta t}{2}[\boldsymbol{f}(\boldsymbol{x}(t_k),\ \boldsymbol{y}(t_k),\ t_k) + \boldsymbol{f}(\boldsymbol{x}(t_{k+1}),\ \boldsymbol{y}(t_{k+1}),\ t_{k+1})] \tag{11.46}$$

$$0 = \boldsymbol{g}(\boldsymbol{x}(t_{k+1}), \boldsymbol{y}(t_{k+1}), t_{k+1}) \tag{11.47}$$

Equations (11.46) and (11.47) represent $n + m$ nonlinear algebraic equations that can be solved by any numerical method such as Newton–Raphson method.

- Calculation of the trajectory sensitivity

Now, we want to compute the trajectory sensitivities $\boldsymbol{x_u}(t)$ and $\boldsymbol{y_u}(t)$. Differentiating from Eqs. (11.18) and (11.19) leads to,

$$\boldsymbol{x}^{0}{}_{\boldsymbol{u}}(t) = \boldsymbol{f_x}(t) \cdot \boldsymbol{x_u}(t) + \boldsymbol{f_y}(t) \cdot \boldsymbol{y_u}(t) + \boldsymbol{f_u}(t) \tag{11.48}$$

$$0 = \boldsymbol{g_x}(t) \cdot \boldsymbol{x_u}(t) + \boldsymbol{g_y}(t) \cdot \boldsymbol{y_u}(t) + \boldsymbol{g_u}(t) \tag{11.49}$$

where $\boldsymbol{f_x}, \boldsymbol{f_y}, \boldsymbol{g_x}, \boldsymbol{g_y}$, and $\boldsymbol{f_u}$ are time-dependent matrices that are obtained along the trajectory obtained from solving Eqs. (11.28) and (11.29). Note that, in calculation of the above matrices, $\boldsymbol{u}$ is inserted as $\boldsymbol{u}_0$. Using trapezoidal numerical integration, Eqs. (11.50) and (11.51) are now obtained from Eqs. (11.48) and (11.49).

$$\begin{aligned}\boldsymbol{x_u}(t_{k+1}) = \boldsymbol{x_u}(t_k) + \frac{\Delta t}{2}\Big[&\boldsymbol{f_x}(t_k) \cdot \boldsymbol{x_u}(t_k) + \boldsymbol{f_y}(t_k) \cdot \boldsymbol{y_u}(t_k) + \boldsymbol{f_u}(t_k) \\ &+ \boldsymbol{f_x}(t_{k+1}) \cdot \boldsymbol{x_u}(t_{k+1}) + \boldsymbol{f_y}(t_{k+1}) \cdot \boldsymbol{y_u}(t_{k+1}) + \boldsymbol{f_u}(t_{k+1})\Big]\end{aligned} \tag{11.50}$$

$$0 = \boldsymbol{g_x}(t_{k+1}) \cdot \boldsymbol{x_u}(t_{k+1}) + \boldsymbol{g_y}(t_{k+1}) \cdot \boldsymbol{y_u}(t_{k+1}) + \boldsymbol{g_u}(t_{k+1}) \tag{11.51}$$

It should be noted that here t_k and t_{k+1} are not the sampling times, and the distance between them is determined based on the integration time step. $\boldsymbol{x_u}(t)$ and $\boldsymbol{y_u}(t)$ are determined at each instant by solving Eqs. (11.50) and (11.51).

To calculate the initial conditions $\boldsymbol{x_u}(t_0)$ and $\boldsymbol{y_u}(t_0)$, the fact that state variables cannot change instantly is used, therefore

$$\boldsymbol{x_u}(t_0) = 0 \tag{11.52}$$

Having $\boldsymbol{x_u}(t_0)$, $\boldsymbol{y_u}(t_0)$ is obtained from Eq. (11.51) at the moment t_0.

$$0 = \boldsymbol{g_x}(t_0) \cdot \boldsymbol{x_u}(t_0) + \boldsymbol{g_y}(t_0) \cdot \boldsymbol{y_u}(t_0) + \boldsymbol{g_u}(t_0) \tag{11.53}$$

A complete description of the trajectory sensitivity calculation is provided in [20].

11.5.3 Cost Function

Given that the goal is to bring the trajectory of the controlled variables closer to the desired trajectory in the prediction period by the minimum actions, the cost function at sampling time t_k can be considered as in Eq. (11.54) [14, 22, 23].

$$J_1 = \int_{t_k}^{t_k + T_{\text{ph}}(t_k)} \left((\hat{\boldsymbol{y}} - \boldsymbol{y}_r)^T \boldsymbol{Q} (\hat{\boldsymbol{y}} - \boldsymbol{y}_r) + (\boldsymbol{u}(t_k^+) - \boldsymbol{u}(t_k^-))^T \boldsymbol{R} (\boldsymbol{u}(t_k^+) - \boldsymbol{u}(t_k^-))\right) dt \tag{11.54}$$

where $\hat{\boldsymbol{y}} = \hat{\boldsymbol{y}}\left(t, \boldsymbol{x}(t_k), \boldsymbol{u}\left(t_k^+\right)\right)$ represents the vector of the predicted variables. In the problem of emergency voltage control, $\hat{\boldsymbol{y}}$ includes the load buses voltage magnitudes. $\boldsymbol{y}_r$ is the desired trajectory vector. $\boldsymbol{Q}$ and $\boldsymbol{R}$ are diagonal matrices that, respectively, determine the relative importance of variables control and priority of control actions.

A method to consider constraints is to add a penalty term to the cost function as Eq. (11.55) [14, 22].

$$J_2 = \int_{t_k}^{t_k + T_{\text{ph}}(t_k)} \left((\hat{\boldsymbol{y}} - \boldsymbol{y}_r)^T \boldsymbol{Q} (\hat{\boldsymbol{y}} - \boldsymbol{y}_r) + \left(\boldsymbol{u}\left(t_k^+\right) - \boldsymbol{u}\left(t_k^-\right)\right)^T \boldsymbol{R} \left(\boldsymbol{u}\left(t_k^+\right) - \boldsymbol{u}\left(t_k^-\right)\right) + P \right) dt \tag{11.55}$$

The penalty factor can be selected in such a way that its increment when violating the constraints causes the main purpose of control to be to minimize any constraint violation. An example of penalty function is given in Eq. (11.56) [14]. In this equation, $y_{C,\,k}$ is a limit that is selected based on the constraint related to the kth controlled variable. For example, for emergency control of voltage, $y_{C,\,k}$ can be 0.9 pu for the load buses voltages. C is the set of constraints in which violation is occurred.

$$P\left(t, \boldsymbol{x}(t_k), \boldsymbol{u}\left(t_k^+\right)\right) = 10\left(1 + \sum_{k \in C} \left(\hat{y}_k - y_{C,k}\right)^2\right) \tag{11.56}$$

11.6 Summary

Despite preventive control, the power system may always encounter contingencies for which proper actions have not been taken before. In this situation, the system needs emergency actions. Therefore, emergency actions are taken when the voltage instability has started and if these actions are not taken in time, the voltage instability may occur in a few moments or minutes. Under voltage load shedding is one of the most widely used emergency actions in which only load shedding is used as an emergency action. On the other hand, the goal of a number of methods is to implement several different actions in a coordinated manner. These methods usually use the concept of predictive model control to determine the optimal actions. Since determination and implementation of emergency actions must be done in real time, reducing the solving time is very important. For this purpose, several approximate methods for predicting the trajectory of variables are presented.

References

1 Otomega, B. and Cutsem, V. (2007). Undervoltage load shedding using distributed controllers. *IEEE Trans. Power Syst.* 22 (4): 1898–1907. https://doi.org/10.1109/PSAMP.2007.4740897.

2 Nikolaidis, V.C., Vournas, C.D., Fotopoulos, G.A., et al. (2007). Automatic load shedding schemes against voltage instability in the hellenic system. *2007 IEEE Power Engineering Society General Meeting*, Tampa, FL (24–28 June 2007), pp. 1–7. https://doi.org/10.1109/PES.2007.386083.

3 Moors, C., Lefebvre, D., and Van Cutsem, T. (2000). Design of load shedding schemes against voltage instability. *2000 IEEE Power Engineering Society Winter Meeting. Conference Proceedings (Cat. No.00CH37077)*, Singapore (23–27 January 2000), pp. 1495–1500. https://doi.org/10.1109/PESW.2000.850200.

4 Jin, L., Kumar, R., and Elia, N. (2010). Model predictive control-based real-time power system protection schemes. *IEEE Trans. Power Syst.* 25 (2): 988–998. https://doi.org/10.1109/TPWRS.2009.2034748.

5 Glavic, M. and Van Cutsem, T. (2006). Some reflections on Model Predictive Control of transmission voltages. *Proceedings of the 2006 38th Annual North American Power Symposium NAPS-2006*, Carbondale, IL (17–19 September 2006), pp. 625–632. https://doi.org/10.1109/NAPS.2006.359637.

6 Adewole, A.C., Tzoneva, R., and Apostolov, A. (2016). Adaptive under-voltage load shedding scheme for large interconnected smart grids based on wide area synchrophasor measurements. *IET Gener. Transm. Distrib.* 10 (8): 1957–1968. https://doi.org/10.1049/iet-gtd.2015.1250.

7 Otomega, B. and Van Cutsem, T. (2009). Local vs. wide-area undervoltage load shedding in the presence of induction motor loads. *2009 IEEE Bucharest PowerTech*, Bucharest, Romania (28 June 2009 to 2 July 2009), pp. 1–7. https://doi.org/10.1109/PTC.2009.5282247.

8 Otomega, B. and Van Cutsem, T. (2011). A load shedding scheme against both short-and long-term voltage instabilities in the presence of induction motors. *2011 IEEE Trondheim PowerTech*, Trondheim, Norway (19–23 June 2011), pp. 1–7. https://doi.org/10.1109/PTC.2011.6019184.

9 Van Cutsem, T., Moors, C. and Lefebvre, D. (2000). Design of load shedding schemes against voltage instability using combinatorial optimization. *2002 IEEE Power Engineering Society Winter Meeting. Conference Proceedings (Cat. No.02CH37309)*, New York (27–31 January 2002), vol. 2, no. c, pp. 1495–1500. https://doi.org/10.1109/PESW.2000.850200.

10 Glavic, M. and Van Cutsem, T. (2010). Adaptive wide-area closed-loop undervoltage load shedding using synchronized measurements. *IEEE PES General Meeting PES 2010*, Minneapolis, MN (25–29 July 2010), pp. 1–8. https://doi.org/10.1109/PES.2010.5589279.

11 Otomega, B., Glavic, M. and Van Cutsem, T. (2014). A two-level emergency control scheme against power system voltage instability. *Control Eng. Pract.* 30: 93–104. https://doi.org/10.1016/j.conengprac.2013.10.007.

12 Karbalaei, F. and Shahbazi, H. (2018). Determining an appropriate partitioning method to reduce the power system dimensions for real time voltage control. *Int. J. Electr. Power Energy Syst.* 100: 58–68. https://doi.org/10.1016/j.ijepes.2018.02.025.

13 Glavic, M. and Van Cutsem, T. (2009). Wide-area detection of voltage instability from synchronized phasor measurements. Part I: principle. *IEEE Trans. Power Syst.* 24 (3): 1408–1416. https://doi.org/10.1109/TPWRS.2009.2023271.

14 Larsson, M., Hill, D.J., and Olsson, G. (2002). Emergency voltage control using search and predictive control. *Int. J. Electr. Power Energy Syst.* 24 (2): 121–130. https://doi.org/10.1016/S0142-0615(01)00017-5.

15 Karbalaei, F. and Shahbazi, H. (2017). A quick method to solve the optimal coordinated voltage control problem based on reduction of system dimensions. *Electr. Power Syst. Res.* 142: 310–319. https://doi.org/10.1016/j.epsr.2016.10.002.

16 Islam, S.R., Sutanto, D. and Muttaqi, K.M. (2015). Coordinated decentralized emergency voltage and reactive power control to prevent long-term voltage instability in a power system. *IEEE Trans. Power Syst.* 30 (5): 2591–2603. https://doi.org/10.1109/TPWRS.2014.2369502.

17 Islam, S.R., Muttaqi, K.M. and Sutanto, D. (2015). A Decentralized Multiagent-based voltage control for catastrophic disturbances in a power system. *IEEE Trans. Ind. Appl.* 51 (2): 1201–1214. https://doi.org/10.1109/TIA.2014.2350072.

18 Huaidong, Y., Jun, S. and Jingjing, B. (2021). A Method for Calculating Electrical Distance Between Nodes Based on Voltage-reactive Power Sensitivity. *Proceedings of the 2021 IEEE Sustainable Power and Energy Conference: Energy Transition for Carbon Neutrality, iSPEC 2021*, pp. 314–318. https://doi.org/10.1109/iSPEC53008.2021.9735538.

19 Ospina, L.D.P. and Van Cutsem, T. (2021). Emergency support of transmission voltages by active distribution networks: A non-intrusive scheme. *IEEE Trans. Power Syst.* 36 (5): 3887–3896. https://doi.org/10.1109/TPWRS.2020.3027949.

20 Hiskens, I.A. and Pai, M.A. (2000). Trajectory sensitivity analysis of hybrid systems. *IEEE Trans. Circuits Syst. I Fundam. Theory Appl.* 47 (2): 204–220. https://doi.org/10.1109/81.828574.

21 Kundur, P., Neal, J.B., and Mark, G.L. (1994). *Power System Stability and Control*, vol. 7. New York: McGraw-Hill.

22 Larrson, M. and Karlsoon, D. (2003). Coordinated system protection scheme against voltage collapse using heuristic search and predictive control. *IEEE Power Eng. Rev.* 18 (3): 1001–1006. https://doi.org/10.1109/MPER.2002.4312296.

23 Wen, J.Y., Wu, Q.H., Turner, D.R. et al. (2004). Optimal coordinated voltage control for power system voltage stability. *IEEE Trans. Power Syst.* 19 (2): 1115–1122. https://doi.org/10.1109/TPWRS.2004.825897.

Index

Voltage Stability in Electrical Power Systems: Concepts, Assessment, and Methods for Improvement, First Edition. Farid Karbalaei, Shahriar Abbasi, and Hamid Reza Shabani.

Published 2023 by John Wiley & Sons, Inc.

u

v

w

Printed and bound by CPI Group (UK) Ltd, Croydon, CR0 4YY

16/04/2025

14658595-0002